Universal Formulas in Integral and Fractional Differential Calculus

Universal Formulas in Integral and Fractional Differential Calculus

Khavtgai Namsrai

Mongolian Academy of Sciences, Mongolia

World Scientific

NEW JERSEY · LONDON · SINGAPORE · BEIJING · SHANGHAI · HONG KONG · TAIPEI · CHENNAI · TOKYO

Published by

World Scientific Publishing Co. Pte. Ltd.

5 Toh Tuck Link, Singapore 596224

USA office: 27 Warren Street, Suite 401-402, Hackensack, NJ 07601

UK office: 57 Shelton Street, Covent Garden, London WC2H 9HE

Library of Congress Cataloging-in-Publication Data

Names: Namsraæi, Khavtgaæin, 1943–

Title: Universal formulas in integral and fractional differential calculus /
   by Khavtgai Namsrai (Mongolian Academy of Sciences, Mongolia).

Description: New Jersey : World Scientific, 2016. | Includes bibliographical references and index.

Identifiers: LCCN 2015040816 | ISBN 9789814675598 (hardcover : alk. paper)

Subjects: LCSH: Differential calculus. | Calculus, Integral. | Mathematics--Formulae.

Classification: LCC QA304 .N25 2016 | DDC 515/.3--dc23

LC record available at http://lccn.loc.gov/2015040816

British Library Cataloguing-in-Publication Data

A catalogue record for this book is available from the British Library.

Dedicated to
my mother Samdan Myadag and father Damdin Khavtgai

Preface

It is well known that the laws of nature and society are formulated by means of higher mathematics, i.e., of language of differentials and integrals.

Already, during the last three centuries and more, classical differential and integral calculations by famous mathematicians and physicists like Isaac Newton (1643-1727), Gottfried W. Leibniz (1646-1716) and others played a vital role in our lives (see, for example, the wonderful textbooks of Ya. B. Zel'dovich and I. M. Yaglam, 1982, E. T. Whittaker and G. N. Watson, 1996).

Our method is based on the complex number analysis by using the complex plane of integration and connected with the Mellin representation for sign variable functions.

Recently, students and researchers studying physics, chemistry, engineering technology, computer science, communications technology and other branches of science know that carrying out differentiation of any complicated mathematical expressions with respect to independent variables of any kind is easy to calculate. However, we have experienced that the procedures of taking antiderivatives or integrations will encounter some difficulties. In the former case, there exists a definite rule of differentiation procedure. Until now, such rules are absent for integration methods, and we work only differently for different or concrete cases of integration.

In this book, we will show that there exist concrete and unified formulas where one can calculate enormous numbers of integrals by hand like the ones that use the Pythagorean Theorem in geometry. It turns out that according to our derived universal formulas there exists an enormous number of equivalent integrals which differ from each other by constants. In our approach, taking integrals becomes a design culture, since by appropriate choices of parameters entering into universal formulas, one can obtain a nice compact solution for integrals.

Chapter 1 deals with the mathematical base of our approach. Here, we use the Mellin representation of sign variable functions and sketch some useful formulas, equations and expressions which will be needed in the following chapters.

In Chapter 2, we will obtain unified formulas for calculating definite integrals which contain circular or trigonometric and power functions. As an example, some standard integrals are calculated by using newly obtained formulas.

Chapter 3 deals with obtaining unified formulas for definite integrals, the integrant of which includes functions of the type $\sin^q(ax^\nu)$, $\cos^m(ax^\nu)$, x^γ and $(p + qx^\rho)^{-\lambda}$.

In Chapter 4, we will derive general formulas for integrals which involve any powers of x, binomial functions of the type $(a + tx^\sigma)$ and trigonometric functions.

Chapter 5 is devoted to obtaining unique formulas for calculations of definite integrals containing $\exp[-ax^\nu]$, power and polynomial functions including trigonometric functions. Some formulas and summations are in standard textbooks (Gradshteyn and Ryzhik, 1980, Wheelon and Robacker, 1954).

In Chapters 6, 7, 8 and 9, we will derive some unified general formulas for the calculation of integrals involving special functions like cylindrical ones including Bessel, Neumann, Struve functions, etc. and two trigonometric functions.

Chapter 10 considers the calculation of fractional derivatives and inverse operators.

This book is devoted to undergraduate and graduate students, and researchers.

I would like to thank Professor G. V. Efimov (JINR, Dubna, RF), Professor Chunli Bai (President of Chinese Academy of Sciences), H. V. Von Geramb (Hamburg, Germany), Shih-Lin Chang (National Synchrotron Radiation Research Center, Taiwan, ROC), M. L. Klein (Temple University, Philadelphia, USA), A. K. Cheetham (University of Cambridge, United Kingdom), C. N. R. Rao (Jawaharla Nehru Centre for Advanced Scientific Research, Bangalore, India), Professors C. S. Lim and Hidenori Sonata (Kobe University, Japan) and my colleagues B. Chadraa, Ts. Baatar and S. Baigalsaikhan for their enthusiastic support and their interest. I am also grateful to my students Ts. Myanganbayar, Ts. Tsogbayar and Ts. Banzragch for their help in the preparation of this text.

Finally, I would like to thank my wife Jadambaa Tserendulam and my children Nyamtseren and Tsedevsuren. At every step of the way, I was showered with their love and support.

Kh. Namsrai
Ulaanbaatar, Mongolia
May 2015

Contents

Chapter 1

Mathematical Preparation

A number theory is beautiful but the complex number theory is more beautiful.

P. A. M. Dirac

1.1 Going to the Complex Number of Integration and the Mellin Representation

Let us consider a very simple and famous function: $\exp[-x]$ for which the following expansion is valid:

$$e^{-x} = 1 - x + \frac{x^2}{2!} - \frac{x^3}{3!} + \ldots = \sum_{n=0}^{\infty} (-1)^n \frac{x^n}{n!}. \tag{1.1}$$

Now we can go to the complex plane ξ and present this sum (1.1) in the integral form:

$$e^{-x} = \frac{1}{2i} \int_{-\beta+i\infty}^{-\beta-i\infty} d\xi \frac{x^\xi}{\sin(\pi\xi)\,\Gamma(1+\xi)}. \tag{1.2}$$

Here the contour of integration over the complex variable ξ is shown in Figure 1.1, where $-1 < \beta < 0$.

1

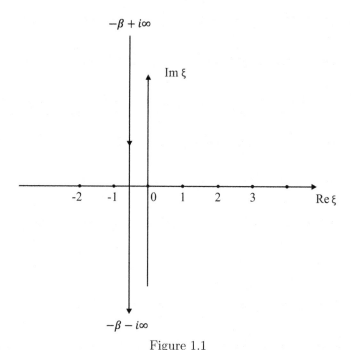

Figure 1.1

Typical form of the integration contour which will play an important role in obtaining explicit formulas for defined integrals.

The integral form of the type (1.2) is called the **Mellin representation** which plays a vital role in the integral calculation. Here the **important principal** is that one can move the integration contour to the left in any desired order through the points $\xi = -1$, -2, etc. It is possible due to the $\Gamma(x)$-gamma function properties, where the expression $1/[\sin \pi \xi \, \Gamma(1+\xi)]$ has no poles at the points $\xi = -1$, $-2,\dots$. Such a shift of the integration contour to the left allows us to get finite integrals at the limit:

$$\lim_{\varepsilon \to 0} \int_{\varepsilon}^{\infty} dx \; x^{\xi+\nu}.$$

On the contrary, displacement of the integration contour to the right gives rise to poles at the points $\xi = n$, $n = 0$, 1, $2,\dots$, because of the sine function $\sin \pi \xi$. In general, the calculation of the Mellin integral of the type (1.2) and taking residue in it will need some background and skills in the theory of complex variable functions. Now we will go into this problem in more detail.

1.2 Theory of Residues and Gamma-Function Properties

1.2.1 *Basic Theorem of the Residue Theory*

Theorem 1.1

Let $f(z)$ be an analytic function in the closed domain $\overline{G}$ with the exception of a finite number of isolated singular points $z_k (k = 1, \ldots, N)$ lying inside domain G. Then

$$\int_{L^+} d\xi f(\xi) = 2\pi i \sum_{k=1}^{N} Res[f(z), z_k], \qquad (1.3)$$

where L^+ is a full boundary of the domain G going through it in the positive direction.

Let a point z_0 be the pole of m-th order of the function $f(z)$. Then the formula for the calculation of residue at the pole of m-th order is given as

$$Res[f(z), z_0] = \frac{1}{(m-1)!} \lim_{z \to z_0} \frac{d^{m-1}}{dz^{m-1}} \left[(z - z_0)^m f(z) \right]. \qquad (1.4)$$

1.2.2 *The L'Hôpital Rule*

Theorem 1.2

If functions $f(z)$ and $g(z)$ have zero of m-th order at the point $z = z_0$, then

$$\lim_{z \to z_0} \frac{f(z)}{g(z)} = \lim_{z \to z_0} \frac{\dfrac{d^m}{dz^m} f(z)}{\dfrac{d^m}{dz^m} g(z)}. \qquad (1.5)$$

Example 1.1

$$\lim_{x \to 0} \frac{\sin x}{x} = \lim_{x \to 0} \frac{(\sin x)'}{x'} = \lim_{x \to 0} \frac{\cos x}{1} = 1.$$

Example 1.2

$$\lim_{z \to n} \frac{z - n}{\sin \pi z} = \lim_{z \to n} \frac{(z - n)'}{(\sin \pi z)'} = \lim_{z \to n} \frac{1}{\pi \cos \pi z} = \frac{1}{\pi}(-1)^n. \qquad (1.6)$$

Example 1.3

$$\lim_{z \to n} \frac{(z - n)^2}{\sin^2 \pi z} = \lim_{z \to n} \frac{2(z - n)}{2 \sin \pi x \, \cos \pi z \, \pi} = \lim_{z \to n} \frac{1}{\pi \cos \pi z} \lim_{z \to n} \frac{z - n}{\sin \pi z}$$

$$= \lim_{z \to n} \frac{1}{\pi \cos \pi z} \lim_{z \to n} \frac{1}{\pi \cos \pi z} = \frac{1}{\pi^2}(-1)^{2n} = \frac{1}{\pi^2}. \qquad (1.7)$$

1.2.3   *Calculation of Residue Encountered in the Mellin Integrals*

Let us calculate residues which are entered into the Mellin integrals. They are:

(1) **The pole of the first order:** According to the formula (1.4), one gets

$$Res_1 = \lim_{\xi \to n} \left[\frac{\xi - n}{\sin \pi \xi} f(\xi) \right].$$

Making use of the limit (1.6) it is easy to find

$$Res_1 = \frac{(-1)^n}{\pi} f(n). \tag{1.8}$$

(2) **The pole of the second order:** Again the definition (1.4) is used. It reads

$$
\begin{aligned}
Res_2 &= \lim_{\xi \to n} \frac{d}{d\xi} \left\{ \frac{(\xi - n)^2}{\sin^2 \pi \xi} f(\xi) \right\} \\
&= \lim_{\xi \to n} \left\{ \left[\frac{2(\xi - n)}{\sin^2 \pi \xi} - \frac{2(\xi - n)^2}{\sin^3 \pi \xi} \pi \cos \pi \xi \right] f(\xi) \right. \\
&\quad \left. + \frac{(\xi - n)^2}{\sin^2 \pi \xi} f'(\xi) \right\}.
\end{aligned}
\tag{1.9}
$$

Here the second limit is given by the formula (1.7). To find the first limit, we use the L'Hôpital rule (1.5). It gives

$$
\begin{aligned}
L_1 &= \lim_{\xi \to n} \left[\frac{2(\xi - n)}{\sin^2 \pi \xi} - \frac{2(\xi - n)^2}{\sin^3 \pi \xi} \pi \cos \pi \xi \right] \\
&= \lim_{\xi \to n} \frac{2(\xi - n)}{\sin \pi \xi} \lim_{\xi \to n} \left[\frac{\sin \pi \xi - (\xi - n) \pi \cos \pi \xi}{\sin^2 \pi \xi} \right].
\end{aligned}
$$

Here the second limit is calculated by using again the L'Hôpital rule. That is

$$l_1 = \lim_{\xi \to n} \frac{\pi \cos \pi \xi - \pi \cos \pi \xi + \pi^2 \sin \pi \xi \, (\xi - n)}{2 \sin \pi \xi \, \cos \pi \xi \, \pi} = \lim_{\xi \to n} \frac{\pi}{2} \frac{\xi - n}{\cos \pi \xi}.$$

Finally, we have

$$L_1 = \pi \lim_{\xi \to n} \frac{(\xi - n)}{\sin \pi \xi} \lim_{\xi \to n} \frac{\xi - n}{\cos \pi \xi} = 0.$$

Thus, the residue (1.9) takes the form

$$Res_2 = \frac{1}{\pi^2} f'(n). \tag{1.10}$$

For completeness we find other residues arising from the pole at the odd number $n = m + \frac{1}{2}$, $m = 0, 1, 2, \ldots$. That is

$$Res_3 = \lim_{\xi \to n} \left\{ \frac{\xi - n}{\cos \pi \xi} f(\xi) \right\} = \lim_{\xi \to n} \left\{ -\frac{1}{\pi \sin \pi \xi} f(\xi) \right\},$$

where

$$\sin \pi \left(m + \frac{1}{2}\right) = \sin \pi m \cos \frac{\pi}{2} + \cos \pi m \sin \frac{\pi}{2} = (-1)^m,$$

so that

$$Res_3 = -\frac{(-1)^m}{\pi} f\left(m + \frac{1}{2}\right). \tag{1.11}$$

1.2.4 *Gamma Function Properties*

As usual, in the Mellin integrals, gamma-functions are always presented. For this reason, we would like to study some properties of this function:

1. $\Gamma(1+x) = x\Gamma(x)$, (1.12)

2. $\Gamma(1+n) = n!$,

   $\Gamma(1) = \Gamma(2) = 1$ for integer number n,

3. $n! = 1 \cdot 2 \cdot 3 \cdot \ldots \cdot n$, $0! = 1$,

4. $(2n+1)!! = 1 \cdot 3 \cdot \ldots \cdot (2n+1)$,

5. $(2n)!! = 2 \cdot 4 \cdot \ldots \cdot (2n)$,

6. $\dbinom{p}{n} = \dfrac{p(p-1)\ldots(p-n+1)}{1 \cdot 2 \cdot \ldots \cdot n}$, $\dbinom{p}{0} = 1$,

7. $\dfrac{(2n+1)!!}{(2n+1)!} = \dfrac{1}{2^n}\dfrac{1}{n!}$ (1.13)

   and $\dfrac{(2n-1)!!}{(2n)!!} = \dfrac{(2n-1)!}{2^{2n-1}(n!)(n-1)!}$.

   The main properties of the gamma-function:

8. $\Gamma(2x) = \dfrac{2^{2x-1}}{\sqrt{\pi}}\Gamma(x)\Gamma\left(x+\dfrac{1}{2}\right)$, (1.14)

9. $\Gamma(x)\Gamma(1-x) = \dfrac{\pi}{\sin \pi x}$, (1.15)

10. $\Gamma\left(\dfrac{1}{2}+x\right)\Gamma\left(\dfrac{1}{2}-x\right) = \dfrac{\pi}{\cos \pi x}$. (1.16)

    Since $\cos 0 = 1$, we find from (1.16), $\Gamma\left(\dfrac{1}{2}\right) = \sqrt{\pi}$.

11. We use the following identities:

    $\Gamma\left(-\dfrac{1}{2}\right) = \dfrac{\left(-\frac{1}{2}\right)\Gamma\left(-\frac{1}{2}\right)}{\left(-\frac{1}{2}\right)} = \dfrac{\Gamma\left(1-\frac{1}{2}\right)}{\left(-\frac{1}{2}\right)} = \dfrac{\Gamma\left(\frac{1}{2}\right)}{\left(-\frac{1}{2}\right)} = \dfrac{\sqrt{\pi}}{\left(-\frac{1}{2}\right)}$.

    So that

    $\Gamma\left(-\dfrac{1}{2}\right) = -2\sqrt{\pi}$. (1.17)

12. For natural n, we have

    $\Gamma\left(n+\dfrac{1}{2}\right) = \dfrac{\sqrt{\pi}}{2^n}(2n-1)!!$, (1.18)

    $\Gamma\left(\dfrac{1}{2}-n\right) = (-1)^n\dfrac{2^n\sqrt{\pi}}{(2n-1)!!}$, (1.19)

    $\dfrac{\Gamma\left(p+n+\frac{1}{2}\right)}{\Gamma\left(p-n+\frac{1}{2}\right)} = \dfrac{(4p^2-1^2)(4p^2-3^2)\ldots[4p^2-(2n-1)^2]}{2^{2n}}$. (1.20)

From (1.14), it follows that

$$\Gamma\left(-\frac{1}{4}\right) = -\frac{4\pi\sqrt{2}}{\Gamma\left(\frac{1}{4}\right)} \tag{1.21}$$

and

$$\Gamma\left(\frac{3}{4}\right) = \frac{\pi\sqrt{2}}{\Gamma\left(\frac{1}{4}\right)}. \tag{1.22}$$

For gamma-functions, there exist some useful relations:

13. $\displaystyle\int\limits_{0}^{\pi} d\theta(\sin\theta)^k = \sqrt{\pi}\,\frac{\Gamma\left(\frac{k+1}{2}\right)}{\Gamma\left(\frac{k+2}{2}\right)}.$

14. $\displaystyle\Gamma(-n+\varepsilon) = \frac{(-1)^n}{n!}\left[\frac{1}{\varepsilon} + \left(1 + \frac{1}{2} + \ldots + \frac{1}{n} - C\right) + O(\varepsilon)\right],$

where C is the Euler constant. The last formula means that for integers n, gamma-function $\Gamma(-n)$ has the single pole of the type of $(1/\varepsilon)$.

1.2.5   Psi-Function $\Psi(x)$

In the Mellin representation for any functions, there exists $\Psi(x)$-function and its definition is

$$\Psi(x) = \frac{d}{dx}\ln\Gamma(x). \tag{1.23}$$

Its functional relations take the forms:

1. $\displaystyle\Psi(x+1) = \Psi(x) + \frac{1}{x},$ (1.24)

2. $\displaystyle\Psi(x+n) = \Psi(x) + \sum_{k=0}^{n-1}\frac{1}{x+k},$ (1.25)

3. $\displaystyle\Psi(n+1) = -C + \sum_{k=1}^{n}\frac{1}{k},$ (1.26)

4. $\Psi(1-z) = \Psi(z) + \pi\cot\pi z,$ (1.27)

5. $\displaystyle\Psi\left(\frac{1}{2}+z\right) = \Psi\left(\frac{1}{2}-z\right) + \pi\tan\pi z.$ (1.28)

Its particular quantities are:

1. $\Psi(1) = -C,$

2. $\displaystyle\Psi\left(\frac{1}{2}\right) = -C - 2\ln 2,$

3. $\Psi\left(\dfrac{1}{4}\right) = -C - \dfrac{\pi}{2} - 3\ln 2,$

4. $\Psi\left(\dfrac{3}{4}\right) = -C + \dfrac{\pi}{2} - 3\ln 2,$

5. $\Psi\left(\dfrac{1}{3}\right) = -C - \dfrac{\pi}{2}\sqrt{\dfrac{1}{3}} - \dfrac{3}{2}\ln 3,$

6. $\Psi\left(\dfrac{2}{3}\right) = -C + \dfrac{\pi}{2}\sqrt{\dfrac{1}{3}} - \dfrac{3}{2}\ln 3,$

7. $\Psi'\left(1\right) = \dfrac{\pi^2}{6},$

8. $\Psi'\left(\dfrac{1}{2}\right) = \dfrac{\pi^2}{2},$

where C is the Euler number

$$C = -\Psi(1) = 0.57721566490\ldots.$$

1.3 Calculation of Integrals in the Form of the Mellin Representation

The next step is to calculate an integral by using the Mellin representation like
(1.2). Let us consider a very simple integral

$$I_1 = \int_0^\infty dx\, e^{-x} = \frac{1}{2i} \int_{-\beta+i\infty}^{-\beta-i\infty} d\xi \frac{1}{\sin \pi\xi\, \Gamma(1+\xi)} \lim_{\varepsilon\to 0} \int_\varepsilon^\infty dx\, x^\xi, \tag{1.29}$$

where

$$i_1 = \lim_{\varepsilon\to 0} \int_\varepsilon^\infty x^\xi dx = -\lim_{\varepsilon\to 0} \frac{\varepsilon^{\xi+1}}{1+\xi}.$$

For this case, the integration contour should be move through the point $\xi = -1$,
i.e, $\beta = -1 - \delta,\ \delta > 0$.

In this case, the contour of integration has the form of Figure 1.2.

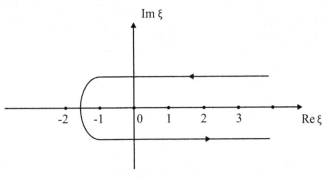

Figure 1.2

Thus,

$$I_1 = \lim_{\varepsilon \to 0} \frac{-1}{2i} \int\limits_{-\beta'+i\infty}^{-\beta'-i\infty} d\xi \frac{\varepsilon^{1+\xi}}{\sin \pi\xi \, \Gamma(2+\xi)},$$

where we have used the gamma-function property

$$\Gamma(1+\xi)(1+\xi) = \Gamma(2+\xi).$$

In this simple case, the residue at the point $\xi = -1$ gives a non zero result

$$Res f(\xi)_{\xi=-1} = \frac{2\pi i}{2i}(-1)\frac{1}{\pi \cos \pi\xi}|_{\xi=-1} = 1,$$

where $\Gamma(1) = 1$.

Finally, we get

$$I_1 = \int\limits_0^\infty dx e^{-x} = 1$$

as it should be. As seen above, our method is very simple and more natural.

1.4   The Mellin Representation of Functions

1.4.1   *The Exponential Function*

$$e^{-ax^\nu} = \frac{1}{2i} \int\limits_{-\beta+i\infty}^{-\beta-i\infty} d\xi \frac{a^\xi x^{\nu\xi}}{\sin \pi\xi \, \Gamma(1+\xi)}, \quad (-1 < \beta < 0). \tag{1.30}$$

1.4.2 The Trigonometric Functions (See Exercises in Section 1.4.8, Chapter 1)

1. $\sin^q(bx^\nu) = \dfrac{1}{2^{q-1}} \dfrac{1}{2i} \displaystyle\int\limits_{\alpha+i\infty}^{-\alpha-i\infty} d\xi \dfrac{(2bx^\nu)^{2\xi}}{\sin \pi\xi \; \Gamma(1+2\xi)} I_q(\xi),$ \hfill (1.31)

where $q = 2,\ 4,\ 6,\ldots,\ 0 < \alpha < 1,$

$$\left.\begin{aligned}
&I_2(\xi) = -1,\ I_4(\xi) = 2^{2\xi} - 4, \\
&I_6(\xi) = -3^{2\xi} + 6\cdot 2^{2\xi} - 15 \\
&\cdots\cdots\cdots\cdots\cdots\cdots\cdots\cdots\cdots\cdots\cdots\cdots\cdots
\end{aligned}\right\}. \qquad (1.32)$$

2. $\sin^m(bx^\nu) = \dfrac{1}{2^{m-1}} \dfrac{1}{2i} \displaystyle\int\limits_{-\beta+i\infty}^{-\beta-i\infty} d\xi \dfrac{(bx^\nu)^{2\xi+1}}{\sin \pi\xi \; \Gamma(2+2\xi)} N_m(\xi).$ \hfill (1.33)

Here $m = 1,\ 3,\ 5,\ 7,\ldots,\ -1 < \beta < 0,$

$$\left.\begin{aligned}
&N_1 = 1,\ N_3(\xi) = 3 - 3^{2\xi+1}, \\
&N_5(\xi) = 5^{2\xi+1} - 5\cdot 3^{2\xi+1} + 10, \\
&N_7(\xi) = -7^{2\xi+1} + 7\cdot 5^{2\xi+1} - 21\cdot 3^{2\xi+1} + 35 \\
&\cdots\cdots\cdots\cdots\cdots\cdots\cdots\cdots\cdots\cdots\cdots\cdots\cdots
\end{aligned}\right\}. \qquad (1.34)$$

3. $\cos^m(bx^\nu) = \dfrac{1}{2^{m-1}} \dfrac{1}{2i} \displaystyle\int\limits_{-\beta+i\infty}^{-\beta-i\infty} d\xi \dfrac{(bx^\nu)^{2\xi}}{\sin \pi\xi \; \Gamma(1+2\xi)} N'_m(\xi),$ \hfill (1.35)

where $m = 1,\ 3,\ 5,\ 7,\ldots,\ -1 < \beta < 0,$

$$\left.\begin{aligned}
&N'_1(\xi) = 1,\ N'_3(\xi) = 3 + 3^{2\xi}, \\
&N'_5(\xi) = 5^{2\xi} + 5\cdot 3^{2\xi} + 10, \\
&N'_7(\xi) = 7^{2\xi} + 7\cdot 5^{2\xi} + 21\cdot 3^{2\xi} + 35 \\
&\cdots\cdots\cdots\cdots\cdots\cdots\cdots\cdots\cdots\cdots\cdots\cdots\cdots
\end{aligned}\right\}. \qquad (1.36)$$

4. $\cos^q(bx^\nu) - 1 = \dfrac{1}{2^{q-1}} \dfrac{1}{2i} \displaystyle\int\limits_{\alpha+i\infty}^{-\alpha-i\infty} d\xi \dfrac{(2bx^\nu)^{2\xi}}{\sin \pi\xi \; \Gamma(1+2\xi)} I'_q(\xi),$ \hfill (1.37)

where $q = 2,\ 4,\ 6,\ldots,\ 0 < \alpha < 1,$

$$\left.\begin{aligned}
&I'_2(\xi) = 1,\ I'_4(\xi) = 2^{2\xi} + 4, \\
&I'_6(\xi) = 3^{2\xi} + 6\cdot 2^{2\xi} + 15 \\
&\cdots\cdots\cdots\cdots\cdots\cdots\cdots\cdots\cdots\cdots\cdots\cdots\cdots
\end{aligned}\right\}. \qquad (1.38)$$

1.4.3 The Cylindrical Functions

1. $J_\rho(ax^\nu) = \dfrac{1}{2i} \displaystyle\int\limits_{-\beta+i\infty}^{-\beta-i\infty} d\xi \dfrac{[ax^\nu/2]^{2\xi+\rho}}{\sin \pi\xi \; \Gamma(1+\xi)\; \Gamma(\rho+\xi+1)},$ \hfill (1.39)

$$(-1 < \beta < 0)$$

2. $J_\rho(ax^\nu)J_\mu(ax^\nu) = \dfrac{1}{2i} \displaystyle\int\limits_{-\beta+i\infty}^{-\beta-i\infty} d\xi \dfrac{\left(\frac{ax^\nu}{2}\right)^{\rho+\mu+2\xi}}{\sin \pi\xi \, \Gamma(\rho+\mu+\xi+1)}$

$$\times \dfrac{\Gamma(\rho+\mu+2\xi+1)}{\Gamma(\rho+\xi+1)\,\Gamma(\mu+\xi+1)}, \tag{1.40}$$

3. $K_0(ax^\nu) = -\dfrac{\pi}{2i} \displaystyle\int\limits_{-\beta+i\infty}^{-\beta-i\infty} d\xi \dfrac{\left(\frac{ax^\nu}{2}\right)^{2\xi}}{\sin^2 \pi\xi \, \Gamma^2(1+\xi)}. \tag{1.41}$

1.4.4   The Struve Function

$$\mathbf{H}_\rho(ax^\nu) = \dfrac{1}{2i} \int\limits_{-\beta+i\infty}^{-\beta-i\infty} d\xi \dfrac{(ax^\nu/2)^{2\xi+\rho+1}}{\sin \pi\xi \, \Gamma\left(\xi+\frac{3}{2}\right)\Gamma\left(\rho+\xi+\frac{3}{2}\right)}. \tag{1.42}$$

1.4.5   The $\beta(x)$-Function

$$\beta(ax^\nu) = \dfrac{1}{2i} \int\limits_{-\beta+i\infty}^{-\beta-i\infty} d\xi \dfrac{1}{\sin \pi\xi \, (\xi+ax^\nu)}. \tag{1.43}$$

1.4.6   The Incomplete Gamma-Function

Definition

$$\gamma(\alpha, x) = \int\limits_0^x e^{-t}t^{\alpha-1}dt, \quad \mathrm{Re}\,\alpha > 0,$$

$$\Gamma(\alpha, x) = \int\limits_x^\infty e^{-t}t^{\alpha-1}dt,$$

$$\gamma(\alpha, ax^\nu) = \dfrac{1}{2i} \int\limits_{-\beta+i\infty}^{-\beta-i\infty} d\xi \dfrac{(ax^\nu)^{\alpha+\xi}}{\sin \pi\xi \, \Gamma(1+\xi)\,(\alpha+\xi)}, \tag{1.44}$$

$$-\alpha < \beta < 0,$$

$$\Gamma(\alpha, ax^\nu) = \Gamma(\alpha) - \dfrac{1}{2i} \int\limits_{-\beta+i\infty}^{-\beta-i\infty} d\xi \dfrac{(ax^\nu)^{\alpha+\xi}}{\sin \pi\xi \, \Gamma(1+\xi)\,(\alpha+\xi)}, \tag{1.45}$$

$$-\alpha < \beta < 0.$$

1.4.7 The Probability Integral and Integrals of Frenel

Definition

$$\Phi(x) = \frac{2}{\sqrt{\pi}} \int_0^x dt\, e^{-t^2},$$

$$S(x) = \frac{2}{\sqrt{2\pi}} \int_0^x dt\, \sin t^2,$$

$$C(x) = \frac{2}{\sqrt{2\pi}} \int_0^x dt\, \cos t^2,$$

$$\Phi(ax^\nu) = -\frac{2}{\sqrt{\pi}} \int_{\alpha+i\infty}^{\alpha-i\infty} d\xi \frac{(ax^\nu)^{2\xi-1}}{\sin \pi\xi\, (2\xi-1)\, \Gamma(\xi)}, \tag{1.46}$$

$$S(ax^\nu) = \frac{2}{\sqrt{2\pi}} \int_{\beta'+i\infty}^{\beta'-i\infty} d\xi \frac{(ax^\nu)^{4\xi+3}}{\sin \pi\xi\, \Gamma(2\xi+2)\, (4\xi+3)}, \tag{1.47}$$

$$C(ax^\nu) = \frac{2}{\sqrt{2\pi}} \int_{\beta+i\infty}^{\beta-i\infty} d\xi \frac{(ax^\nu)^{4\xi+1}}{\sin \pi\xi\, \Gamma(1+2\xi)\, (4\xi+1)}, \tag{1.48}$$

where $\dfrac{1}{2} < \alpha < 1,\ -\dfrac{3}{4} < \beta' < 0,\ -\dfrac{1}{4} < \beta < 0.$

1.4.8 Exercises

Calculate expressions $I_q(\xi)$, $N_m(\xi)$, $N'_m(\xi)$ and $I'_q(\xi)$ for any numbers q and m in the equations (1.31), (1.33), (1.35) and (1.37) in this chapter.

Solutions are giving by the following formulas:

$$I_q(\xi) = \sum_{k=0}^{q/2-1} (-1)^{q/2-k} \left(\frac{q}{2} - k\right)^{2\xi} \binom{q}{k}, \tag{1.49}$$

where $q = 2,\ 4,\ 6,\ldots$

$$N_m(\xi) = \sum_{k=0}^{\frac{m+1}{2}-1} (-1)^{\frac{m+1}{2}+k-1} \binom{m}{k} (m-2k)^{1+2\xi}, \tag{1.50}$$

where $m = 1,\ 3,\ 5,\ 7,\ldots.$

$$N'_m(\xi) = \sum_{k=0}^{\frac{m+1}{2}-1} \binom{m}{k} (m-2k)^{2\xi}, \tag{1.51}$$

where $m = 1, \ 3, \ 5, \ 7, \ldots$

$$I_q'(\xi) = \sum_{k=0}^{q/2-1} \left(\frac{q}{2} - k\right)^{2\xi} \binom{q}{k},$$ (1.52)

where $q = 2, \ 4, \ 6, \ldots$.

    Here

$$\binom{q}{k} = \frac{q(q-1)\cdots(q-k+1)}{1\cdot 2\cdots\cdots k}, \qquad \binom{q}{0} = 1.$$

Chapter 2

Calculation of Integrals Containing Trigonometric and Power Functions

2.1 Derivation of General Unified Formulas

2.1.1 *The First General Formula*

First of all, we derive a general formula for the integrals of the type:

$$N_1(q, b, \nu, \gamma < 0) = \int_0^\infty dx\, x^\gamma \sin^q(bx^\nu). \tag{2.1}$$

By using the Mellin representation (1.31), one gets

$$N_1 = \frac{1}{2^{q-1}} \frac{1}{2i} \int_{\alpha+i\infty}^{\alpha-i\infty} d\xi \frac{(2b)^{2\xi}}{\sin \pi\xi} \frac{I_q(\xi)}{\Gamma(1+2\xi)} \Lambda_1(\gamma, \nu), \tag{2.2}$$

where

$$\Lambda_1(\gamma, \nu) = \lim_{\varepsilon \to 0} \int_\varepsilon^\infty dx\, x^{\gamma+2\nu\xi} = -\lim_{\varepsilon \to 0} \frac{\varepsilon^{\gamma+1+2\nu\xi}}{\gamma+1+2\nu\xi}.$$

Due to the limit $\varepsilon \to 0$, the integral (2.2) is calculated by means of the residue at the one point $\xi = -(\gamma+1)/(2\nu)$. The result reads

$$N_1 = \frac{1}{2^{q-1}} I_q\left(\xi = -\frac{\gamma+1}{2\nu}\right) \frac{\sqrt{\pi}}{2\nu} b^{-\frac{\gamma+1}{\nu}} \frac{\Gamma\left(\dfrac{\gamma+1}{2\nu}\right)}{\Gamma\left(\dfrac{1}{2} - \dfrac{\gamma+1}{2\nu}\right)}. \tag{2.3}$$

Here $I_q(\xi)$ is determined by expressions (1.32). This is the first general unified formula aimed to calculate an enormous number of integrals by an appropriate choice of the parameters q, γ, b, ν. It is important to notice that the upper and lower bounds (ranges of changing) of parameters γ, ν are established from the original integrals (2.1). Two cases $\gamma < 0$ and $\gamma > 0$ are studied differently.

2.1.2   The Second General Formula

Let us consider the second integral:

$$N_2(m, b, \nu, \gamma < 0) = \int\limits_0^\infty dx\, x^\gamma \sin^m(bx^\nu), \tag{2.4}$$

where $\sin^m(bx^\nu)$ is presented by the Mellin representation (1.33). As previously, similar calculation gives the second general formula

$$
\begin{aligned}
N_2 = \frac{1}{2^{m-1}} N_m &\left(\xi = -\frac{1}{2}\left[1 + \frac{\gamma+1}{\nu}\right] \right) \left(\frac{2}{b}\right)^{\frac{\gamma+1}{\nu}} \\
&\times \frac{\sqrt{\pi}}{2\nu} \frac{\Gamma\left[\dfrac{1}{2}\left(1 + \dfrac{\gamma+1}{\nu}\right)\right]}{\Gamma\left(1 - \dfrac{\gamma+1}{2\nu}\right)},
\end{aligned} \tag{2.5}
$$

where $m = 1,\ 3,\ 5,\ 7,\ldots$ and $N_m(\xi)$ is given by expression (1.34).

2.1.3   The Third General Formula

Now we calculate integral with cosine function.

$$N_3(m, \gamma, b, \nu) = \int\limits_0^\infty dx\, x^\gamma \cos^m(bx^\nu),$$

$$
N_3 = \frac{1}{2^{m-1}} N'_m \left(\xi = -\frac{\gamma+1}{2\nu} \right) \frac{\sqrt{\pi}}{2\nu} \left(\frac{2}{b}\right)^{\frac{\gamma+1}{\nu}} \frac{\Gamma\left(\dfrac{\gamma+1}{2\nu}\right)}{\Gamma\left(\dfrac{1}{2} - \dfrac{\gamma+1}{2\nu}\right)}, \tag{2.6}
$$

where $m = 1,\ 3,\ 5,\ 7,\ldots$ and $N'_m(\xi)$ is defined by the expressions (1.36).

2.1.4   The Fourth General Formula

It defines the following integral

$$N_4(q, \gamma, b, \nu) = \int\limits_0^\infty dx\, x^\gamma \left[\cos^q(bx^\nu) - 1\right].$$

Similar calculation gives

$$N_4 = \frac{1}{2^{q-1}} I_q' \left(\xi = -\frac{\gamma+1}{2\nu} \right) \frac{\sqrt{\pi}}{2\nu} b^{-\frac{\gamma+1}{\nu}} \frac{\Gamma\left(\dfrac{\gamma+1}{2\nu}\right)}{\Gamma\left(\dfrac{1}{2} - \dfrac{\gamma+1}{2\nu}\right)}. \tag{2.7}$$

Here $q = 2,\ 4,\ 6, \ldots$ and $I_q'(\xi)$ are given by the formulas (1.38).

2.2 Calculation of Concrete Particular Integrals Involving $x^{-\gamma}$ and Sine Functions

Now we are able to calculate any concrete integrals by means of these four general unified formulas.

From the unified formula (2.5) we obtain:

(1)

$$i_1 = \int\limits_0^\infty dx \frac{\sin(bx)}{x} = \frac{\sqrt{\pi}}{2} \Gamma\left(\frac{1}{2}\right) = \frac{\pi}{2}, \tag{2.8}$$

where $\gamma = -1,\ m = 1,\ \nu = 1$.

(2) Assuming $\gamma = 0,\ \nu = 2,\ m = 1$, then

$$i_2 = \int\limits_0^\infty dx \sin(bx^2) = \left(\frac{2}{b}\right)^{1/2} \frac{\sqrt{\pi}}{4} \frac{\Gamma\left(\frac{3}{4}\right)}{\Gamma\left(\frac{3}{4}\right)} = \sqrt{\frac{\pi}{2b}} \frac{1}{2}. \tag{2.9}$$

(3) Let $\gamma = -2,\ \nu = 2,\ m = 1$, then

$$i_3 = \int\limits_0^\infty dx \frac{\sin(bx^2)}{x^2} = \frac{\sqrt{\pi}}{4} \left(\frac{b}{2}\right)^{1/2} \frac{\Gamma\left(\frac{1}{4}\right)}{\Gamma\left(\frac{3}{4}\right)} = \sqrt{\frac{\pi b}{2}}. \tag{2.10}$$

(4) Let $\gamma = 0,\ \nu = 1,\ m = 1$, then

$$i_4 = \int\limits_0^\infty dx \sin(bx) = \frac{\sqrt{\pi}}{2} \left(\frac{2}{b}\right) \frac{\Gamma(1)}{\Gamma\left(\frac{1}{2}\right)} = \frac{1}{b}, \tag{2.11}$$

where $\Gamma\left(\frac{1}{2}\right) = \sqrt{\pi}$.

(5) Put $q = 2,\ \gamma = -4,\ \nu = 2,\ b \to a^2$, then from the unified formula (2.3) it follows

$$i_5 = \int\limits_0^\infty dx \frac{\sin^2(a^2x^2)}{x^4} = -\frac{\sqrt{\pi}}{8} a^3 \frac{\Gamma\left(-\frac{3}{4}\right)}{\Gamma\left(\frac{5}{4}\right)},$$

where

$$\Gamma\left(-\frac{3}{4}\right) = \frac{(-3/4)}{(-3/4)} \Gamma\left(-\frac{3}{4}\right) = -\frac{4}{3} \Gamma\left(1 - \frac{3}{4}\right) = -\frac{4}{3} \Gamma\left(\frac{1}{4}\right),$$

$$\Gamma\left(\frac{5}{4}\right) = \Gamma\left(1 + \frac{1}{4}\right) = \frac{1}{4}\Gamma\left(\frac{1}{4}\right).$$

So that

$$i_5 = \frac{2}{3}\sqrt{\pi}a^3. \tag{2.12}$$

(6) Assuming $m = 3$, $\gamma = -2$, $b \to a^2$, $\nu = 2$ and $N_3 = 3 - 3^{2\xi+1}$, one gets from the formula (2.5)

$$i_6 = \int_0^\infty dx \frac{\sin^3(a^2 x^2)}{x^2} = \frac{\sqrt{\pi}}{4}\frac{a}{\sqrt{2}}$$

$$\times \frac{\Gamma\left(\frac{1}{4}\right)}{\Gamma\left(\frac{5}{4}\right)}\frac{(3 - \sqrt{3})}{4} = \frac{\sqrt{2\pi}}{8}(3 - \sqrt{3})a. \tag{2.13}$$

Now let us calculate the following two integrals by using the Mellin representations (1.31) and (1.37) directly.

(7)

$$i_7 = \int_0^\infty \frac{dx}{x^4}(\sin x^2 - x^2 \cos x^2)$$

$$= \frac{1}{2i}\int_{-\alpha+i\infty}^{-\alpha-i\infty} d\xi \frac{1}{\sin \pi\xi \, \Gamma(1 + 2\xi)}\left[\frac{1}{1 + 2\xi} - 1\right]$$

$$\times \lim_{\varepsilon \to 0}\int_\varepsilon^\infty dx x^{4\xi-2} = -\frac{\pi}{4}\frac{1}{\sin\frac{\pi}{4}\,\Gamma\left(1 + \frac{1}{2}\right)}$$

$$\times \left(\frac{1}{1 + \frac{1}{2}} - 1\right) = \frac{1}{3}\sqrt{\frac{\pi}{2}}, \tag{2.14}$$

where

$$\sin\frac{\pi}{4} = \frac{1}{\sqrt{2}}, \quad \Gamma\left(\frac{1}{2}\right) = \sqrt{\pi}.$$

(8) The second integral is given by

$$i_8 = \int_0^\infty dx x^{-8}\left[\sin^2 x^2 - x^4 \cos x^2\right]$$

$$= \int_0^\infty dx x^{-8}\left[\sin^2 x^2 - x^4 - x^4(\cos^2 x^2 - 1)\right]$$

$$= -\frac{1}{2i}\lim_{\varepsilon \to 0}\left\{\int_{\alpha'+i\infty}^{\alpha'-i\infty} d\xi \int_\varepsilon^\infty dx x^{4\xi-8} + \int_{\alpha+i\infty}^{\alpha-i\infty} d\xi \int_\varepsilon^\infty dx x^{4\xi-4}\right\}$$

$$\times \frac{2^{2\xi-1}}{\sin \pi\xi \; \Gamma(1+2\xi)} = \frac{\pi}{4}\left[\frac{1}{\sin \frac{7}{4}\pi \; \Gamma\left(1+\frac{7}{2}\right)} \; \frac{2^{\frac{5}{2}}}{}\right.$$

$$+ \left. \frac{1}{\sin \frac{3\pi}{4} \; \Gamma\left(1+\frac{3}{2}\right)} \; \frac{2^{\frac{1}{2}}}{}\right] = \frac{38}{105}\sqrt{\pi}, \tag{2.15}$$

where

$$\sin \frac{3}{4}\pi = \sin\left(\pi - \frac{\pi}{4}\right) = \sin \pi \; \cos\frac{\pi}{4} - \cos \pi \; \sin\frac{\pi}{4} = \frac{1}{\sqrt{2}},$$

$$\sin \frac{7}{4}\pi = \sin\left(2\pi - \frac{\pi}{4}\right) = \sin 2\pi \; \cos\frac{\pi}{4} - \cos 2\pi \; \sin\frac{\pi}{4} = -\frac{1}{\sqrt{2}}$$

and

$$\Gamma\left(1+\frac{7}{2}\right) = \frac{7}{2}\frac{5}{2}\frac{3}{2}\frac{1}{2}\sqrt{\pi}.$$

(9) Let $m = 3$, $b \to a$, $\nu = 1$, $\gamma = -\delta$, then from (2.5) it follows:

$$i_9 = \int\limits_0^\infty dx\frac{\sin^3 ax}{x^\delta} = \frac{\sqrt{\pi}}{2}a^{-(1-\delta)}2^{-\delta+1}\frac{\Gamma\left(\frac{1}{2}(2-\delta)\right)}{\Gamma\left(1-\frac{1-\delta}{2}\right)}$$

$$\times \frac{1}{4}\left[3 - 3^{-2\frac{(1+1-\delta)}{2}+1}\right]$$

$$= \frac{1}{4}(3 - 3^{\delta-1})a^{\delta-1}\cos\frac{\pi\delta}{2}\;\Gamma(1-\delta), \tag{2.16}$$

where $a > 0$, $0 < \operatorname{Re} \delta < 2$ and the following relations are used:

$$\Gamma\left(\frac{1}{2}+\frac{\delta}{2}\right) = \frac{\pi}{\cos\frac{\pi\delta}{2}\;\Gamma\left(\frac{1}{2}-\frac{\delta}{2}\right)},$$

$$\Gamma\left(1-\frac{\delta}{2}\right)\Gamma\left(\frac{1}{2}-\frac{\delta}{2}\right) = \sqrt{\pi}\;2^\delta\;\Gamma(1-\delta).$$

(10) From the formula (2.16), one gets

$$i_{10} = \int\limits_0^\infty dx\frac{\sin^3 ax}{x} = \frac{\pi}{4}\operatorname{sign} a, \tag{2.17}$$

where we have used the limit

$$\lim_{\nu\to 1}\left[\cos\frac{\pi\nu}{2}\;\Gamma(1-\nu)\right] = \lim_{\nu\to 1}\cos\frac{\pi\nu}{2}\;\frac{\pi}{\sin\pi\nu\;\Gamma(\nu)}$$

$$= \lim_{\nu\to 1}\frac{\cos\frac{\nu\pi}{2}\;\pi}{2\sin\frac{\pi\nu}{2}\;\cos\frac{\pi\nu}{2}} = \frac{\pi}{2}.$$

(11) Let $m = 3$, $b \to a$, $\gamma = -2$ be in (2.5), then one gets

$$i_{11} = \int\limits_0^\infty dx\frac{\sin^3 ax}{x^2} = -\frac{a}{4}\lim_{\nu\to 2}\left[(3 - 3^{\nu-1})\;\Gamma(1-\nu)\right]$$

$$= -\frac{a}{4}\lim_{\nu\to 2}\left[\frac{(3 - 3^{\nu-1})}{\Gamma(\nu)\;\sin\pi\nu}\;\pi\right] = -\frac{\pi a}{4}\frac{1}{\Gamma(2)}\lim_{\nu\to 2}\left[\frac{3 - 3^{\nu-1}}{\sin\pi\nu}\right].$$

Now we use the L'Hôpital rule as above, then

$$i_{11} = -\frac{\pi a}{4}\frac{(-3^{-1})\;3^2\;\ln 3}{\pi \cos\pi\nu}\bigg|_{\nu=2} = \frac{3}{4}a\ln 3. \tag{2.18}$$

(12) Let us calculate the integral

$$
i_{12} = \int\limits_0^\infty dx \frac{\sin^3 ax}{x^3} = \left(-\frac{6}{4}\right) a^2 \lim_{\nu \to 3}\left[\cos\frac{\nu\pi}{2}\,\Gamma(1-\nu)\right]
$$

$$
= -\frac{3}{2}a^2 \lim_{\nu \to 3}\left[\cos\frac{\nu\pi}{2}\,\frac{\pi}{\sin\pi\nu\,\Gamma(\nu)}\right] = \frac{3}{8}a^2\pi,
\tag{2.19}
$$

where $a > 0$.

(13) Let

$$
i_{13} = \int\limits_0^\infty dx \frac{\sin^4 ax}{x^2},
$$

where $q = 4$, $b \to a$, $\nu = 1$, $\gamma = -2$,

$$
I_4 = 2^{2\xi} - 4.
$$

Then, from (2.3) it follows

$$
i_{13} = \frac{1}{2^3}(-2)\frac{\sqrt{\pi}}{2}a\frac{\Gamma(-\frac{1}{2})}{\Gamma(1)} = \frac{a\pi}{4},
\tag{2.20}
$$

where

$$
\Gamma\left(-\frac{1}{2}\right) = -2\sqrt{\pi}, \quad a > 0.
$$

(14) In the general formula (2.3), we put $q = 4$, $b \to a$, $\nu = 1$, $\gamma = -3$, $I_4 = 2^{2\xi} - 4$. Then, we have

$$
i_{14} = \int\limits_0^\infty dx \frac{\sin^4 ax}{x^3} = \frac{1}{8}\frac{\sqrt{\pi}}{2}a^2 \lim_{\varepsilon \to 1}\frac{\Gamma(-\varepsilon)}{\Gamma\left(\frac{3}{2}\right)}\left(2^{2\varepsilon} - 4\right),
$$

where

$$
\lim_{\varepsilon \to 1}\left(2^{2\varepsilon} - 4\right)\Gamma(-\varepsilon) = -\frac{\pi}{\Gamma(2)}\lim_{\varepsilon \to 1}\frac{2^{2\varepsilon} - 4}{\sin\pi\varepsilon}.
$$

Here we again use the L'Hôpital rule and get

$$
\lim_{\varepsilon \to 1}\left(2^{2\varepsilon} - 4\right)\Gamma(-\varepsilon) = -\pi\lim_{\varepsilon \to 1}\frac{2^{2\varepsilon}\ln 4}{\cos\pi\varepsilon} = 8\ln 2.
$$

Finally, we find

$$
i_{14} = \int\limits_0^\infty dx \frac{\sin^4 ax}{x^3} = a^2\ln 2.
\tag{2.21}
$$

(15) Similarly, for the case $q = 4$, $b \to a$, $\nu = 1$, $\gamma = -4$,

$$I_4 = 2^{2\xi} - 4,$$

one gets

$$i_{15} = \int\limits_0^\infty dx \frac{\sin^4 ax}{x^4} = \frac{\sqrt{\pi}}{16} a^3 \left(2^3 - 4\right) \frac{\Gamma\left(-\frac{3}{2}\right)}{\Gamma(2)} = \frac{1}{3}\pi a^3, \qquad (2.22)$$

where

$$\Gamma\left(-\frac{3}{2}\right) = \frac{(-3/2)}{(-3/2)}\Gamma\left(-\frac{3}{2}\right) = -\frac{2}{3}\Gamma\left(1 - \frac{3}{2}\right)$$

$$= -\frac{2}{3}\Gamma\left(-\frac{1}{2}\right) = \frac{4}{3}\sqrt{\pi}.$$

(16) Now we use the formula (2.5), and put in it $m = 5$, $\nu = 1$, $\gamma = -2$, $b \to a$,

$$N_5 = 5^{2\xi+1} - 5 \cdot 3^{2\xi+1} + 10,$$

then

$$i_{16} = \int\limits_0^\infty dx \frac{\sin^5 ax}{x^2} = \frac{\sqrt{\pi}a}{2^6} \frac{1}{\Gamma\left(\frac{3}{2}\right)} \lim_{\varepsilon \to 0} \Gamma(\varepsilon) \left[5^{2\varepsilon+1} - 5 \cdot 3^{2\varepsilon+1} + 10\right],$$

where

$$\lim_{\varepsilon \to 0} \Gamma(\varepsilon) \left[5^{2\varepsilon+1} - 5 \cdot 3^{2\varepsilon+1} + 10\right] = 10\left[3\ln 3 - \ln 5\right].$$

So that

$$i_{16} = \frac{5}{16} a(3\ln 3 - \ln 5). \qquad (2.23)$$

(17) Similar calculation for $m = 5$, $\nu = 1$, $\gamma = -3$,

$$N_5 = 5^{2\xi+1} - 5 \cdot 3^{2\xi+1} + 10$$

gives the integral

$$i_{17} = \int\limits_0^\infty dx \frac{\sin^5 ax}{x^3} = \frac{\sqrt{\pi}}{2^7} a^2 \frac{\Gamma\left(-\frac{1}{2}\right)}{\Gamma(2)} \left(5^2 - 5 \cdot 3^2 + 10\right) = \frac{5}{32}\pi a^2. \qquad (2.24)$$

(18) Assuming $m = 5$, $\gamma = -4$, $\nu = 1$, $b \to a$,

$$N_5 = 5^{2\xi+1} - 5 \cdot 3^{2\xi+1} + 10,$$

one gets the integral

$$i_{18} = \int\limits_0^\infty dx \frac{\sin^5 ax}{x^4} = \frac{\sqrt{\pi}}{2^8} a^3 \frac{1}{\Gamma\left(\frac{5}{2}\right)} \lim_{\varepsilon \to 1} \Gamma(-\varepsilon) \left(5^{2\varepsilon+1} - 5 \cdot 3^{2\varepsilon+1} + 10\right).$$

Thus,

$$i_{18} = \frac{5}{96} a^3 (25\ln 5 - 27\ln 3). \qquad (2.25)$$

(19) Similar calculation for $\gamma = -5$, $m = 5$, $\nu = 1$, $b \to a$ reads

$$i_{19} = \int_0^\infty dx \frac{\sin^5 ax}{x^5} = \frac{\sqrt{\pi}}{2^9} a^4 \frac{\Gamma\left(-\frac{3}{2}\right)}{\Gamma(3)} \left(5^4 - 5 \cdot 3^4 + 10\right) = \frac{115}{384} \pi a^4. \qquad (2.26)$$

(20) Now we use the formula (2.3). Assuming $q = 6$, $\gamma = -2$, $\nu = 1$, $b \to a$

$$I_6 = -3^{2\xi} + 6 \cdot 2^{2\xi} - 15,$$

one gets

$$i_{20} = \int_0^\infty dx \frac{\sin^6 ax}{x^2} = \frac{\sqrt{\pi}}{2^6} (-6) a \frac{\Gamma\left(-\frac{1}{2}\right)}{\Gamma(1)} = \frac{3}{16} \pi a, \qquad (2.27)$$

where $a > 0$.

(21) Assuming $q = 6$, $\gamma = -3$, $\nu = 1$, $b \to a$, one gets

$$i_{21} = \int_0^\infty dx \frac{\sin^6 ax}{x^3} = \frac{\sqrt{\pi}}{2^6} \frac{1}{\Gamma\left(\frac{3}{2}\right)} \lim_{\varepsilon \to 1} \Gamma(-\varepsilon) \left(-3^{2\varepsilon} + 6 \cdot 2^{2\varepsilon} - 15\right)$$

$$= \frac{3}{16} a^2 (8 \ln 2 - 3 \ln 3), \qquad (2.28)$$

where the limit is given by

$$\lim_{\varepsilon \to 1} \Gamma(-\varepsilon) \left(-3^{2\varepsilon} + 6 \cdot 2^{2\varepsilon} - 15\right) = -\pi \lim_{\varepsilon \to 1} \left(\frac{-3^{2\varepsilon} + 6 \cdot 2^{2\varepsilon} - 15}{\sin \pi \varepsilon}\right)$$

$$= 2(24 \ln 2 - 9 \ln 3).$$

Here, we have used the gamma function relation

$$\Gamma(-\varepsilon)\Gamma(1 + \varepsilon) = \frac{\pi}{-\sin \pi \varepsilon}.$$

(22) In the general formula (2.3), we put its parameters $q = 6$, $\gamma = -5$, $\nu = 1$, $b \to a$, then one gets the following integral:

$$i_{22} = \int_0^\infty dx \frac{\sin^6 ax}{x^5} = \frac{\sqrt{\pi}}{2^6} a^4 \frac{1}{\Gamma\left(\frac{5}{2}\right)} \lim_{\varepsilon \to 2} \Gamma(-\varepsilon) \left(-3^{2\varepsilon} + 6 \cdot 2^{2\varepsilon} - 15\right).$$

Here the limit takes the form

$$\lim_{\varepsilon \to 2} \Gamma(-\varepsilon) \left(-3^{2\varepsilon} + 6 \cdot 2^{2\varepsilon} - 15\right) = 3(27 \ln 3 - 32 \ln 2).$$

Therefore

$$i_{22} = \frac{1}{16} a^4 (27 \ln 3 - 32 \ln 2). \qquad (2.29)$$

(23) Putting $q = 6$, $\gamma = -6$, $\nu = 1$, $b \to a$ in (2.3), we have

$$i_{23} = \int_0^\infty dx \frac{\sin^6 ax}{x^6} = \frac{\sqrt{\pi}}{2^6} a^5 \frac{\Gamma\left(-\frac{5}{2}\right)}{\Gamma(3)} \left(-3^5 + 6 \cdot 2^5 - 15\right) = \frac{11}{30} \pi a^5, \qquad (2.30)$$

where $a > 0$ and according to the identities:

$$\left(-\frac{5}{2}\right) \Gamma\left(-\frac{5}{2}\right) = \Gamma\left(1 - \frac{5}{2}\right) = \Gamma\left(-\frac{3}{2}\right),$$

$$\left(-\frac{3}{2}\right) \Gamma\left(-\frac{3}{2}\right) = \Gamma\left(1 - \frac{3}{2}\right) = \Gamma\left(-\frac{1}{2}\right),$$

we have

$$\Gamma\left(-\frac{5}{2}\right) = \frac{(-5/2)(-3/2)}{(-5/2)(-3/2)} \Gamma\left(-\frac{5}{2}\right) = -\frac{8}{15}\sqrt{\pi}.$$

(24) The formula (2.5) with $\gamma = -1/2$, $m = 1$, $\nu = 1$, $b \to a$, $N_1(\xi) = 1$ reads

$$i_{24} = \int_0^\infty dx \frac{\sin ax}{\sqrt{x}} = \frac{\sqrt{\pi}}{2} \left(\frac{2}{a}\right)^{1/2} \frac{\Gamma\left(\frac{3}{4}\right)}{\Gamma\left(1 - \frac{1}{4}\right)} = \sqrt{\frac{\pi}{2a}}. \qquad (2.31)$$

(25) Assuming $m = 1$, $b \to a$, $\gamma = 0$, $N_1 = 1$ in the formula (2.5), we have

$$i_{25} = \int_0^\infty dx \sin(ax^\nu) = \frac{\sqrt{\pi}}{2\nu} \left(\frac{2}{a}\right)^{1/\nu} \frac{\Gamma\left(\frac{1}{2} + \frac{1}{\nu}\right)}{\Gamma\left(1 - \frac{1}{2\nu}\right)},$$

where

$$\bullet\; \Gamma\left(1 - \frac{1}{2\nu}\right) = \frac{\pi}{\sin\left(\frac{\pi}{2\nu}\right)\Gamma\left(\frac{1}{2\nu}\right)},$$

$$\bullet\; \Gamma\left(\frac{1}{2\nu}\right) \Gamma\left(\frac{1}{2} + \frac{1}{2\nu}\right) = \sqrt{\pi}\Gamma\left(\frac{1}{\nu}\right) 2^{1 - \frac{1}{\nu}}.$$

So that

$$i_{25} = \frac{\Gamma(1/\nu)}{\nu\, a^{1/\nu}} \sin\left(\frac{\pi}{2\nu}\right). \qquad (2.32)$$

(26) The unified formula (2.5) with $\gamma = 0$, $m = 7$, $\nu = 1$, $b \to a$,

$$N_7(\xi) = -7^{2\xi+1} + 7 \cdot 5^{2\xi+1} - 21 \cdot 3^{2\xi+1} + 35, \quad \xi = -\frac{1}{2}\left(1 + \frac{\gamma+1}{\nu}\right)$$

gives

$$i_{26} = \int_0^\infty dx \sin^7(ax) = \frac{\sqrt{\pi}}{a} \frac{\Gamma(1)}{\Gamma\left(\frac{1}{2}\right)} \frac{1}{2^6}$$

$$\times \left(-7^{-1} + 7 \cdot 5^{-1} - 21 \cdot 3^{-1} + 35\right) = \frac{16}{35} \frac{1}{a}. \qquad (2.33)$$

(27) The formula (2.5) with $\gamma = -4$, $m = 7$, $\nu = 2$, $b \to a$ reads

$$i_{27} = \int_0^\infty dx \frac{\sin^7(ax^2)}{x^4} = \frac{\sqrt{\pi}}{4} \left(\frac{a}{2}\right)^{3/2} \frac{\Gamma\left(-\frac{1}{4}\right)}{\Gamma\left(1+\frac{3}{4}\right)} \frac{1}{2^6} N_7,$$

where

$$N_7 = -7^{3/2} + 7 \cdot 5^{3/2} - 21 \cdot 3^{3/2} + 35 = -7 \left[9\sqrt{3} + \sqrt{7} - 5(\sqrt{5}+1)\right].$$

So that

$$i_{27} = \sqrt{\frac{\pi}{2}} \, a^{3/2} \frac{7}{32} \left[9\sqrt{3} + \sqrt{7} - 5(\sqrt{5}+1)\right]. \tag{2.34}$$

(28) Let $\gamma = -7$, $m = 7$, $\nu = 3$, $b \to a$, one gets

$$i_{28} = \int_0^\infty dx \frac{\sin^7(ax^3)}{x^7} = \frac{\sqrt{\pi}}{6} \left(\frac{a}{2}\right)^2 \frac{\Gamma\left(-\frac{1}{2}\right)}{\Gamma(2)} \frac{1}{2^6} N_7,$$

where

$$N_7 = -7^2 + 7 \cdot 5^2 - 21 \cdot 3^2 + 35 = -28.$$

So that

$$i_{28} = \frac{7}{192} \pi a^2. \tag{2.35}$$

(29) Parameters $\gamma = -7$, $m = 7$, $\nu = 2$, $b \to a$ in the formula (2.5) read:

$$i_{29} = \int_0^\infty dx \frac{\sin^7(ax^2)}{x^7} = \frac{\sqrt{\pi}}{2^{11}} \left(\frac{a}{2}\right)^3 \frac{1}{\Gamma\left(1+\frac{3}{2}\right)}$$

$$\times \lim_{\varepsilon \to 1} \Gamma(-\varepsilon) \left(-7^{2\varepsilon+1} + 7 \cdot 5^{2\varepsilon+1} - 21 \cdot 3^{2\varepsilon+1} + 35\right).$$

This limit takes the form

$$14[125 \ln 5 - 49 \ln 7 - 81 \ln 3].$$

Thus,

$$i_{29} = \frac{7}{768} a^3 \left[125 \ln 5 - 49 \ln 7 - 81 \ln 3\right]. \tag{2.36}$$

(30) The formula (2.5) with the parameters $\gamma = -\frac{1}{2}$, $\nu = \frac{1}{2}$, $b \to a$, $m = 7$ gives

$$i_{30} = \int_0^\infty \frac{dx}{\sqrt{x}} \sin^7(a\sqrt{x}) = 2\frac{\sqrt{\pi}}{a} \frac{\Gamma(1)}{\Gamma\left(\frac{1}{2}\right)} \frac{1}{2^6} N(-2),$$

where

$$N(-2) = -7^{-1} + 7 \cdot 5^{-1} - 21 \cdot 3^{-1} + 35 = \frac{1}{35} 2^{10}$$

so that

$$i_{30} = \frac{32}{35} \frac{1}{a}. \tag{2.37}$$

(31) Similarly, assuming $\gamma = -\frac{1}{2}$, $\nu = -\frac{1}{2}$, $b \to a$, $m = 7$, one gets

$$i_{31} = \int\limits_0^\infty \frac{dx}{\sqrt{x}} \sin^7\left(a\frac{1}{\sqrt{x}}\right) = -\sqrt{\pi}\frac{a}{2}\frac{1}{\Gamma\left(\frac{3}{2}\right)}\frac{1}{2^6}\lim_{\varepsilon\to 0}\Gamma(\varepsilon)N_7(\varepsilon).$$

Due to the L'Hôpital rule (1.5), this limit takes the form

$$\lim_{\varepsilon\to 0}\Gamma(\varepsilon)\left[-7^{2\varepsilon+1} + 7\cdot 5^{2\varepsilon+1} - 21\cdot 3^{2\varepsilon+1} + 35\right]$$
$$= 2(-7\ln 7 + 35\ln 5 - 63\ln 3).$$

Finally, we have

$$i_{31} = \frac{7}{32}a\left[\ln 7 - 5\ln 5 + 9\ln 3\right]. \tag{2.38}$$

(32) From the formula (2.3), it follows ($\gamma = -2$, $\nu = 1$, $b \to a$, $q = 2$):

$$i_{32} = \int\limits_0^\infty dx\frac{\sin^2 ax}{x^2} = \frac{\sqrt{\pi}}{4}a\frac{\Gamma\left(-\frac{1}{2}\right)}{\Gamma(1)}(-1) = \frac{\pi a}{2}. \tag{2.39}$$

(33) The choice of the parameters $m = 7$, $b \to a$, $\nu = -\frac{1}{4}$, $\gamma = -\frac{1}{8}$ entering in (2.5) leads to the integral:

$$i_{33} = \int\limits_0^\infty dx x^{-1/8}\sin^7(ax^{-1/4}) = -\frac{\sqrt{\pi}}{2^5}\left(\frac{a}{2}\right)^{7/2}\frac{\Gamma\left(-\frac{5}{4}\right)}{\Gamma\left(1+\frac{7}{4}\right)}N_7,$$

where

- $N_7 = -7^{\frac{7}{2}} + 7\cdot 5^{\frac{7}{2}} - 21\cdot 3^{\frac{7}{2}} + 35 = 7\left(-49\sqrt{7} + 125\sqrt{5} - 81\sqrt{3} + 5\right),$

- $\dfrac{\Gamma\left(-\frac{5}{4}\right)}{\Gamma\left(1+\frac{7}{4}\right)} = \dfrac{2^8}{3\cdot 5\cdot 7}.$

Thus,

$$i_{33} = \sqrt{\frac{\pi}{2}}\frac{1}{15}a^{7/2}\left(49\sqrt{7} + 81\sqrt{3} - 125\sqrt{5} - 5\right). \tag{2.40}$$

2.3 Integrals Involving $x^{-\gamma}$, Sine and Cosine Functions

From the general formula (2.6) it follows:

(34) $\gamma = -\frac{1}{2}$, $\nu = 1$, $b \to p$, $m = 1$:

$$i_{34} = \int\limits_0^\infty dx\frac{1}{\sqrt{x}}\cos(px) = \frac{\sqrt{\pi}}{2}\left(\frac{2}{p}\right)^{1/2}\frac{\Gamma\left(\frac{1}{4}\right)}{\Gamma\left(\frac{1}{4}\right)} = \sqrt{\frac{\pi}{2p}}. \tag{2.41}$$

(35) $\gamma = -\frac{1}{2}$, $\nu = 1$, $b \to p$, $m = 3$:

$$i_{35} = \int\limits_0^\infty dx \frac{1}{\sqrt{x}} \cos^3(px) = \frac{\sqrt{\pi}}{2} \left(\frac{2}{p}\right)^{1/2} \frac{1}{2^2} N_3'$$

$$= \frac{1}{2^2} \sqrt{\frac{\pi}{2p}} \left(3 + \frac{1}{\sqrt{3}}\right). \tag{2.42}$$

(36) $\gamma = -\frac{1}{2}$, $\nu = 1$, $b \to p$, $m = 5$:

$$i_{36} = \int\limits_0^\infty dx \frac{1}{\sqrt{x}} \cos^5(px) = \frac{\sqrt{\pi}}{2} \left(\frac{2}{p}\right)^{1/2} \frac{1}{2^4} \left(10 + 5\frac{1}{\sqrt{3}} + \frac{1}{\sqrt{5}}\right)$$

$$= \frac{1}{2^4} \sqrt{\frac{\pi}{2p}} \left(10 + 5\frac{1}{\sqrt{3}} + \frac{1}{\sqrt{5}}\right). \tag{2.43}$$

(37) $\gamma = -\frac{1}{2}$, $\nu = 1$, $b \to p$, $m = 7$:

$$i_{37} = \int\limits_0^\infty dx \frac{1}{\sqrt{x}} \cos^7(px) = \frac{1}{2^6} \sqrt{\frac{\pi}{2p}} \left(35 + 21\frac{1}{\sqrt{3}} + 7\frac{1}{\sqrt{5}} + \frac{1}{\sqrt{7}}\right). \tag{2.44}$$

Thus, by the induction rule, one gets:

(38) $\gamma = -\frac{1}{2}$, $\nu = 1$, $b \to p$, $m = 2n + 1$, $n = 0, 1, 2, 3, \ldots$:

$$i_{38} = \int\limits_0^\infty dx \frac{1}{\sqrt{x}} \cos^{2n+1}(px)$$

$$= \frac{1}{2^{2n}} \sqrt{\frac{\pi}{2p}} \sum_{k=0}^n \binom{2n+1}{n+k+1} \frac{1}{\sqrt{2k+1}}, \tag{2.45}$$

where

$$\binom{p}{n} = \frac{p(p-1)\ldots(p-(n-1))}{1 \cdot 2 \cdot \ldots \cdot n}, \qquad \binom{p}{0} = 1.$$

For example, when $n = 1$ we have

$$\sum_{k=0}^1 \binom{3}{2+k} \frac{1}{\sqrt{2k+1}} = \binom{3}{2} \frac{1}{\sqrt{1}} + \binom{3}{3} \frac{1}{\sqrt{3}} = 3 + \frac{1}{\sqrt{3}},$$

where

$$\binom{3}{2} = \frac{3 \cdot 2}{2} = 3, \qquad \binom{3}{3} = \frac{3 \cdot 2 \cdot 1}{1 \cdot 2 \cdot 3} = 1.$$

If $n = 2$, one gets

$$\sum_{k=0}^2 \binom{5}{3+k} \frac{1}{\sqrt{2k+1}} = \binom{5}{3} \frac{1}{\sqrt{1}} + \binom{5}{4} \frac{1}{\sqrt{3}}$$

$$+ \binom{5}{5} \frac{1}{\sqrt{5}} = 10 + 5\frac{1}{\sqrt{3}} + \frac{1}{\sqrt{5}}.$$

Because of

$$\binom{5}{3} = \frac{5 \cdot 4 \cdot 3}{1 \cdot 2 \cdot 3} = 10, \qquad \binom{5}{4} = \frac{5 \cdot 4 \cdot 3 \cdot 2}{1 \cdot 2 \cdot 3 \cdot 4} = 5,$$

and

$$\binom{5}{5} = \frac{5 \cdot 4 \cdot 3 \cdot 2 \cdot 1}{1 \cdot 2 \cdot 3 \cdot 4 \cdot 5} = 1$$

and etc.

(39) From the general formula (2.7), one gets:

$$i_{39} = \int_0^\infty dx \frac{1 - \cos(ax)}{x^2} = -\frac{\sqrt{\pi}}{2} \left(\frac{a}{2}\right) \frac{\Gamma\left(-\frac{1}{2}\right)}{\Gamma(1)} = \frac{\pi a}{2}. \tag{2.46}$$

(40) By the induction rule, it is easy to show that from (2.5) it follows

$$i_{40} = \int_0^\infty \frac{dx}{x} \sin^{2n+1} x = \frac{\pi}{2} \frac{(2n-1)!!}{(2n)!!}, \tag{2.47}$$

where definitions of symbols $(2n-1)!!$ and $(2n)!!$ are given in Section 1.2.4.

(41) The following integral with $\gamma = -5$, $\nu = 1$ and $m = 3$ has the form

$$i_{41} = \int_0^\infty dx \frac{x^3 - \sin^3 x}{x^5} = -\frac{\sqrt{\pi}}{27} \frac{\Gamma\left(-\frac{3}{2}\right)}{\Gamma(3)} (-13 \cdot 3 \cdot 2) = \frac{13}{32}\pi. \tag{2.48}$$

(42) Similarly, from (2.3) it follows

$$i_{42} = \int_0^\infty dx \frac{\sin^2(ax) - \sin^2(bx)}{x} = -\frac{\sqrt{\pi}}{4} \frac{1}{\Gamma\left(\frac{1}{2}\right)}$$

$$\times \lim_{\varepsilon \to 0} \Gamma(\varepsilon) \left(a^{-\varepsilon} - b^{-\varepsilon}\right) = \frac{1}{4} \ln\left(\frac{a}{b}\right). \tag{2.49}$$

(43) Moreover, it follows from (2.5) and (2.6) that

$$i_{43} = \int_0^\infty dx \frac{\sin(ax) - ax\cos(ax)}{x^3} = \frac{\sqrt{\pi}}{4} a^2 \Gamma\left(-\frac{1}{2}\right)(-1) = \frac{\pi a^2}{4},$$

$$\tag{2.50}$$

$a > 0$, where $\gamma = -3$, $\nu = 1$, $b \to a$, $m = 1$.

(44) If we choose the parameters in (2.5) and (2.6) as $\gamma = -2$, $\nu = 1$, $m = 1$, $q = 1$ and $b = 1$. Then

$$i_{44} = \int_0^\infty dx \frac{\sin x - x\cos x}{x^2} = \frac{\sqrt{\pi}}{4} \frac{1}{\Gamma\left(\frac{3}{2}\right)} 2 = 1. \tag{2.51}$$

(45) If we put in (2.6) $\gamma = -1$, $\nu = 1$, $m = 1$, then

$$i_{45} = \int\limits_0^\infty dx \frac{\cos(ax) - \cos(bx)}{x} = \frac{\sqrt{\pi}}{2} \frac{1}{\Gamma\left(\frac{1}{2}\right)}$$

$$\times \lim_{\varepsilon \to 0} \Gamma(\varepsilon) \left(a^{-\varepsilon} - b^{-\varepsilon}\right) = \frac{1}{2} \ln \frac{b}{a}. \tag{2.52}$$

(46) Putting $\gamma = -2$, $m = 1$ in (2.5), one gets

$$i_{46} = \int\limits_0^\infty dx \frac{a \sin(bx) - b \sin(ax)}{x^2} = \frac{\sqrt{\pi}}{4} a$$

$$\times \lim_{\varepsilon \to 0} \Gamma(\varepsilon) \left(b^\varepsilon\, a - b\, a^\varepsilon\right) = \frac{1}{2} ab\, \ln \frac{a}{b}. \tag{2.53}$$

(47) Let $\gamma = -2$, $\nu = 1$, $m = 1$ be in (2.6), then

$$i_{47} = \int\limits_0^\infty dx \frac{\cos(ax) - \cos(bx)}{x^2} = \frac{\sqrt{\pi}}{4} \frac{\Gamma\left(-\frac{1}{2}\right)}{\Gamma(1)}(a - b) = \frac{b - a}{2}\, \pi,$$

$$\tag{2.54}$$

where $a \geq 0$, $b \geq 0$.

(48) We put $\gamma = \mu - 1$, $b \to a$, $\nu = 1$, $m = 1$ in (2.6) and get

$$i_{48} = \int\limits_0^\infty dx\, x^{\mu-1} \cos(ax) = \frac{\sqrt{\pi}}{2} \left(\frac{2}{a}\right)^\mu \frac{\Gamma\left(\frac{\mu}{2}\right)}{\Gamma\left(\frac{1}{2} - \frac{\mu}{2}\right)},$$

where $a > 0$, $0 < \operatorname{Re} \mu < 1$;

$$\Gamma\left(\frac{\mu}{2}\right) \Gamma\left(\frac{1}{2} + \frac{\mu}{2}\right) = \sqrt{\pi}\, \frac{\Gamma(\mu)}{2^{\mu-1}}.$$

Thus,

$$i_{48} = \frac{\Gamma(\mu)}{a^\mu} \cos \frac{\pi \mu}{2}. \tag{2.55}$$

(49) From (2.6) with $\gamma = -\frac{1}{2}$, $b \to a$, $\nu = 1$, $m = 1$, one gets

$$i_{49} = \int\limits_0^\infty dx \frac{\cos(ax)}{\sqrt{x}} = \frac{\sqrt{\pi}}{2} \left(\frac{2}{a}\right)^{1/2} \frac{\Gamma\left(\frac{1}{4}\right)}{\Gamma\left(\frac{1}{4}\right)} = \sqrt{\frac{\pi}{2a}}. \tag{2.56}$$

(50) Similarly, if $m = 1$, $\gamma = 0$, $b \to a$, $\nu = p$ in (2.6), then

$$i_{50} = \int\limits_0^\infty dx \cos(ax^p) = \frac{\sqrt{\pi}}{2p} \left(\frac{a}{2}\right)^{-\frac{1}{p}} \frac{\Gamma\left(\frac{1}{2p}\right)}{\Gamma\left(\frac{1}{2} - \frac{1}{2p}\right)},$$

where $a > 0$, $p > 1$;

$$\Gamma\left(\frac{1}{2p}\right) \Gamma\left(\frac{1}{2} + \frac{1}{2p}\right) = \sqrt{\pi}\, \frac{\Gamma\left(\frac{1}{p}\right)}{2^{\frac{1}{p}-1}}.$$

So that

$$i_{50} = \frac{1}{p} \frac{1}{a^{1/p}} \cos \frac{\pi}{2p}\, \Gamma\left(\frac{1}{p}\right). \tag{2.57}$$

(51) If $q = 4$, $\gamma = 0$, $\nu = 2$, $b \to a$,

$$I'_4 = 2^{2\xi} + 4$$

in (2.7), then

$$i_{51} = \int_0^\infty dx \left[\cos^4(ax^2) - \cos^4(bx^2) \right]$$

$$= \int_0^\infty dx \left[\cos^4(ax^2) - 1 + 1 - \cos^4(bx^2) \right]$$

$$= \frac{\sqrt{\pi}}{2^6} \frac{\Gamma\left(\frac{1}{4}\right)}{\Gamma\left(\frac{1}{2} - \frac{1}{4}\right)} \left(\sqrt{2} + 8\right) \left(a^{-1/2} - b^{-1/2}\right).$$

Therefore

$$i_{51} = \frac{1}{64} \left(\sqrt{2} + 8\right) \left(\sqrt{\frac{\pi}{a}} - \sqrt{\frac{\pi}{b}}\right). \qquad (2.58)$$

(52) We choose $\gamma = -2$, $\nu = 1$, $m = 1$ in the formula (2.6) and obtain

$$i_{52} = \int_0^\infty dx \frac{\cos(ax) - \cos(bx)}{x^2} = \frac{\sqrt{\pi}}{4} \frac{\Gamma\left(-\frac{1}{2}\right)}{\Gamma(1)} (a - b)$$

$$= \frac{b - a}{2} \pi. \qquad (2.59)$$

(53) By means of the formula (2.3) with $q = 4$, $\nu = 2$, $\gamma = 0$, $b \to a$, one gets

$$i_{53} = \int_0^\infty dx \left[\sin^4(ax^2) - \sin^4(bx^2) \right]$$

$$= \frac{1}{2^5} \sqrt{\pi} \left(\frac{1}{\sqrt{2}} - 4\right) \frac{\Gamma\left(\frac{1}{4}\right)}{\Gamma\left(\frac{1}{2} - \frac{1}{4}\right)} \left(a^{-1/2} - b^{-1/2}\right).$$

Thus,

$$i_{53} = \frac{1}{64} \left(8 - \sqrt{2}\right) \left(\sqrt{\frac{\pi}{b}} - \sqrt{\frac{\pi}{a}}\right). \qquad (2.60)$$

(54) Using (2.7) with $q = 2$, $\nu = 2$, $\gamma = 0$, one gets

$$i_{54} = \int_0^\infty dx \left(\cos^2(ax^2) - \cos^2(bx^2) \right) \qquad (2.61)$$

$$= \int_0^\infty dx \left[(\cos^2(ax^2) - 1) + (1 - \cos^2(bx^2)) \right]$$

$$= \frac{\sqrt{\pi}}{8} \frac{\Gamma\left(\frac{1}{4}\right)}{\Gamma\left(\frac{1}{2} - \frac{1}{4}\right)} \left(a^{-1/2} - b^{-1/2}\right) = \frac{1}{8} \left(\sqrt{\frac{\pi}{a}} - \sqrt{\frac{\pi}{b}}\right).$$

(55) Similarly, from (2.3) with $q = 2$, $\gamma = 0$, $\nu = 2$, one obtains

$$i_{55} = \int_0^\infty dx \left[\sin^2(ax^2) - \sin^2(bx^2) \right] = -\frac{\sqrt{\pi}}{8} \frac{\Gamma\left(\frac{1}{4}\right)}{\Gamma\left(\frac{1}{2} - \frac{1}{4}\right)}$$

$$\times \left(a^{-1/2} - b^{-1/2} \right) = \frac{1}{8} \left(\sqrt{\frac{\pi}{b}} - \sqrt{\frac{\pi}{a}} \right). \tag{2.62}$$

(56) From (2.5) with $\gamma = -7$, $\nu = 3$, $m = 7$, $b \to a$, one gets

$$i_{56} = \int_0^\infty dx x^{-7} \sin^7(ax^3) = \frac{\sqrt{\pi}}{3 \cdot 2^9} a^2 \frac{\Gamma\left(-\frac{1}{2}\right)}{\Gamma(2)} N_7,$$

where

$$N_7 = -7^2 + 7 \cdot 5^2 - 21 \cdot 3^2 + 35.$$

So that

$$i_{56} = \frac{7}{192} \pi a^2. \tag{2.63}$$

(57) Assuming $q = 4$, $\gamma = 1$, $\nu = 1$ in (2.7), one gets

$$i_{57} = \int_0^\infty dx \frac{\cos^4(bx) - 1}{x} = \frac{\sqrt{\pi}}{4} \frac{\Gamma(0)}{\Gamma\left(\frac{1}{2}\right)} = \infty. \tag{2.64}$$

(58) Similarly for $q = 4$, $\gamma = -2$, $\nu = 1$, we have

$$i_{58} = \int_0^\infty dx \frac{\cos^4(bx) - 1}{x^2} = \frac{3}{8} \sqrt{\pi} b \frac{\Gamma\left(-\frac{1}{2}\right)}{\Gamma(1)} = -\frac{3}{4} \pi b, \tag{2.65}$$

where

$$I_4' = 2^{2\xi} + 4 = 6.$$

(59) Moreover, for $q = 4$, $\gamma = -3$, $\nu = 1$, one gets

$$i_{59} = \int_0^\infty dx \frac{\cos^4(bx) - 1}{x^3} = \frac{\sqrt{\pi}}{2} b^2 \frac{\Gamma(-1)}{\Gamma\left(\frac{3}{2}\right)} = \infty, \tag{2.66}$$

where

$$I_4' = 2^{2\xi} + 4 = 2^2 + 4 = 8.$$

(60) For $q = 4$, $\gamma = -4$, $\nu = 1$, $2\xi = 3$

$$I_4' = 2^{2\xi} + 4 = 2^3 + 4 = 12,$$

one gets

$$i_{60} = \int_0^\infty dx \frac{\cos^4(bx) - 1}{x^4} = \frac{1}{2^3} 12 \frac{\sqrt{\pi}}{2} b^3 \frac{\Gamma\left(-\frac{3}{2}\right)}{\Gamma(2)} = \pi b^3. \tag{2.67}$$

(61) Let $q = 4$, $\gamma = -1$, $\nu = 2$ in (2.7), then

$$i_{61} = \int_0^\infty dx \frac{\cos^4(bx^2) - 1}{x} = \infty.$$

(62) For $q = 4$, $\gamma = -2$, $\nu = 2$, $2\xi = \frac{1}{2}$,

$$I_4' = 2^{2\xi} + 4 = 4 + \sqrt{2},$$

one gets

$$i_{62} = \int_0^\infty dx \frac{\cos^4(bx^2) - 1}{x^2} = \frac{1}{2^3}\left(\sqrt{2} + 4\right)\frac{\sqrt{\pi}}{4} b^{1/2}$$

$$\times \frac{\Gamma\left(-\frac{1}{4}\right)}{\Gamma\left(\frac{1}{2} + \frac{1}{4}\right)} = -\frac{1}{8}\sqrt{\pi b}\left(4 + \sqrt{2}\right). \tag{2.68}$$

(63) Assuming $q = 4$, $\gamma = -3$, $\nu = 2$, where $2\xi = 1$,

$$I_4' = 2^{2\xi} + 4 = 6,$$

we have

$$i_{63} = \int_0^\infty dx \frac{\cos^4(bx^2) - 1}{x^3} = \frac{1}{2^3} 6 \frac{\sqrt{\pi}}{4} b \frac{\Gamma\left(-\frac{1}{2}\right)}{\Gamma(1)} = -\frac{3}{8}\pi b. \tag{2.69}$$

(64) Let $q = 4$, $\gamma = -4$, $\nu = 2$, where

$$I_4' = 2^{2\xi} + 4 = 2\left(\sqrt{2} + 2\right),$$

one gets

$$I_{64} = \int_0^\infty dx \frac{\cos^4(bx^2) - 1}{x^4} = \frac{1}{2^3} 2\left(\sqrt{2} + 2\right)\frac{\sqrt{\pi}}{4} b^{3/2}$$

$$\times \frac{\Gamma\left(-\frac{3}{4}\right)}{\Gamma\left(\frac{5}{4}\right)} = -\frac{1}{3}\left(\sqrt{2} + 2\right)\sqrt{\pi}\, b^{3/2}\Gamma\left(\frac{1}{4}\right). \tag{2.70}$$

(65) In the case $q = 4$, $\gamma = -1$, $\nu = 3$, one gets

$$i_{65} = \int_0^\infty dx \frac{\cos^4(bx^3) - 1}{x} = \infty.$$

(66) For the case $q = 4$, $\gamma = -2$, $\nu = 3$, the general formula (2.7) reads

$$i_{66} = \int_0^\infty dx \frac{\cos^4(bx^3) - 1}{x^2} = -\frac{1}{8}\left(\sqrt[3]{2} + 4\right)\sqrt{\pi}\, b^{1/3}\frac{\Gamma\left(\frac{5}{6}\right)}{\Gamma\left(\frac{2}{3}\right)}, \tag{2.71}$$

where

$$\frac{\Gamma\left(\frac{5}{6}\right)}{\Gamma\left(\frac{1}{6}\right)} = \frac{\Gamma^2\left(\frac{2}{3}\right)}{\Gamma^2\left(\frac{1}{3}\right)}\frac{1}{\sqrt[3]{2}}.$$

(67) Let $q = 4$, $\gamma = -3$, $\nu = 3$ for which

$$2\xi = \frac{2}{3}, \quad I_4' = 2^{2\xi} + 4 = 2^{2/3} + 4,$$

one gets

$$i_{67} = \int\limits_0^\infty dx \frac{\cos^4(bx^3) - 1}{x^3} = \frac{1}{2^3}\left(\sqrt[3]{4} + 4\right) \frac{\sqrt{\pi}}{6} b^{2/3} \frac{\Gamma\left(-\frac{1}{3}\right)}{\Gamma\left(\frac{5}{6}\right)},$$

where

$$\bullet \; \Gamma\left(\frac{1}{6}\right) = \sqrt{\pi}\, \Gamma\left(\frac{1}{3}\right) \frac{1}{\Gamma\left(\frac{2}{3}\right)}\, 2^{2/3},$$

$$\bullet \; \Gamma\left(\frac{5}{6}\right) = \sqrt{\pi}\, \Gamma\left(\frac{2}{3}\right) \frac{1}{\Gamma\left(\frac{1}{3}\right)}\, 2^{1/3}.$$

So that

$$\bullet \; \frac{\Gamma\left(\frac{5}{6}\right)}{\Gamma\left(\frac{1}{6}\right)} = \frac{\Gamma^2\left(\frac{2}{3}\right)}{\Gamma^2\left(\frac{1}{3}\right)} \frac{1}{\sqrt[3]{2}},$$

$$\bullet \; \Gamma\left(\frac{2}{3}\right) = \frac{2^{2\frac{1}{3}-1}}{\sqrt{\pi}} \Gamma\left(\frac{1}{3}\right) \Gamma\left(\frac{5}{6}\right).$$

Thus,

$$i_{67} = -\frac{1}{16}\left(\sqrt[3]{4} + 4\right) b^{2/3} \frac{1}{\sqrt[3]{2}} \Gamma\left(\frac{1}{3}\right). \tag{2.72}$$

(68) Assuming $q = 4$, $\gamma = -4$, $\nu = 3$, where $2\xi = 1$, $2^{2\xi} + 4 = 6$, one gets

$$i_{68} = \int\limits_0^\infty dx \frac{\cos^4(bx^3) - 1}{x^4} = \frac{1}{2^3}\, 6\, \frac{\sqrt{\pi}}{6}\, b\, \frac{\Gamma\left(-\frac{1}{2}\right)}{\Gamma(1)}$$

$$= -\frac{1}{4}\pi b. \tag{2.73}$$

(69) Now let us calculate concrete integrals involving $\cos^6(bx^\nu) - 1$, where

$$I_6' = 3^{2\xi} + 6 \cdot 2^{2\xi} + 15.$$

Assuming $q = 6$, $\gamma = -1$, then for any ν, all such type integrals are equal to infinity. It is seen from the formula (2.7). Thus,

$$i_{69} = \int\limits_0^\infty dx \frac{\cos^6(bx^\nu) - 1}{x} = \infty. \tag{2.74}$$

(70) Let $q = 6$, $\gamma = -2$, $\nu = 1$, where

$$2\xi = 1, \quad I_6' = 30,$$

one gets

$$i_{70} = \int\limits_0^\infty dx \frac{\cos^6(bx) - 1}{x^2} = \frac{1}{2^5}\, 30\, \frac{\sqrt{\pi}}{2}\, b\, \frac{\Gamma\left(-\frac{1}{2}\right)}{\Gamma(1)} = -\frac{15}{16}\pi b. \tag{2.75}$$

(71) The choice $q = 6$, $\gamma = -2$, $\nu = 1/2$ in (2.7), where

$$2\xi = 2, \quad I_6' = 48,$$

reads

$$i_{71} = \int\limits_0^\infty dx \frac{\cos^6\left(b\sqrt{x}\right) - 1}{x^2} = \frac{1}{2^5}\, 48\, \frac{\sqrt{\pi}}{1}\, b^2\, \frac{\Gamma(-1)}{\Gamma\left(\frac{3}{2}\right)}$$

$$= \frac{1}{2^5}\, 48\, \frac{\sqrt{\pi}}{1}\, b^2\, \frac{\Gamma(-1)}{\Gamma\left(\frac{3}{2}\right)} = \infty. \tag{2.76}$$

(72) For $q = 6$, $\gamma = -2$, $\nu = 2$, where

$$2\xi = \frac{1}{2}, \quad I_6' = \sqrt{3} + 6\sqrt{2} + 15,$$

one gets

$$i_{72} = \int\limits_0^\infty dx \frac{\cos^6(bx^2) - 1}{x^2}$$

$$= \frac{1}{2^5}\left(\sqrt{3} + 6\sqrt{2} + 15\right) \frac{\sqrt{\pi}}{4}\, b^{1/2}\, \frac{\Gamma\left(-\frac{1}{4}\right)}{\Gamma\left(\frac{3}{4}\right)}$$

$$= -\frac{\sqrt{\pi b}}{32}\left(\sqrt{3} + 6\sqrt{2} + 15\right). \tag{2.77}$$

(73) For the case $\gamma = -2$, $q = 6$, $\nu = -\frac{1}{2}$, where

$$2\xi = -2, \quad I_4' = \frac{299}{18},$$

one gets from (2.7)

$$i_{73} = \int\limits_0^\infty dx \frac{\cos^6\left(\frac{b}{\sqrt{x}}\right) - 1}{x^2}$$

$$= \frac{1}{2^5}\, \frac{299}{18}\, \frac{\sqrt{\pi}}{(-1)}\, b^{-2}\, \frac{\Gamma(1)}{\Gamma\left(-\frac{1}{2}\right)} = \frac{299}{9}\, \frac{1}{2^7}\, b^{-2}. \tag{2.78}$$

(74) It is easy to show that

$$i_{74} = \int\limits_0^\infty dx \frac{\cos^6(bx) - 1}{x^3} = \infty, \tag{2.79}$$

$$i_{74}' = \int\limits_0^\infty dx \frac{\cos^6\left(bx^{1/2}\right) - 1}{x^3} = \infty. \tag{2.80}$$

(75) Let $q = 6$, $\gamma = -3$, $\nu = 2$, where

$$2\xi = 1, \quad I_6' = 30,$$

then from (2.7) we have

$$i_{75} = \int_0^\infty dx \frac{\cos^6(bx^2) - 1}{x^3}$$

$$= \frac{1}{2^5} \, 30 \, \frac{\sqrt{\pi}}{4} \, b \, \frac{\Gamma\left(-\frac{1}{2}\right)}{\Gamma(1)} = -\frac{15}{32}\pi b. \tag{2.81}$$

(76) Assuming $q = 6$, $\gamma = -4$, $\nu = 2$, where

$$2\xi = \frac{3}{2}, \quad I_6' = 3\left(\sqrt{3} + 4\sqrt{2} + 5\right),$$

one gets from (2.7)

$$i_{76} = \int_0^\infty dx \frac{\cos^6(bx^2) - 1}{x^4}$$

$$= \frac{1}{2^5} \, 3\left(\sqrt{3} + 4\sqrt{2} + 5\right) \, \frac{\sqrt{\pi}}{4} \, b^{3/2} \, \frac{\Gamma\left(-\frac{3}{4}\right)}{\Gamma\left(\frac{1}{2} + \frac{3}{4}\right)}$$

$$= -\frac{\sqrt{\pi}}{8}\left(\sqrt{3} + 4\sqrt{2} + 5\right) \, b^{3/2}. \tag{2.82}$$

(77) It is obvious that

$$i_{77} = \int_0^\infty dx \frac{\cos^6(bx^2) - 1}{x^5} = \infty. \tag{2.83}$$

(78) Now let us calculate the integral

$$i_{78} = \int_0^\infty dx \frac{\cos^6(bx^2) - 1}{x^6} = \frac{\sqrt{\pi}}{10} \, b^{5/2} \left(3\sqrt{3} + 8\sqrt{2} + 5\right), \tag{2.84}$$

where we have used the gamma-function properties

$$\bullet \; \Gamma\left(-\frac{1}{4}\right) = \left(-\frac{5}{4}\right)\Gamma\left(-\frac{5}{4}\right),$$

$$\bullet \; \left(-\frac{1}{4}\right)\Gamma\left(-\frac{1}{4}\right) = \Gamma\left(\frac{3}{4}\right).$$

(79) Let $q = 6$, $\gamma = -\frac{1}{4}$, $\nu = 1$, where

$$2\xi = -\frac{3}{4}, \quad I_6' = 3^{-\frac{3}{4}} + 6 \cdot 2^{-\frac{3}{4}} + 15,$$

one gets from the formula (2.7)

$$i_{79} = \int_0^\infty dx \frac{\cos^6(bx) - 1}{\sqrt[4]{x}}$$

$$= \frac{1}{2^5}\left(\frac{1}{\sqrt[4]{27}} + \frac{6}{\sqrt[4]{8}} + 15\right) \frac{\sqrt{\pi}}{2} \, b^{-3/2} \, \frac{\Gamma\left(-\frac{3}{8}\right)}{\Gamma\left(\frac{7}{8}\right)},$$

where

$$\bullet \left(-\frac{3}{8}\right) \Gamma\left(-\frac{3}{8}\right) = \Gamma\left(\frac{5}{8}\right),$$

$$\bullet \frac{\Gamma\left(\frac{5}{8}\right)}{\Gamma\left(\frac{7}{8}\right)} = \frac{2^{3/4}}{\sqrt{\pi}} \Gamma\left(\frac{1}{4}\right) \cos\frac{3\pi}{8}.$$

So that

$$i_{79} = -\frac{1}{24}\left(\frac{1}{\sqrt[4]{27}} + \frac{6}{\sqrt[4]{8}} + 15\right)\left(\frac{2}{b}\right)^{3/4} \Gamma\left(\frac{1}{4}\right) \cos\frac{3\pi}{8}. \tag{2.85}$$

(80) We would like to calculate some concrete integrals by using the formula (2.6). Assuming $m = 3$, $\gamma = -\frac{1}{2}$, $\nu = 1$ in this formula (2.6), where

$$2\xi = -\frac{1}{2}, \quad N_3' = 3 + 3^{-1/2} = 3 + \frac{1}{\sqrt{3}},$$

one gets

$$i_{80} = \int_0^\infty dx \frac{\cos^3(bx)}{\sqrt{x}}$$

$$= \frac{1}{2^2}\left(3 + \frac{1}{\sqrt{3}}\right) \frac{\sqrt{\pi}}{2} \left(\frac{b}{2}\right)^{-1/2} \frac{\Gamma\left(\frac{1}{4}\right)}{\Gamma\left(\frac{1}{2} - \frac{1}{4}\right)}$$

$$= \frac{1}{8}\left(3 + \frac{1}{\sqrt{3}}\right) \sqrt{\frac{2\pi}{b}}. \tag{2.86}$$

(81) It is obvious that

$$i_{81} = \int_0^\infty dx \frac{\cos^3(bx) - 1}{x} = \infty. \tag{2.87}$$

(82) Putting $m = 3$, $\gamma = -2$, $\nu = 1$ in (2.6), where

$$2\xi = 1, \quad N_3' = 6,$$

one gets

$$i_{82} = \int_0^\infty dx \frac{\cos^3(bx) - 1}{x^2}$$

$$= \frac{1}{2^2} 6 \frac{\sqrt{\pi}}{2} \left(\frac{b}{2}\right) \frac{\Gamma\left(-\frac{1}{2}\right)}{\Gamma(1)} = -\frac{3}{4}\pi b. \tag{2.88}$$

(83) Let $m = 3$, $\gamma = -4$, $\nu = 2$ in (2.6), where

$$2\xi = \frac{3}{2}, \quad N_3' = 3 + 3\sqrt{3} = 3\left(1 + \sqrt{3}\right),$$

one gets

$$i_{83} = \int_0^\infty dx \frac{\cos^3(bx^2) - 1}{x^4} = \frac{1}{2} 3\left(1 + \sqrt{3}\right) \frac{\sqrt{\pi}}{4}$$

$$\times \left(\frac{b}{2}\right)^{3/2} \frac{\Gamma\left(-\frac{3}{4}\right)}{\Gamma\left(\frac{5}{4}\right)} = -\frac{b}{2} \sqrt{\frac{\pi b}{2}}\left(1 + \sqrt{3}\right). \tag{2.89}$$

(84) The choice $m = 3$, $\gamma = -\frac{3}{2}$, $\nu = 2$, where

$$2\xi = \frac{1}{4}, \quad N_3' = 3 + \sqrt[4]{3}$$

leads to the integral

$$i_{84} = \int\limits_0^\infty dx \frac{\cos^3(bx^2) - 1}{x^{3/2}} = \frac{1}{2^2}\left(3 + \sqrt[4]{3}\right)\frac{\sqrt{\pi}}{4}\left(\frac{b}{2}\right)^{1/4}\frac{\Gamma\left(-\frac{1}{8}\right)}{\Gamma\left(\frac{5}{8}\right)},$$

where

$$\left(-\frac{1}{8}\right)\Gamma\left(-\frac{1}{8}\right) = \Gamma\left(\frac{7}{8}\right)$$

and

$$\frac{\Gamma\left(\frac{7}{8}\right)}{\Gamma\left(\frac{5}{8}\right)} = \frac{\sqrt{\pi}}{2^{3/4}\,\Gamma\left(\frac{1}{4}\right)\cos\frac{3\pi}{8}}.$$

So that

$$i_{84} = -\frac{\pi}{4}\left(3 + \sqrt[4]{3}\right)\sqrt[4]{b}\,\frac{1}{\Gamma\left(\frac{1}{4}\right)\cos\frac{3\pi}{8}}. \tag{2.90}$$

(85) The parameters $\gamma = -\frac{1}{4}$, $\nu = 1$, $m = 7$ with

$$2\xi = -\frac{3}{4}, \quad N_7' = \frac{1}{\sqrt[4]{343}} + \frac{7}{\sqrt[4]{125}} + \frac{21}{\sqrt[4]{27}} + 35$$

in the formula (2.6) lead to

$$i_{85} = \int\limits_0^\infty dx \frac{\cos^7(bx)}{x^{1/4}} = \frac{1}{2^6}\left(\frac{1}{\sqrt[4]{343}} + \frac{7}{\sqrt[4]{125}} + \frac{21}{\sqrt[4]{27}} + 35\right)$$

$$\times \frac{\sqrt{\pi}}{2}\left(\frac{b}{2}\right)^{-3/4}\frac{\Gamma\left(\frac{3}{8}\right)}{\Gamma\left(\frac{7}{8}\right)},$$

where

$$\frac{\Gamma\left(\frac{3}{8}\right)}{\Gamma\left(\frac{7}{8}\right)} = \frac{\sqrt{\pi}\,\Gamma\left(\frac{3}{4}\right)}{2^{-1/4}\,\Gamma^2\left(\frac{7}{8}\right)},$$

$$\Gamma\left(\frac{7}{8}\right) = \frac{\pi}{\Gamma\left(\frac{1}{8}\right)\sin\frac{\pi}{8}}.$$

Thus,

$$i_{85} = \frac{1}{2^6}\left[\frac{1}{\sqrt[4]{343}} + \frac{7}{\sqrt[4]{125}} + \frac{21}{\sqrt[4]{27}} + 35\right]$$

$$\times \frac{\sqrt{2}}{\sqrt[4]{b^3}}\frac{\sin^2\left(\frac{\pi}{8}\right)}{\Gamma\left(\frac{1}{4}\right)}\Gamma^2\left(\frac{1}{8}\right). \tag{2.91}$$

(86) It is obvious that

$$i_{86} = \int\limits_0^\infty dx \frac{\cos^7(bx^4) - 1}{x^9} = \infty. \tag{2.92}$$

(87) The integral with the parameters $\gamma = -\frac{1}{2}$, $\nu = -\frac{1}{2}$, $m = 7$ is easy to calculate

$$i_{87} = \int\limits_0^\infty dx \frac{\cos^7\left(b\frac{1}{\sqrt{x}}\right)}{\sqrt{x}}$$

$$= \frac{1}{2^6} \, 7 \cdot 4 \cdot 5 \frac{\sqrt{\pi}}{(-1)} \left(\frac{b}{2}\right) \frac{\Gamma\left(-\frac{1}{2}\right)}{\Gamma(1)} = \frac{35}{16}\pi b. \tag{2.93}$$

(88) From the formula (2.6), it follows that

$$i_{88} = \int\limits_0^\infty dx \frac{\cos^7\left(bx^{-1/3}\right) - 1}{\sqrt[3]{x}} = \infty. \tag{2.94}$$

(89) Now let us calculate the following integral arising from the formula (2.6) with the parameters: $\gamma = -\frac{3}{2}$, $\nu = -\frac{1}{3}$, $m = 7$, where

$$2\xi = -\frac{3}{2}, \quad N_7' = \frac{1}{7\sqrt{7}} + \frac{7}{5\sqrt{5}} + \frac{21}{3\sqrt{3}} + 35,$$

$$i_{89} = \int\limits_0^\infty dx \frac{\cos^7\left(bx^{-1/3}\right) - 1}{x^{3/2}} = \frac{1}{2^6} N_7' \frac{\sqrt{\pi}}{2\left(-\frac{1}{3}\right)} \left(\frac{b}{2}\right)^{-3/2} \frac{\Gamma\left(\frac{3}{4}\right)}{\Gamma\left(\frac{5}{4}\right)},$$

where

$$\bullet \, \Gamma\left(\frac{5}{4}\right) = \Gamma\left(1 + \frac{1}{4}\right) = \frac{1}{4}\Gamma\left(\frac{1}{4}\right),$$

$$\bullet \, \Gamma\left(\frac{3}{4}\right) = \frac{\pi}{\sin\frac{\pi}{4}} \frac{1}{\Gamma\left(\frac{1}{4}\right)} = \frac{\pi\sqrt{2}}{\Gamma\left(\frac{1}{4}\right)}.$$

So that

$$i_{89} = -\frac{3}{8}\left(\frac{1}{7\sqrt{7}} + \frac{7}{5\sqrt{5}} + \frac{7}{\sqrt{3}} + 35\right)\left(\frac{\pi}{b}\right)^{3/2} \frac{1}{\Gamma^2\left(\frac{1}{4}\right)}. \tag{2.95}$$

(90) The integral arising from the formula (2.6) with parameters $\gamma = -7$, $\nu = 3$, $m = 7$ is diverged.

$$i_{90} = \int\limits_0^\infty dx \frac{\cos^7(bx^3) - 1}{x^7} = \infty. \tag{2.96}$$

(91) The formula (2.6) with the parameters $\gamma = -7$, $\nu = -3$, $m = 7$ leads to the integral:

$$i_{91} = \int\limits_0^\infty dx \frac{\cos^7(bx^{-3}) - 1}{x^7} = \frac{1}{2^6} \frac{2161 \cdot 2^6}{3 \cdot 5^2 \cdot 7^2}$$

$$\times \frac{\sqrt{\pi}}{(-6)}\left(\frac{b}{2}\right)^{-2} \frac{\Gamma(1)}{\Gamma\left(-\frac{1}{2}\right)} = \frac{2161}{3^2 \cdot 5^2 \cdot 7^2}\frac{1}{b^2}. \tag{2.97}$$

(92) Finally, we want to calculate the following integral with the parameters $\gamma = -11$, $\nu = -\frac{5}{2}$, $m = 7$. That is

$$i_{92} = \int_0^\infty dx \frac{\cos^7(bx^{-5/2}) - 1}{x^{11}} = \frac{2^6}{2^6} \frac{22329151}{3^3 \cdot 5^4 \cdot 7^4}$$

$$\times \frac{\sqrt{\pi}}{(-5)} \left(\frac{b}{2}\right)^{-4} \frac{\Gamma(2)}{\Gamma\left(-\frac{3}{2}\right)} = -\frac{22329151}{3^2 \cdot 5^5 \cdot 7^4} \frac{4}{b^4}. \tag{2.98}$$

(93) Now we derive some integrals from (2.6) with $m = 5$, where

$$N_5' = 5^{2\xi} + 5 \cdot 3^{2\xi} + 10.$$

Let $\gamma = -\frac{1}{2}$, $\nu = 1$, $m = 5$, then

$$i_{93} = \int_0^\infty dx \frac{\cos^5(bx)}{\sqrt{x}}$$

$$= \frac{1}{2^4} \left(\frac{1}{\sqrt{5}} + \frac{5}{\sqrt{3}} + 10\right) \frac{\sqrt{\pi}}{2} \left(\frac{b}{2}\right)^{-1/2} \frac{\Gamma\left(\frac{1}{4}\right)}{\Gamma\left(\frac{1}{2} - \frac{1}{4}\right)}$$

$$= \frac{1}{32} \left(\frac{1}{\sqrt{5}} + \frac{5}{\sqrt{3}} + 10\right) \sqrt{\frac{2\pi}{b}}. \tag{2.99}$$

(94) The integral with parameters $\gamma = -\frac{1}{2}$, $\nu = \frac{1}{2}$, $m = 5$ goes to zero

$$i_{94} = \int_0^\infty dx \frac{\cos^5(b\sqrt{x})}{\sqrt{x}} = 0. \tag{2.100}$$

(95) But the following integral is easily calculated.

$$i_{95} = \int_0^\infty dx \frac{\cos^5(bx^{-1/2})}{\sqrt{x}}$$

$$= \frac{1}{2^4} 30 \frac{\sqrt{\pi}}{(-1)} \left(\frac{b}{2}\right) \frac{\Gamma\left(-\frac{1}{2}\right)}{\Gamma(1)} = \frac{15}{8} \pi b. \tag{2.101}$$

(96) Let $\gamma = -\frac{3}{2}$, $\nu = 1$, $m = 5$, then

$$i_{96} = \int_0^\infty dx \frac{\cos^5(bx) - 1}{x^{3/2}} = -\frac{1}{8} \left(\sqrt{5} + 5\sqrt{3} + 10\right) \sqrt{\frac{\pi b}{2}}. \tag{2.102}$$

(97) If we put $\gamma = -\frac{3}{2}$, $\nu = \frac{1}{2}$, $m = 5$ in the main formula (2.6), then

$$i_{97} = \int_0^\infty dx \frac{\cos^5(b\sqrt{x}) - 1}{x^{3/2}}$$

$$= \frac{1}{2^4} 30 \sqrt{\pi} \left(\frac{b}{2}\right) \frac{\Gamma\left(-\frac{1}{2}\right)}{\Gamma(1)} = -\frac{15}{8} \pi b. \tag{2.103}$$

(98) Let us calculate integral arising from the formula (2.6) with the parameters $\gamma = -\frac{5}{2}$, $\nu = 1$, $m = 5$. That is

$$i_{98} = \int\limits_0^\infty dx \frac{\cos^5(bx) - 1}{x^{5/2}}$$

$$= \frac{1}{2^4} \, 5 \left(\sqrt{5} + 3\sqrt{3} + 2 \right) \frac{\sqrt{\pi}}{2} \left(\frac{b}{2} \right)^{3/2} \frac{\Gamma\left(-\frac{3}{4}\right)}{\Gamma\left(\frac{5}{4}\right)}$$

$$= -\frac{5}{12} \left(\sqrt{5} + 3\sqrt{3} + 2 \right) b \sqrt{\frac{\pi b}{2}}. \tag{2.104}$$

(99) Let $\gamma = -3$, $\nu = 1$, $m = 5$, then

$$i_{99} = \int\limits_0^\infty dx \frac{\cos^5(bx) - 1}{x^3} = \infty.$$

(100) Let $\gamma = -3$, $\nu = 2$, $m = 5$, then

$$i_{100} = \int\limits_0^\infty dx \frac{\cos^5(bx^2) - 1}{x^3} = -\frac{15}{32} \pi b. \tag{2.105}$$

(101) Let $\gamma = -\frac{7}{2}$, $\nu = 5$, $m = 5$, then

$$i_{101} = \int\limits_0^\infty dx \frac{\cos^5(bx^5) - 1}{x^{7/2}}$$

$$= \frac{1}{2^4} \left(\sqrt{5} + 5\sqrt{3} + 10 \right) \frac{\sqrt{\pi}}{10} \left(\frac{b}{2} \right)^{1/2} \frac{\Gamma\left(-\frac{1}{4}\right)}{\Gamma\left(\frac{3}{4}\right)}$$

$$= -\frac{1}{40} \left(\sqrt{5} + 5\sqrt{3} + 10 \right) \sqrt{\frac{\pi b}{2}}. \tag{2.106}$$

Finally, let us calculate some integrals:

$$T_1 = \int\limits_0^\infty dx \frac{\sin^{12}(bx^{-6})}{x^{13}}, \qquad T_2 = \int\limits_0^\infty dx \frac{\sin^{15}(bx^{40})}{x^{21}},$$

$$T_3 = \int\limits_0^\infty dx \frac{\cos^{19}(bx^{-25})}{x^{51}}, \qquad T_4 = \int\limits_0^\infty dx \frac{\cos^{20}(bx^{50}) - 1}{x^{51}},$$

$$T_5 = \int\limits_0^\infty dx \frac{\sin^{20}(bx^{-30})}{x^{61}}.$$

By using the general formulas (2.3), (2.5), (2.6) and (2.7), one gets

$$T_1 = \frac{1}{3 \cdot b^2 \cdot 2^{14}} I_{12}(-1), \qquad T_2 = \frac{1}{5 \cdot 2^{16}} \sqrt{\frac{\pi b}{2}} N_{15} \left(-\frac{1}{4} \right),$$

$$T_3 = \frac{1}{25 \cdot b^2 \cdot 2^{18}} N'_{19}(-1), \qquad T_4 = -\frac{\pi b}{25} \frac{1}{2^{20}} I'_{20} \left(\frac{1}{2} \right),$$

$$T_5 = \frac{1}{15 \cdot b^2 \cdot 2^{22}} \, I_{20}(-1),$$

where

$$I_{12}(-1) = \sum_{k=0}^{5} (-1)^{6-k} (6-k)^{-2} \binom{12}{k}$$

$$= -\frac{413413}{2^3 \cdot 3 \cdot 5^2} = -\frac{7^2 \cdot 11 \cdot 13 \cdot 59}{2^3 \cdot 3 \cdot 5^2},$$

$$N_{15}\left(-\frac{1}{4}\right) = \sum_{k=0}^{7} (-1)^{7+k} (15 - 2k)^{1/2} \binom{15}{k}$$

$$= 7 \cdot 11 \cdot 13 \left(3\sqrt{5} - 5\sqrt{3}\right) - \sqrt{15}$$

$$+ 3 \cdot 5 \left[2^3 \cdot 5 \cdot 13 + \sqrt{13} - 7 \left(\sqrt{11} + 13\sqrt{7}\right)\right],$$

$$N'_{19}(-1) = \sum_{k=0}^{9} (19 - 2k)^{-2} \binom{19}{k}$$

$$= \frac{2^5}{5^2 \cdot 7 \cdot 9} \frac{111391852837}{11^2 \cdot 13^2 \cdot 17^2 \cdot 19^2},$$

$$I'_{20}\left(\frac{1}{2}\right) = \sum_{k=0}^{9} (10 - k) \binom{20}{k} = 2 \cdot 3 \cdot 5 \cdot 7 \cdot 53 \cdot 83$$

and

$$I_{20}(-1) = \sum_{k=0}^{9} (-1)^{10-k} (10 - k)^{-2} \binom{20}{k} = -\frac{19 \cdot 37 \cdot 646528247}{2^3 \cdot 7^2 \cdot 9^2 \cdot 10^2}.$$

Last expressions are calculated by using the formulas (1.49)-(1.52) in Chapter 1.

Integrals Involving x^γ, $(p + tx^\rho)^{-\lambda}$, Sine and Cosine Functions

3.1 Derivation of General Unified Formulas for this Class of Integrals

3.1.1 *The Fifth General Formula*

For integrals of the type

$$N_5(q, b, p, t, \gamma, \rho, \lambda, \nu) = \int\limits_0^\infty dx \frac{x^\gamma}{[p + tx^\rho]^\lambda} \sin^q(bx^\nu),$$ (3.1)

one can derive the following general unified formula

$$
\begin{aligned}
N_5 = {} & \frac{1}{\rho} \frac{1}{\Gamma(\lambda)} \frac{1}{p^\lambda} \left(\frac{p}{t}\right)^{\frac{\gamma+1}{\rho}} \frac{1}{2^{q-1}} \frac{1}{2i} \int\limits_{\alpha+i\infty}^{\alpha-i\infty} d\xi \frac{(2b)^{2\xi}}{\sin \pi\xi\, \Gamma(1 + 2\xi)} \\
& \times I_q(\xi) \left(\frac{p}{t}\right)^{2\nu\xi/\rho} \Gamma\left(\frac{\gamma+1}{\rho} + \frac{2\nu\xi}{\rho}\right) \Gamma\left(\lambda - \frac{\gamma+1+2\nu\xi}{\rho}\right),
\end{aligned}
$$ (3.2)

where $q = 2, 4, 6, \ldots$, $0 < \alpha < 1$,

$$I_2 = -1, \quad I_4 = \left(2^{2\xi} - 4\right), \quad I_6 = -3^{2\xi} + 6 \cdot 2^{2\xi} - 15, \ldots.$$

Ranges (upper and lower bounds) of changing parameters $\gamma, \rho, \lambda, \nu$, are established from the original integrals (3.1). The cases $\gamma > 0$, $\gamma < 0$ are studied differently. Here, we have used the following integral

$$
\begin{aligned}
\Omega_1 = {} & \int\limits_0^\infty dx \frac{x^{\gamma+1+2\nu\xi-1}}{[p + tx^\rho]^\lambda} = \frac{1}{\Gamma(\lambda)} \frac{1}{\rho} \frac{1}{p^\lambda} \left(\frac{p}{t}\right)^{\frac{\gamma+1+2\nu\xi}{\rho}} \\
& \times \Gamma\left(\frac{\gamma+1}{\rho} + \frac{2\nu\xi}{\rho}\right) \Gamma\left(\lambda - \frac{\gamma+1+2\nu\xi}{\rho}\right).
\end{aligned}
$$ (3.3)

3.1.2   The Sixth General Formula

A similar unified formula exists

$$
\begin{aligned}
N_6 &= \frac{1}{\rho}\,\frac{1}{\Gamma(\lambda)}\,\frac{1}{p^\lambda}\left(\frac{p}{t}\right)^{\frac{\gamma+\nu+1}{\rho}}\frac{1}{2^{m-1}}\,\frac{1}{2i}\int\limits_{-\beta+i\infty}^{-\beta-i\infty}d\xi\frac{b^{2\xi+1}}{\sin\pi\xi\,\Gamma(2+2\xi)}\\
&\times N_m(\xi)\left(\frac{p}{t}\right)^{2\nu\xi/\rho}\Gamma\left(\frac{\gamma+1+\nu+2\nu\xi}{\rho}\right)\Gamma\left(\lambda-\frac{\gamma+\nu+1+2\nu\xi}{\rho}\right)
\end{aligned}
$$

(3.4)

for the following integrals

$$
N_6(m,b,p,t,\gamma,\rho,\lambda,\nu) = \int\limits_0^\infty dx\frac{x^\gamma}{[p+tx^\rho]^\lambda}\sin^m(bx^\nu).
$$

(3.5)

Here

$$
m = 1,\ 3,\ 5,\ldots,\quad -1 < \beta < 0
$$

and $N_m(\xi)$ is given by the formula (1.34) in Chapter 1.

3.1.3   The Seventh General Formula

For integrals

$$
N_7(m,b,p,t,\gamma,\rho,\lambda,\nu) = \int\limits_0^\infty dx\frac{x^\gamma}{[p+tx^\rho]^\lambda}\cos^m(bx^\nu),
$$

(3.6)

we have the following general formula

$$
\begin{aligned}
N_7 &= \frac{1}{\rho}\,\frac{1}{\Gamma(\lambda)}\,\frac{1}{p^\lambda}\left(\frac{p}{t}\right)^{\frac{\gamma+1}{\rho}}\frac{1}{2^{m-1}}\,\frac{1}{2i}\int\limits_{-\beta+i\infty}^{-\beta-i\infty}d\xi\frac{b^{2\xi}}{\sin\pi\xi\,\Gamma(1+2\xi)}\\
&\times N'_m(\xi)\left(\frac{p}{t}\right)^{2\nu\xi/\rho}\Gamma\left(\frac{\gamma+1}{\rho}+\frac{2\nu\xi}{\rho}\right)\Gamma\left(\lambda-\frac{\gamma+1+2\nu\xi}{\rho}\right),
\end{aligned}
$$

(3.7)

where $m = 1,\ 3,\ 5,\ldots,\ -1 < \beta < 0$ and $N'_m(\xi)$-functions are defined by the expressions (1.36) in Chapter 1.

3.1.4   *The Eighth General Formula*

This formula is given by

$$
N_8 = \frac{1}{\rho} \frac{1}{\Gamma(\lambda)} \frac{1}{p^\lambda} \left(\frac{p}{t}\right)^{\frac{\gamma+1}{\rho}} \frac{1}{2^{q-1}} \frac{1}{2i} \int\limits_{\alpha+i\infty}^{\alpha-i\infty} d\xi \frac{(2b)^{2\xi}}{\sin \pi\xi\, \Gamma(1 + 2\xi)}
$$

$$
\times I'_q(\xi) \left(\frac{p}{t}\right)^{2\nu\xi/\rho} \Gamma\left(\frac{\gamma+1}{\rho} + \frac{2\nu\xi}{\rho}\right) \Gamma\left(\lambda - \frac{\gamma+1+2\nu\xi}{\rho}\right) \tag{3.8}
$$

for the integrals

$$
N_8(q, b, p, t, \gamma, \rho, \lambda, \nu) = \int\limits_{0}^{\infty} dx \frac{x^\gamma}{[p + tx^\rho]^\lambda} [\cos^q(bx^\nu) - 1]. \tag{3.9}
$$

Here $q = 2, 4, 6, \ldots$ and $I'_q(\xi)$ are given by the expressions (1.38) in Chapter 1.

3.2   Calculation of Concrete Integrals

The calculation of particular integrals by means of the formulas (3.2), (3.4), (3.7) and (3.8) will encounter some difficulties with respect to the given integrals in the previous Chapter 2. But this procedure of taking out integrals does not give rise to any problems and we have nicely calculated them.

(1) The choice

$$
\gamma = 1, \; b \to a, \; p \to \beta^2, \; m = 1, \; \rho = 2,
$$

$$
\lambda = 1, \; m = 1, \; \nu = 1, \; t = 1, \; N_1(\xi) = 1
$$

in the formula (3.4) leads to an integral

$$
i_{102} = \int\limits_{0}^{\infty} dx \frac{x \sin(ax)}{\beta^2 + x^2} = \frac{1}{2} \frac{1}{\Gamma(1)} \beta a
$$

$$
\times \frac{1}{2i} \int\limits_{-\beta+i\infty}^{-\beta-i\infty} d\xi \frac{(a\beta)^{2\xi}}{\sin \pi\xi\, \Gamma(2 + 2\xi)} \Gamma\left(\frac{3}{2} + \xi\right) \Gamma\left(1 - \frac{3}{2} - \xi\right),
$$

where

$$
\bullet\; \Gamma\left(\frac{3}{2} + \xi\right) \Gamma\left(1 - \frac{3}{2} - \xi\right) = \frac{\pi}{\sin\left(\frac{3}{2} + \xi\right)\pi},
$$

$$
\bullet\; \sin \pi\left(\frac{3}{2} + \xi\right) = \sin \frac{3\pi}{2} \cos \pi\xi + \sin \pi\xi\, \cos \frac{3\pi}{2} = -\cos \pi\xi.
$$

So that

$$i_{102} = -\beta \, a \, \frac{\pi}{2i} \int\limits_{-\beta+i\infty}^{-\beta-i\infty} d\xi \frac{(a^2\beta^2)^\xi}{\sin 2\pi\xi \; \Gamma(2+2\xi)}.$$

After changing the integration variable

$$2\xi \to y - 1$$

and taking into account the identity:

$$\sin \pi(y-1) = \sin \pi y \, \cos \pi - \sin \pi \, \cos \pi y = -\sin \pi y,$$

one gets

$$i_{102} = \frac{\pi}{4i} \int\limits_{-\beta+i\infty}^{-\beta-i\infty} dy \frac{(a\beta)^y}{\sin \pi y \; \Gamma(1+y)} = \frac{\pi}{2} \sum_{n=0}^{\infty} (-1)^n \frac{(a\beta)^n}{n!}.$$

Thus,

$$i_{102} = \frac{\pi}{2} e^{-a\beta}, \quad a > 0, \; \operatorname{Re} \beta > 0. \tag{3.10}$$

(2) Assuming

$$\gamma = -1, \; b \to a, \; p \to \beta^2, \; m = 1,$$

$$\rho = 2, \; \lambda = 1, \; t = 1, \; \nu = 1, \; N_1 = 1$$

in the formula (3.4), one gets

$$i_{103} = \int\limits_{0}^{\infty} dx \frac{\sin(ax)}{x(\beta^2+x^2)} = \frac{a}{2\beta} \frac{1}{2i} \int\limits_{-\beta+i\infty}^{-\beta-i\infty} d\xi \frac{(a\beta)^{2\xi}}{\sin \pi\xi \; \Gamma(2+2\xi)}$$

$$\times \Gamma\left(\frac{1}{2}+\xi\right) \Gamma\left(\frac{1}{2}-\xi\right),$$

where

$$\Gamma\left(\frac{1}{2}+\xi\right) \Gamma\left(\frac{1}{2}-\xi\right) = \frac{\pi}{\cos \pi\xi}.$$

Again putting the integration variable

$$2\xi \to y - 1, \; \beta \to \beta', \; \frac{1}{2} < \beta' < 1,$$

one gets

$$i_{103} = \frac{\pi}{2\beta^2} \left(1 - e^{-a\beta}\right). \tag{3.11}$$

(3) We choose the parameters:

$$\gamma = -1, \ b \to a, \ p \to b^2, \ m = 1,$$

$$t = -1, \ \rho = 2, \ \lambda = 1, \ \nu = 1, \ N_1 = 1$$

in the formula (3.4) and obtain

$$i_{104} = \int\limits_0^\infty dx \frac{\sin(ax)}{x(\beta^2 - x^2)} = \frac{1}{2}\frac{1}{b^2}\left(\frac{b^2}{-1}\right)^{1/2}$$

$$\times \frac{1}{2i}\int\limits_{-\beta+i\infty}^{-\beta-i\infty} d\xi \frac{a^{2\xi+1}}{\sin \pi\xi \ \Gamma(2+2\xi)}\left(\frac{b^2}{-1}\right)^\xi \Gamma\left(\frac{1}{2}+\xi\right)\Gamma\left(\frac{1}{2}-\xi\right)$$

$$= \frac{\pi a}{bi}\frac{1}{2i}\int\limits_{-\beta+i\infty}^{-\beta-i\infty} d\xi \frac{(ab/i)^{2\xi}}{\sin 2\pi\xi \ \Gamma(2+2\xi)}.$$

Making use of the integration variable

$$\xi \to x - \frac{1}{2}, \ \beta \to \beta', \ \frac{1}{2} < \beta' < 1$$

and taking into account the following relations

- $\sin \pi(x - \frac{1}{2}) = \sin \pi x \ \cos \frac{\pi}{2} - \cos \pi x \ \sin \frac{\pi}{2} = -\cos \pi x,$

- $\cos \pi\left(x - \frac{1}{2}\right) = \cos \pi x \ \cos \frac{\pi}{2} + \sin \pi x \ \sin \frac{\pi}{2} = \sin \pi x,$

one gets

$$i_{104} = -\frac{\pi}{2bi}\frac{a}{2i}\int\limits_{-\beta'+i\infty}^{-\beta'-i\infty} dx \frac{(ab/i)^{2x-1}}{\sin \pi x \ \cos \pi x \ \Gamma(1+2x)},$$

where

$$\frac{1}{2} < \beta' < 1.$$

Now taking the residue at the point $x = n$, one obtains

$$i_{104} = -\frac{\pi}{2bi} a \frac{2\pi i}{2i}\frac{1}{\pi}\sum_{n=1}^\infty \frac{(ab/i)^{2n-1}}{(2n)!}$$

$$= -\frac{\pi}{2bi} a \frac{2\pi i}{2i\pi}\frac{i}{ab}\left[\cosh\left(\frac{ab}{i}\right) - 1\right].$$

Finally, we have

$$i_{104} = \int\limits_0^\infty dx \frac{\sin(ax)}{x(\beta^2 - x^2)} = \frac{\pi}{2b^2}\left[1 - \cos(ab)\right], \qquad (3.12)$$

where we have used the relation:

$$\cosh(-ix) = \cosh(ix) = \cos x.$$

(4) Using the main formula (3.7) and parameters:

$$m = 1, \ \gamma = 0, \ \nu = 1, \ b \to a, \ p \to \beta^2,$$

$$m = 1, \ \rho = 2, \ t = 1, \ \lambda = 1,$$

we have

$$i_{105} = \int_0^\infty dx \frac{\cos(ax)}{\beta^2 + x^2} = \frac{\pi}{2\beta} \frac{1}{2i} \int_{-\beta+i\infty}^{-\beta-i\infty} d\xi \frac{(a\beta)^{2\xi}}{\sin \pi\xi \ \cos \pi\xi \ \Gamma(1 + 2\xi)},$$

where we have used the relation:

$$\Gamma\left(\frac{1}{2} + \xi\right) \Gamma\left(\frac{1}{2} - \xi\right) = \frac{\pi}{\cos \pi\xi}.$$

Making use of the change $2\xi \to x$, one gets

$$i_{105} = \frac{\pi}{2\beta} \frac{1}{2i} \int_{-\beta+i\infty}^{-\beta-i\infty} \frac{dx}{\sin(\pi x)} \frac{(a\beta)^x}{\Gamma(1 + x)}$$

$$= \frac{\pi}{2\beta} \sum_{n=0}^\infty \frac{(-1)^n}{n!} (a\beta)^n = \frac{\pi}{2\beta} e^{-a\beta}. \tag{3.13}$$

(5) The case $m = 1, \ \gamma = 0, \ \nu = 1, \ b \to a, \ p \to b^2, \ t = -1, \ \rho = 2, \ \lambda = 1$ in the formula (3.7) reads

$$i_{106} = \int_0^\infty dx \frac{\cos(ax)}{b^2 - x^2}$$

$$= \frac{\pi}{2bi} \frac{1}{2i} \int_{-\beta+i\infty}^{-\beta-i\infty} d\xi \frac{(ab/i)^{2\xi}}{\sin \pi\xi \ \cos \pi\xi \ \Gamma(1 + 2\xi)},$$

where we have used the relation

$$\Gamma\left(\frac{1}{2} + \xi\right) \Gamma\left(\frac{1}{2} - \xi\right) = \frac{\pi}{\cos \pi\xi}.$$

Next we change variable $x + \frac{1}{2} = \xi$ and

$$\bullet \ \sin \pi \left(x + \frac{1}{2}\right) = \cos \pi x,$$

$$\bullet \ \cos \pi \left(x + \frac{1}{2}\right) = -\sin \pi x.$$

Then

$$i_{106} = \frac{\pi}{2bi}(-1) \frac{1}{2i} \int_{-\beta+i\infty}^{-\beta-i\infty} dx \frac{(ab/i)^{2x+1}}{\sin \pi x \ \cos \pi x \ \Gamma(2 + 2x)}.$$

Calculate residue at the point $x = n, \ n = 0, \ 1, \ 2, \ldots$, and obtain

$$i_{106} = \frac{\pi}{-2bi} \sinh\left(\frac{ab}{i}\right) = \frac{\pi}{2b} \sin(ab). \tag{3.14}$$

(6) We put $\gamma = 1$, $m = 1$, $\nu = 1$, $\lambda = 1$, $\rho = 2$, $p \to b^2$, $b \to a$, $t = -1$ in the main formula (3.4) and obtain:

$$i_{107} = \int_0^\infty dx \frac{x \, \sin(ax)}{b^2 - x^2}$$

$$= \frac{1}{2} \frac{\pi b}{-i} \frac{a}{2i} \int_{-\beta+i\infty}^{-\beta-i\infty} d\xi \frac{(ab/i)^{2\xi}}{\sin \pi \xi \, \Gamma(2 + 2\xi)} \frac{1}{\sin\left(\frac{3}{2} + \xi\right)\pi}.$$

The change of the variable $\xi \to x - \frac{1}{2}$ reads

$$i_{107} = -\pi \frac{1}{2i} \int_{-\beta+i\infty}^{-\beta-i\infty} dx \frac{(ab/i)^{2x}}{\sin(2\pi x) \, \Gamma(1 + 2x)}.$$

Taking the residue at the point $x = n$, one gets

$$i_{107} = -\frac{\pi}{2} \sum_{n=0}^\infty \frac{(ab/i)^{2n}}{(2n)!} = -\frac{\pi}{2} \cosh\left(\frac{ab}{i}\right) = -\frac{\pi}{2} \cos(ab). \qquad (3.15)$$

(7) From the formulas (3.13) and (3.14), we have

$$i_{108} = \int_0^\infty dx \frac{\cos(ax)}{b^4 - x^4} = \frac{1}{2b^2} \int_0^\infty dx \cos(ax)$$

$$\times \left[\frac{1}{b^2 - x^2} + \frac{1}{b^2 + x^2}\right]$$

$$= \frac{1}{2b^2} \left[\frac{\pi}{2b} \sin(ab) + \frac{\pi}{2b} e^{-ab}\right]$$

$$= \frac{\pi}{4b^3} \left[e^{-ab} + \sin(ab)\right]. \qquad (3.16)$$

(8) Also from the integrals (3.10) and (3.15), one gets

$$i_{109} = \int_0^\infty dx \frac{x \, \sin(ax)}{b^4 - x^4} = \frac{1}{2b^2} \left\{ \int_0^\infty dx \frac{x \, \sin(ax)}{b^2 - x^2} \right.$$

$$\left. + \int_0^\infty dx \frac{x \, \sin(ax)}{b^2 - x^2} \right\} = \frac{\pi}{4b^2} \left[e^{-ab} - \cos(ab)\right]. \qquad (3.17)$$

(9) From the integral (3.15), it is easy to calculate

$$i_{110} = \int_0^\infty dx \frac{x^2 \, \cos(ax)}{b^2 - x^2} = \frac{\partial}{\partial a} i_{107} = \frac{\pi b}{2} \sin(ab). \qquad (3.18)$$

(10) Also, we have

$$i_{111} = \int_0^\infty dx \frac{x^3 \, \sin(ax)}{b^2 - x^2} = -\frac{\partial^2}{\partial a^2} i_{107} = -\frac{\pi b^2}{2} \cos(ab). \qquad (3.19)$$

(11) We now try to calculate the integral

$$i_{112} = \int\limits_0^\infty dx \frac{\cos(ax)}{b^4 + x^4},$$

which arises from the main formula (3.7) with the parameters:

$$m = 1, \ \gamma = 0, \ \nu = 1, \ b \to a,$$

$$p \to b^4, \ t = 1, \ \rho = 4, \ \lambda = 1.$$

By using the identity

$$\frac{1}{x^4 + b^4} = \frac{1}{(-2ib^2)} \left\{ \frac{1}{x^2 + ib^2} - \frac{1}{x^2 - ib^2} \right\},$$

this integral is divided by two parts:

$$i_{112} = [i_{112a} + i_{112b}] \ \frac{1}{-2ib^2},$$

where

$$i_{112a} = \int\limits_0^\infty dx \frac{\cos(ax)}{x^2 + ib^2}, \quad i_{112b} = \int\limits_0^\infty dx \frac{\cos(ax)}{x^2 - ib^2}$$

for which we use the main formula (3.7). Thus,

$$i_{112a} = \frac{1}{2} \frac{1}{ib^2} (ib^2)^{1/2} \frac{1}{2i} \int\limits_{-\beta+i\infty}^{-\beta-i\infty} d\xi \frac{a^{2\xi} \ (ib^2)^{\xi}}{\sin \pi\xi \ \Gamma(1 + 2\xi)}$$

$$\times \ \Gamma \left(\frac{1}{2} + \xi \right) \Gamma \left(\frac{1}{2} - \xi \right)$$

$$= \frac{\pi}{2} \frac{1}{(ib^2)^{1/2}} \frac{1}{2i} \int\limits_{-\beta+i\infty}^{-\beta-i\infty} d\xi \frac{i^{\xi} \ (ab)^{2\xi}}{\sin \pi\xi \ \cos \pi\xi \ \Gamma(1 + 2\xi)}.$$

From the integral i_{105}, it follows directly

$$i_{112a} = \frac{\pi}{2b\sqrt{i}} \ e^{-\sqrt{i}ab}$$

and

$$i_{112b} = \int\limits_0^\infty dx \frac{\cos(ax)}{x^2 - ib^2} = \frac{\pi}{2b\sqrt{-i}} \ e^{-\sqrt{-i}ab},$$

where

$$\sqrt{i} = \frac{1}{\sqrt{2}} (1 + i), \qquad \sqrt{-i} = \frac{1}{\sqrt{2}} (1 - i).$$

So that

1)

$$e^{-\sqrt{i}\,ab} = e^{-(1+i)ab/\sqrt{2}} = e^{-ab/\sqrt{2}}\, e^{-iab/\sqrt{2}},$$

2)

$$e^{-\sqrt{-i}\,ab} = e^{-(1-i)ab/\sqrt{2}} = e^{-ab/\sqrt{2}}\, e^{iab/\sqrt{2}}.$$

Here

$$\frac{1}{\sqrt{i}} = \frac{\sqrt{2}}{1+i}\frac{1-i}{1-i} = \frac{1}{\sqrt{2}}(1-i), \quad \frac{1}{\sqrt{-i}} = \frac{1}{\sqrt{2}}(1+i).$$

Finally, we have

$$i_{112} = \frac{\pi}{2b}\frac{1}{(-2ib^2)}\, e^{-ab/\sqrt{2}} \left[\frac{1}{\sqrt{2}}(1-i)\, e^{-iab/\sqrt{2}} - \frac{1}{\sqrt{2}}(1+i)\, e^{iab/\sqrt{2}}\right],$$

where expressions in the square brackets are given by

$$\begin{aligned}
[\ ] &= \frac{1}{\sqrt{2}}(1-i)\left(\cos\frac{ab}{\sqrt{2}} - i\sin\frac{ab}{\sqrt{2}}\right) \\
&\quad - \frac{1}{\sqrt{2}}(1+i)\left(\cos\frac{ab}{\sqrt{2}} + i\sin\frac{ab}{\sqrt{2}}\right) \\
&= -i\sqrt{2}\left(\cos\frac{ab}{\sqrt{2}} + \sin\frac{ab}{\sqrt{2}}\right).
\end{aligned}$$

Thus,

$$i_{112} = \int\limits_0^\infty dx\frac{\cos(ax)}{b^4 + x^4} = \frac{\pi}{4b^3}\sqrt{2}\, e^{-ab/\sqrt{2}}\left(\cos\frac{ab}{\sqrt{2}} + \sin\frac{ab}{\sqrt{2}}\right). \qquad (3.20)$$

(12) Notice that from this integral, the following series integrals arise:

$$i_{113} = \int\limits_0^\infty dx\frac{x\,\sin(ax)}{b^4 + x^4} = -\frac{\partial}{\partial a}i_{112} = \frac{\pi}{2}\frac{1}{b^2}\sin\frac{ab}{\sqrt{2}}\, e^{-ab/\sqrt{2}}, \qquad (3.21)$$

$$i_{114} = \int\limits_0^\infty dx\frac{x^2\,\cos(ax)}{b^4 + x^4} = \frac{\partial}{\partial a}i_{113}$$

$$= \frac{\pi}{4b}\sqrt{2}\, e^{-ab/\sqrt{2}}\left(\cos\frac{ab}{\sqrt{2}} - \sin\frac{ab}{\sqrt{2}}\right), \qquad (3.22)$$

$$i_{115} = \int\limits_0^\infty dx\frac{x^3\,\sin(ax)}{b^4 + x^4} = -\frac{\partial}{\partial a}i_{114} = \frac{\pi}{2}\cos\frac{ab}{\sqrt{2}}\, e^{-ab/\sqrt{2}}. \qquad (3.23)$$

(13) By means of the integral i_{105} one can calculate the following type of integrals:

$$i_{116} = \int\limits_0^\infty dx\frac{\cos(ax)}{(\beta^2 + x^2)(\gamma^2 + x^2)}, \qquad (3.24)$$

where we can use the identity:

$$\frac{1}{(\beta^2 + x^2)(\gamma^2 + x^2)} = \frac{1}{(\beta^2 - \gamma^2)}\left\{\frac{1}{\gamma^2 + x^2} - \frac{1}{\beta^2 + x^2}\right\}.$$

(14) Let us calculate the integral

$$i_{117} = \int_0^\infty dx \frac{x^3 \cos(a^4 x^4)}{[b^8 + x^8]^4}$$

arising from the main formula (3.7) with the choice of the parameters:

$$\gamma = 3, \; m = 1, \; b \to a^4, \; \nu = 4, \; p \to b^8, \; t = 1, \; \rho = 8, \; \lambda = 4.$$

Thus,

$$i_{117} = \frac{1}{8} \frac{1}{\Gamma(4)} \frac{1}{b^{28}}$$

$$\times \frac{1}{2i} \int_{-\beta+i\infty}^{-\beta-i\infty} d\xi \frac{(a^4 b^4)^{2\xi}}{\sin \pi \xi \; \Gamma(1 + 2\xi)} \Gamma\left(\frac{1}{2} + \xi\right) \Gamma\left(\frac{7}{2} - \xi\right), \qquad (3.25)$$

where

$$\Gamma\left(\frac{7}{2} - \xi\right) = \left(\frac{15}{8} - \frac{23}{4}\xi + \frac{9}{2}\xi^2 - \xi^3\right) \Gamma\left(\frac{1}{2} - \xi\right).$$

From this last expression, we see that we must calculate the terms of the type

$$\xi \, (a^4 b^4)^{2\xi}, \; \xi^2 (a^4 b^4)^{2\xi}, \; \xi^3 (a^4 b^4)^{2\xi}$$

with powers of ξ^n in the Mellin integral. In this case, the following substitutions are valid:

$$\xi \to \frac{1}{2} X \frac{\partial}{\partial X},$$

$$\xi^2 \to \frac{1}{4} X^2 \frac{\partial^2}{\partial X^2} + \frac{1}{4} X \frac{\partial}{\partial X}, \qquad (3.26)$$

$$\xi^3 \to \frac{1}{8} \left\{ X^3 \frac{\partial^3}{\partial X^3} + 3X^2 \frac{\partial^2}{\partial X^2} + X \frac{\partial}{\partial X} \right\}.$$

If we have terms like

$$\xi \, Y^\xi, \; \xi^2 \, Y^\xi, \ldots$$

in the Mellin representation, then instead of (3.26) we must use the other substitutions:

$$\xi \to Y \frac{\partial}{\partial Y},$$

$$\xi^2 \to Y^2 \frac{\partial^2}{\partial Y^2} + Y \frac{\partial}{\partial Y}, \qquad (3.27)$$

....

Making use of the substitution (3.26), the integral (3.25) is calculated as

$$i_{117} = \frac{\pi}{48 \, b^{28}} \left\{ \frac{15}{8} - \frac{23}{8}(-X) + \frac{9}{8}[X^2 - X] \right.$$

$$\left. - \frac{1}{8} \left[-X^3 + 3X^2 - X \right] \right\} e^{-X},$$

where

$$X = a^4 b^4, \quad e^{-X} = e^{-a^4 b^4}.$$

Finally, we have

$$i_{117} = \frac{\pi}{384 \, b^{28}} \left[15 + 15(ab)^4 + 6(ab)^8 + (ab)^{12} \right] e^{-a^4 b^4}. \tag{3.28}$$

(15) The main formula (3.7) with the parameters

$$\gamma = 0, \ m = 3, \ b \to a, \ \nu = 1, \ p \to b^2, \ t = 1, \ \rho = 2, \ \lambda = 2$$

reads

$$i_{118} = \int_0^\infty dx \frac{\cos^3(ax)}{(b^2 + x^2)^2} = \frac{1}{4} \frac{1}{2 \, b^3}$$

$$\times \frac{1}{2i} \int_{-\beta + i\infty}^{-\beta - i\infty} d\xi \frac{(ab)^{2\xi}}{\sin \pi\xi \, \Gamma(1 + 2\xi)} (3 + 3^{3\xi})$$

$$\times \frac{\Gamma\left(\frac{1}{2} + \xi\right) \Gamma\left(2 - \frac{1}{2} - \xi\right)}{\Gamma(2)}, \tag{3.29}$$

where

$$\Gamma\left(\frac{3}{2} - \xi\right) = \left(\frac{1}{2} - \xi\right) \Gamma\left(\frac{1}{2} - \xi\right),$$

and

$$(3 + 3^{2\xi}) \left(\frac{1}{2} - \xi\right) = \frac{3}{2} + \frac{1}{2} 3^{2\xi} - 3\xi - \xi \, 3^{2\xi}.$$

Again using the substitution (3.26), it is easy to derive

$$i_{118} = \frac{\pi}{16 \, b^3} \left\{ 3(1 + ab) \, e^{-ab} + (1 + 3ab) \, e^{-3ab} \right\}. \tag{3.30}$$

(16) Using the integrals (3.14) and (3.15), it is easy to obtain the following two integrals

$$i_{119} = \int_{-\infty}^\infty dx \frac{\sin(ax)}{\beta - x} = 2 \int_0^\infty dy \frac{y \, \sin(ay)}{\beta^2 - y^2} = -\pi \cos(a\beta) \tag{3.31}$$

and

$$i_{120} = \int_{-\infty}^\infty dx \frac{\cos(ax)}{\beta - x} = 2\beta \int_0^\infty dx \frac{\cos(ax)}{\beta^2 - x^2} = \pi \sin(a\beta). \tag{3.32}$$

(17) From the main formula (3.4), one gets

$$i_{121} = \int\limits_0^\infty dx \frac{x\,\sin(ax)}{[b^2 + x^2]^3},$$

where $\gamma = 1$, $m = 1$, $b \to a$, $p \to b^2$, $t = 1$, $\rho = 2$, $\lambda = 3$. This integral is calculated as follows:

$$i_{121} = \frac{1}{4}\frac{a}{b^3}\frac{1}{2i}\int\limits_{-\beta+i\infty}^{-\beta-i\infty} d\xi \frac{(ab)^{2\xi}}{\sin \pi\xi\,\Gamma(2+2\xi)}\Gamma\left(\frac{3}{2}+\xi\right)\Gamma\left(3-\frac{3}{2}-\xi\right),$$

where

$$\Gamma\left(\frac{3}{2}-\xi\right) = \Gamma\left(3 - \frac{3}{2}-\xi\right) = -\left(\frac{1}{4}-\xi^2\right)\Gamma\left(1-\frac{3}{2}-\xi\right),$$

and

$$\Gamma\left(\frac{3}{2}+\xi\right)\Gamma\left(1-\frac{3}{2}-\xi\right) = \frac{\pi}{\sin\left(\frac{3}{2}+\xi\right)\pi} = -\frac{\pi}{\cos\pi\xi}.$$

So that given integral takes the form

$$i_{121} = \frac{\pi a}{2\,b^3}\frac{1}{2i}\int\limits_{-\beta+i\infty}^{-\beta-i\infty} d\xi \frac{(ab)^{2\xi}}{\sin 2\pi\xi\,\Gamma(2+2\xi)}\left(\frac{1}{4}-\xi^2\right). \qquad (3.33)$$

Next, making use of the integration variable

$$2\xi = y - 1, \quad d\xi = \frac{1}{2}dy, \quad \sin\pi(y-1) = -\sin\pi y,$$

and the substitution (3.27), one gets

$$i_{121} = \frac{\pi a}{16\,b^3}(1+ab)\,e^{-ab}. \qquad (3.34)$$

(18) From the integral i_{105} it is easy to derive:

1) $\quad i_{122} = \int\limits_0^\infty dx \frac{x\,\sin(ax)}{[b^2+x^2]} = -\frac{\partial}{\partial a}i_{105} = \frac{\pi}{2}\,e^{-ab}, \qquad (3.35)$

2) $\quad i_{123} = \int\limits_0^\infty dx \frac{x\,\sin(ax)}{[b^2+x^2]^2} = -\frac{1}{2b}\frac{\partial}{\partial b}i_{105} = \frac{\pi a}{4b}\,e^{-ab}. \qquad (3.36)$

So that, in the contrary, we have

$$i_{121} = -\frac{1}{4b}\frac{\partial}{\partial b}i_{123} = -\frac{1}{4b}\frac{\partial}{\partial b}\left[\frac{\pi a}{4b}\,e^{-ab}\right] = \frac{\pi a}{16\,b^3}(1+ab)\,e^{-ab}$$

as it should be.

(19) The following integral

$$i_{124} = \int_0^\infty dx \frac{x \, \sin(ax)}{[b^2 + x^2]^3}$$

is easily calculated if we use the integral i_{121}. That is

$$i_{124} = -\frac{1}{6b} \frac{\partial}{\partial b} i_{121} = -\frac{1}{6b} \frac{\partial}{\partial b} \left[\frac{\pi a}{16b^3} (1 + ab) \, e^{-ab} \right]$$

$$= \frac{\pi}{96} \frac{a}{b^5} \left[3(1 + ab) + a^2 b^2 \right] e^{-ab}. \tag{3.37}$$

(20) The main formula (3.4) with

$$\gamma = 0, \; \nu = 1, \; m = 1, \; b \to a, \; p \to \beta^2, \; t = 1, \; \rho = 2, \; \lambda = 1/2,$$

gives the following integral

$$i_{125} = \int_0^\infty dx \frac{\sin(ax)}{\sqrt{\beta^2 + x^2}} = \frac{1}{2} \frac{1}{2i}$$

$$\times \int_{-\beta+i\infty}^{-\beta-i\infty} d\xi \frac{(a\beta)^{2\xi+1}}{\sin \pi\xi \, \Gamma(2 + 2\xi)} \frac{\Gamma(1 + \xi)\Gamma\left(-\frac{1}{2} - \xi\right)}{\Gamma\left(\frac{1}{2}\right)}, \tag{3.38}$$

where

$$\Gamma\left(-\frac{1}{2} - \xi\right) = \Gamma\left(-\frac{1}{2} - \xi\right) \frac{\left(-\frac{1}{2} - \xi\right)}{\left(-\frac{1}{2} - \xi\right)} = \frac{\Gamma\left(\frac{1}{2} - \xi\right)}{\left(-\frac{1}{2} - \xi\right)}$$

$$= -\frac{\pi}{\Gamma\left(\frac{3}{2} + \xi\right)} \frac{1}{\cos \pi\xi},$$

and

$$\Gamma(1 + \xi) = \frac{\sqrt{\pi} \, \Gamma[2(1 + \xi)]}{2^{2\xi+1}} \frac{1}{\Gamma\left(\frac{3}{2} + \xi\right)}.$$

So that

$$i_{125} = -\frac{\pi}{2} \frac{1}{2i} \int_{-\beta+i\infty}^{-\beta-i\infty} d\xi \frac{\left(\frac{a\beta}{2}\right)^{2\xi+1}}{\sin \pi\xi \, \cos \pi\xi \, \Gamma^2\left(\frac{3}{2} + \xi\right)}. \tag{3.39}$$

Here it is worth noticing that due to having the non-analytic function $(\beta^2 + x^2)^{-1/2}$ in the integral (3.38), we must calculate residues at integer $\xi = n$, $(n = 0, 1, 2, \ldots)$ and half-integer points: $\xi = m + \frac{1}{2}$ $(m = 0, 1, 2, \ldots)$. Thus, the first case leads to the integral $[(\sin \pi\xi)' = \pi \cos \pi\xi]$

$$i_{125}^1 = -\frac{\pi}{2} \sum_{n=0}^\infty \frac{(a\beta/2)^{2n+1}}{\Gamma^2\left(\frac{3}{2} + n\right)} = -\frac{\pi}{2} L_0(a\beta), \tag{3.40}$$

where the function $L_0(x)$ is defined by the series

$$L_0(x) = \sum_{n=0}^{\infty} \frac{(x/2)^{2n+1}}{\Gamma^2\left(\frac{3}{2}+n\right)} \tag{3.41}$$

is called the modified Struve function of the zero order.

For the second case, we carry out the change of the variable $\xi = x - 1$ and calculate the residue at the point $x = m + \frac{1}{2}$, where the result reads $[(\cos \pi \xi)' = -\pi \sin \pi \xi]$

$$i_{125}^2 = \frac{\pi}{2} \sum_{m=0}^{\infty} \frac{(a\beta/2)^{2m}}{(m!)^2} = \frac{\pi}{2} I_0(a\beta). \tag{3.42}$$

Here the function $I_0(x)$ is called the modified Bessel function of the first kind. Finally, we have

$$i_{125} = i_{125}^1 + i_{125}^2 = \frac{\pi}{2}\left[I_0(a\beta) - L_0(a\beta)\right]. \tag{3.43}$$

(21) The integral

$$i_{126} = \int_0^{\infty} dx \frac{\cos(ax)}{\sqrt{\beta^2 + x^2}}$$

arising from the main formula (3.7) with substitution:

$$\gamma = 0, \ \nu = 1, \ m = 1, \ b \to a, \ p \to \beta^2, \ t = 1, \ \rho = 2, \ \lambda = 1/2$$

is easily calculated. The result is

$$i_{126} = \frac{1}{2\sqrt{\pi}} \frac{1}{2i} \int_{-\beta+i\infty}^{-\beta-i\infty} d\xi \frac{(a\beta)^{2\xi}}{\sin \pi \xi \, \Gamma(1+2\xi)} \Gamma\left(\frac{1}{2}+\xi\right) \Gamma(-\xi), \tag{3.44}$$

where

$$\bullet \ \Gamma(-\xi) = -\frac{\pi}{\sin \pi \xi \, \Gamma(1+\xi)},$$

$$\bullet \ \Gamma\left(\frac{1}{2}+\xi\right) = \left[\frac{2^{2\xi-1}}{\sqrt{\pi}}\right]^{-1} \frac{\Gamma(2\xi)}{\Gamma(\xi)}.$$

Thus, we have

$$i_{126} = -\frac{\pi}{4i} \int_{-\beta+i\infty}^{-\beta-i\infty} d\xi \frac{(a\beta/2)^{2\xi}}{\sin^2 \pi \xi \, \Gamma^2(1+\xi)}. \tag{3.45}$$

After the calculation of the residue at the pole $\xi = n$ of the second order by the formula (1.10) in Chapter 1, one gets

$$i_{126} = -\ln \frac{a\beta}{2} \sum_{n=0}^{\infty} \frac{(a\beta/2)^{2n}}{(n!)^2}$$

$$+ \sum_{n=0}^{\infty} \frac{(a\beta/2)^{2n}}{(n!)^2} \Psi(1+n) = K_0(a\beta). \tag{3.46}$$

Here the function $K_0(a\beta)$ defined by this series is called the modified Bessel function of the second kind and $\Psi(x)$ is the Psi-function defined in Section 1.2.5 of Chapter 1.

(22) The formula (3.4) with the parameters:

$$\gamma = 1, \ \nu = 1, \ m = 1, \ b \to a, \ p \to \beta^2, \ t = 1, \ \rho = 2, \ \lambda = 3/2$$

leads to the integral:

$$i_{127} = \int\limits_{0}^{\infty} dx \frac{x \, \sin(ax)}{(\beta^2 + x^2)^{3/2}} = \frac{a}{\sqrt{\pi}} \frac{1}{2i}$$

$$\times \int\limits_{-\beta+i\infty}^{-\beta-i\infty} d\xi \frac{(a\beta)^{2\xi}}{\sin \pi\xi \ \Gamma(2 + 2\xi)} \Gamma\left(\frac{3}{2} + \xi\right) \Gamma(-\xi), \qquad (3.47)$$

where

$$\bullet \ \Gamma(-\xi) = -\frac{\pi}{\sin \pi\xi} \frac{1}{\Gamma(1 + \xi)},$$

$$\bullet \ \Gamma\left(1 + \frac{1}{2} + \xi\right) = \left(\frac{1}{2} + \xi\right) \Gamma\left(\frac{1}{2} + \xi\right)$$

$$= \left(\frac{1}{2} + \xi\right) \frac{\sqrt{\pi} \ \Gamma(2\xi) \ 2^{-2\xi+1}}{\Gamma(\xi)},$$

$$\bullet \ \Gamma(2 + 2\xi) = (1 + 2\xi) \ 2\xi \ \Gamma(2\xi).$$

Thus,

$$i_{127} = -\frac{\pi}{2i} \frac{a}{2} \int\limits_{-\beta+i\infty}^{-\beta-i\infty} d\xi \frac{(a\beta/2)^{2\xi}}{\sin^2 \pi\xi \ \Gamma^2(1 + \xi)}$$

$$= a \left\{ -\ln\frac{a\beta}{2} \sum_{n=0}^{\infty} \frac{(a\beta/2)^{2n}}{(n!)^2} \right.$$

$$\left. + \sum_{n=0}^{\infty} \frac{(a\beta/2)^{2n}}{(n!)^2} \Psi(1 + n) \right\} = aK_0(a\beta). \qquad (3.48)$$

(23) From the main formula (3.4), where

$$\gamma = 0, \ \nu = 1, \ m = 1, \ p = \beta^2, \ t = 1, \ \rho = 2, \ \lambda = \frac{1}{2} - \delta,$$

it follows

$$i_{128} = \int\limits_{0}^{\infty} dx \frac{\sin(ax)}{(\beta^2 + x^2)^{1/2-\delta}} = \frac{1}{2} \frac{\beta^{2\delta}}{\Gamma\left(\frac{1}{2} - \delta\right)} \frac{1}{2i}$$

$$\times \int\limits_{-\beta+i\infty}^{-\beta-i\infty} d\xi \frac{(a\beta)^{2\xi+1}}{\sin \pi\xi \ \Gamma(2 + 2\xi)} \Gamma(1 + \xi) \Gamma\left(-\frac{1}{2} - \delta - \xi\right),$$

where

$$\bullet \; \Gamma[2(1+\xi)] = \frac{2^{2(1+\xi)-1}}{\sqrt{\pi}} \Gamma(1+\xi)\Gamma\left(\frac{3}{2}+\xi\right),$$

$$\bullet \; \Gamma\left(-\frac{1}{2}-\delta-\xi\right) = \frac{\left(-\frac{1}{2}-\delta-\xi\right)\Gamma\left(-\frac{1}{2}-\delta-\xi\right)}{\left(-\frac{1}{2}-\delta-\xi\right)} = \frac{\Gamma\left(\frac{1}{2}-\delta-\xi\right)}{\left(-\frac{1}{2}-\delta-\xi\right)}.$$

Thus,

$$i_{128} = -\frac{\beta^{2\delta}}{\Gamma\left(\frac{1}{2}-\delta\right)} \frac{\pi\sqrt{\pi}}{4i} \int\limits_{-\beta+i\infty}^{-\beta-i\infty} d\xi \frac{(a\beta/2)^{2\xi+1}}{\sin \pi\xi \; \cos \pi(\delta+\xi)}$$

$$\times \frac{1}{\Gamma\left(\frac{3}{2}+\xi\right)\Gamma\left(\frac{3}{2}+\delta+\xi\right)}. \tag{3.49}$$

(a) The calculation of the residue at the integers $\xi = n$, where

$$\cos \pi(\delta+n) = \cos \pi\delta \; \cos \pi n - \sin \pi\delta \; \sin \pi n = (-1)^n \cos \pi\delta$$

and

$$(\sin \pi\xi)' = \pi \cos \pi\xi \Big|_{\xi=n} = \pi(-1)^n$$

reads

$$i_{128}^1 = -\frac{\pi\sqrt{\pi}}{2\Gamma\left(\frac{1}{2}-\delta\right)} \frac{\beta^{2\delta}}{\cos \pi\delta} \sum_{n=0}^{\infty} \frac{(a\beta/2)^{2n+1+\delta-\delta}}{\Gamma\left(\frac{3}{2}+n\right)\Gamma\left(\frac{3}{2}+\delta+n\right)}$$

$$= -\frac{\sqrt{\pi}}{2}\Gamma\left(\frac{1}{2}+\delta\right)\left(\frac{2\beta}{a}\right)^{\delta} L_\delta(a\beta), \tag{3.50}$$

where $L_\delta(x)$ is the modified Struve function.

(b) After the calculation of the residue at the half integers

$$\pi(\delta+\xi) = \left(m-\frac{1}{2}\right)\pi,$$

where

$$[\cos \pi(\delta+\xi)]'\Big|_{\xi=m-\frac{1}{2}-\delta} = -\pi \sin\left(m-\frac{1}{2}\right)\pi$$

$$= -\pi\left[\sin \pi m \; \cos \frac{\pi}{2} - \cos \pi m \; \sin \frac{\pi}{2}\right] = (-1)^m \; \pi$$

and

$$\sin \pi\xi \to \sin \pi\left(m-\frac{1}{2}-\delta\right)$$

$$= \sin \pi m \; \cos\left(\frac{1}{2}+\delta\right)\pi - \cos \pi m \; \sin\left(\frac{1}{2}+\delta\right)\pi$$

$$= -(-1)^m\left[\sin \frac{\pi}{2} \; \cos \delta\pi + \cos \frac{\pi}{2} \; \sin \delta\pi\right]$$

$$= -(-1)^m \cos \pi\delta,$$

one gets

$$i_{128}^{2} = \frac{\pi}{\Gamma\left(\frac{1}{2} - \delta\right)\,\cos\pi\delta}\,\frac{\sqrt{\pi}}{2}\,\beta^{2\delta}\sum_{m=0}^{\infty}\frac{(a\beta/2)^{2m-\delta-\delta}}{\Gamma(m+1)\Gamma(m+1-\delta)},$$

where

$$\beta^{2\delta}\left(\frac{a\beta}{2}\right)^{-\delta} = \left(\frac{2\beta}{a}\right)^{\delta}.$$

Therefore

$$i_{128}^{2} = \frac{\sqrt{\pi}}{2}\,\left(\frac{2\beta}{a}\right)^{\delta}\,\Gamma\left(\frac{1}{2} + \delta\right)\sum_{m=0}^{\infty}\frac{(a\beta/2)^{2m-\delta}}{m!\,\Gamma(m+1-\delta)}.$$

Here

$$I_{-\delta} = \sum_{m=0}^{\infty}\frac{(a\beta/2)^{2m-\delta}}{m!\,\Gamma(m+1-\delta)}.$$

Finally, we have

$$i_{128} = i_{128}^{1} + i_{128}^{2}$$

$$= \frac{\sqrt{\pi}}{2}\,\Gamma\left(\frac{1}{2} + \delta\right)\left(\frac{2\beta}{a}\right)^{\delta}\left[I_{-\delta}(a\beta) - L_{\delta}(a\beta)\right]. \qquad (3.51)$$

(24) The integral

$$i_{129} = \int_{0}^{\infty}dx\,\frac{\cos(ax)}{[b^{2} + x^{2}]^{2}}$$

arising from the main formula (3.7) with the substitution:

$$\gamma = 0,\ \nu = 1,\ m = 1,\ p = b^{2},\ t = 1,\ \rho = 2,\ \lambda = 2,\ b \to a$$

is calculated as follows

$$i_{129} = \frac{1}{2}\,\frac{1}{b^{3}}\,\frac{1}{2i}\int_{-\beta+i\infty}^{-\beta-i\infty}d\xi\,\frac{(ab)^{2\xi}}{\sin\pi\xi\,\Gamma(1+\xi)}\,\Gamma\left(\frac{1}{2} + \xi\right)\Gamma\left(2 - \frac{1}{2} - \xi\right), \qquad (3.52)$$

where

$$\Gamma\left(\frac{3}{2} - \xi\right) = \left(\frac{1}{2} - \xi\right)\Gamma\left(\frac{1}{2} - \xi\right).$$

Again making use of the substitution (3.26) and after displacement of the contour integration to the right, one gets

$$i_{129} = \frac{\pi}{4b^{3}}(1 + ab)\,e^{-ab}. \qquad (3.53)$$

(25) From this integral, the following series of integrals are obtained. They are:

$$i_{130} = \int_0^\infty dx \frac{x\,\sin(ax)}{[b^2 + x^2]^2} = -\frac{\partial}{\partial a} i_{129}$$

$$= -\frac{\partial}{\partial a}\left\{\frac{\pi}{4b^3}(1 + ab)\,e^{-ab}\right\} = \frac{\pi a}{4b}\,e^{-ab} \tag{3.54}$$

and

$$i_{131} = \int_0^\infty dx \frac{x^3\,\sin(ax)}{[b^2 + x^2]^2} = -\frac{\partial^2}{\partial a^2} i_{130}$$

$$= -\frac{\partial^2}{\partial a^2}\left\{\frac{\pi a}{4b}\,e^{-ab}\right\} = \frac{\pi}{4}[2 - ab]\,e^{-ab}. \tag{3.55}$$

(26) From the main formulas (3.4) and (3.7), one can calculate the following integral:

$$i_{132} = \int_0^\infty dx \frac{\cos(ax^2) - \sin(ax^2)}{x^4 + b^4} = \frac{1}{4b^3}\frac{1}{2i}$$

$$\times \int_{-\beta+i\infty}^{-\beta-i\infty} d\xi \frac{(ab^2)^{2\xi}}{\sin \pi\xi}\left[\frac{\Gamma\left(\frac{1}{4} + \xi\right)\,\Gamma\left(1 - \frac{1}{4} - \xi\right)}{\Gamma(1 + 2\xi)}\right.$$

$$\left. - b^2\,\frac{\Gamma\left(\frac{3}{4} + \xi\right)\,\Gamma\left(1 - \frac{3}{4} - \xi\right)}{\Gamma(2 + 2\xi)}\right], \tag{3.56}$$

where

$$\bullet\,\Gamma\left(1 - \frac{1}{4} - \xi\right) = \frac{\pi}{\Gamma\left(\frac{1}{4} + \xi\right)\,\sin \pi\left(\frac{1}{4} + \xi\right)},$$

$$\bullet\,\sin \pi\left(\frac{1}{4} + \xi\right) = \frac{1}{\sqrt{2}}(\cos \pi\xi + \sin \pi\xi),$$

and

$$\bullet\,\Gamma\left(1 - \frac{3}{4} - \xi\right) = \frac{\pi}{\Gamma\left(\frac{3}{4} + \xi\right)\,\sin \pi\left(\frac{3}{4} + \xi\right)},$$

$$\bullet\,\sin \pi\left(\frac{3}{4} + \xi\right) = \frac{1}{\sqrt{2}}(\cos \pi\xi - \sin \pi\xi).$$

In the first term of (3.56), we can do substitution

$$2\xi \to 2\xi' + 1,$$

where

$$\sin \pi\xi \to \sin \pi\left(\xi' + \frac{1}{2}\right) = \cos \pi\xi',$$

$$\Gamma(1 + 2\xi) \to \Gamma(2 + 2\xi'),$$

$$\sin \pi\left(\frac{1}{4} + \xi\right) \to \sin\left(\frac{3}{4} + \xi'\right).$$

Next, we come back again to change the notation $\xi' \to \xi$ in this term. After such a procedure, we go to the new integration variable

$$2 + 2\xi \to 1 + x,$$

where

$$\cos \pi\xi \to \cos \pi\left(\frac{x}{2} - \frac{1}{2}\right) = \sin \frac{\pi x}{2},$$

$$\sin \pi\xi \to \sin \pi\left(\frac{x}{2} - \frac{1}{2}\right) = -\cos \frac{\pi x}{2},$$

$$\sin \pi\left(\frac{3}{4} + \xi\right) = \frac{1}{\sqrt{2}}\left(\sin \frac{\pi x}{2} + \cos \frac{\pi x}{2}\right).$$

The resulting integral takes the form

$$i_{132} = \frac{\pi\sqrt{2}}{4b^3} \frac{1}{2i} \int_{-\beta+i\infty}^{-\beta-i\infty} dx \frac{(ab^2)^x}{\sin \pi x \, \Gamma(1 + x)} = \frac{\pi}{2\sqrt{2}\, b^3} e^{-ab^2}. \qquad (3.57)$$

(27) From the integral (3.57), it follows

$$i_{133} = \int_0^\infty dx \frac{\sin(ax^2) + \cos(ax^2)}{x^4 + b^4} x^2 = -\frac{\partial}{\partial a} i_{132} = \frac{\pi}{2\sqrt{2}} \frac{1}{b} e^{-ab^2} \qquad (3.58)$$

and

$$i_{134} = \int_0^\infty dx \frac{\cos(ax^2) + \sin(ax^2)}{[x^4 + b^4]^2} x^2 = -\frac{1}{4b^3} \frac{\partial}{\partial b} i_{133}$$

$$= \frac{\pi}{4\sqrt{2}} \frac{1}{b^3} e^{-ab^2}\left(a + \frac{1}{2b^2}\right). \qquad (3.59)$$

Also, we have

$$i_{135} = \int_0^\infty dx \frac{\cos(ax^2) - \sin(ax^2)}{[x^4 + b^4]^2} x^4 = \frac{\partial}{\partial a} i_{134} = \frac{\pi}{4\sqrt{2}} \frac{1}{b^3}$$

$$\times \frac{\partial}{\partial a}\left\{e^{-ab^2}\left[a + \frac{1}{2b^2}\right]\right\} = \frac{\pi}{4\sqrt{2}} \frac{1}{b}$$

$$\times e^{-ab^2}\left(-a + \frac{1}{2b^2}\right). \qquad (3.60)$$

Chapter 4

Derivation of General Formulas for Integrals Involving Powers of x, $(a + bx)$-Type Binomials and Trigonometric Functions

4.1 Derivation of General Formulas

4.1.1 9th General Formula

$$N_9 = \int\limits_0^1 dx\, x^\delta (1 - x^\sigma)^\mu \, \sin^q \left[bx^\nu (1 - x^\sigma)^\lambda \right] = \frac{1}{2^{q-1}} \frac{1}{2i\sigma}$$

$$\times \int\limits_{\alpha+i\infty}^{\alpha-i\infty} d\xi \frac{(2b)^{2\xi}}{\sin \pi\xi \, \Gamma(1 + 2\xi)} \, I_q(\xi) \tag{4.1}$$

$$\times B\left(\frac{1 + \delta + 2\nu\xi}{\sigma}, \quad 1 + \mu + 2\xi\lambda \right),$$

$$q = 2, \ 4, \ 6, \ldots,$$

where

$$B(x, y) = \frac{\Gamma(x) \, \Gamma(y)}{\Gamma(x + y)} \tag{4.2}$$

is called the Euler integral of the first kind or Beta-function. As before, the functions $I_q(\xi)$ are given by expressions (1.32) in Chapter 1.

4.1.2   10th *General Formula*

$$
N_{10} = \int_0^1 dx\, x^\delta (1 - x^\sigma)^\mu \, \sin^m \left[bx^\nu (1 - x^\sigma)^\lambda \right] = \frac{1}{2^{m-1}} \frac{1}{2i\sigma}
$$

$$
\times \int_{-\beta+i\infty}^{-\beta-i\infty} d\xi \frac{b^{2\xi+1}}{\sin \pi\xi \, \Gamma(2+2\xi)} \, N_m(\xi) \tag{4.3}
$$

$$
\times \, B \left(\frac{\delta + \nu + 2\nu\xi + 1}{\sigma}, \quad 1 + \mu + \lambda + 2\xi\lambda \right),
$$

$$
m = 1,\ 3,\ 5,\ldots,
$$

where functions $N_m(\xi)$ are defined by the expressions (1.34) in Chapter 1.

4.1.3   11th *General Formula*

$$
N_{11} = \int_0^1 dx\, x^\delta (1 - x^\sigma)^\mu \left\{ \cos^q \left[bx^\nu (1 - x^\sigma)^\lambda \right] - 1 \right\}
$$

$$
= \frac{1}{2^{q-1}} \frac{1}{2i\sigma} \int_{\alpha+i\infty}^{\alpha-i\infty} d\xi \frac{(2b)^{2\xi}}{\sin \pi\xi \, \Gamma(1+2\xi)} \, I_q'(\xi) \tag{4.4}
$$

$$
\times B \left(\frac{\delta + 2\nu\xi + 1}{\sigma}, \quad 1 + \mu + 2\xi\lambda \right),
$$

where $q = 2,\ 4,\ 6,\ldots$ and $I_q'(\xi)$-functions are derived from (1.38) in Chapter 1.

4.1.4   12th *General Formula*

$$
N_{12} = \int_0^1 dx\, x^\delta (1 - x^\sigma)^\mu \, \cos^m \left[bx^\nu (1 - x^\sigma)^\lambda \right] = \frac{1}{2^{m-1}}
$$

$$
\times \frac{1}{2i\sigma} \int_{-\beta+i\infty}^{-\beta-i\infty} d\xi \frac{b^{2\xi}}{\sin \pi\xi \, \Gamma(1+2\xi)} \, N_m'(\xi) \tag{4.5}
$$

$$
\times B \left(\frac{\delta + 2\nu\xi + 1}{\sigma}, \quad 1 + \mu + 2\xi\lambda \right).
$$

Here $m = 1,\ 3,\ 5,\ldots$ and $N'_m(\xi)$ are defined by (1.36) in Chapter 1.

One can obtain generalization formulas of these four formulas. They are:

4.1.5   13th *General Formula*

$$N_{13} = \int_a^c dx (x-a)^\delta (c-x)^\mu \, \sin^q \left[b(x-a)^\nu (c-x)^\lambda \right]$$

$$= \frac{1}{2^{q-1}} \frac{1}{2i} \int_{\alpha+i\infty}^{\alpha-i\infty} d\xi \frac{(2b)^{2\xi}}{\sin \pi\xi \, \Gamma(1+2\xi)} \, I_q(\xi)$$

$$\times (c-a)^{1+\delta+\mu+2\nu\xi+2\lambda\xi} \, B\left(\delta + 2\nu\xi + 1,\ 1 + \mu + 2\xi\lambda\right),$$

(4.6)

where $q = 2,\ 4,\ 6,\ldots,\ c > a$.

4.1.6   14th *General Formula*

$$N_{14} = \int_a^c dx (x-a)^\delta (c-x)^\mu \, \sin^m \left[b(x-a)^\nu (c-x)^\lambda \right]$$

$$= \frac{1}{2^{m-1}} \frac{1}{2i} \int_{-\beta+i\infty}^{-\beta-i\infty} d\xi \frac{b^{2\xi+1}}{\sin \pi\xi \, \Gamma(2+2\xi)} \, N_m(\xi)$$

$$\times (c-a)^{1+\delta+\nu+2\nu\xi+\mu+\lambda+2\lambda\xi}$$

$$\times B\left(1 + \delta + \nu + 2\nu\xi,\ 1 + \mu + \lambda + 2\xi\lambda\right),$$

(4.7)

where $m = 1,\ 3,\ 5,\ldots,\ c > a$.

4.1.7　15th *General Formula*

$$N_{15}$$

$$= \int_a^c dx(x-a)^\delta (c-x)^\mu \left\{ \cos^q \left[b(x-a)^\nu (c-x)^\lambda \right] - 1 \right\}$$

$$= \frac{1}{2^{q-1}} \frac{1}{2i} \int_{\alpha+i\infty}^{\alpha-i\infty} d\xi \frac{(2b)^{2\xi}}{\sin \pi \xi \; \Gamma(1+2\xi)} \; I'_q(\xi) \tag{4.8}$$

$$\times (c-a)^{1+\delta+2\nu\xi+\mu+2\lambda\xi}$$

$$\times B\left(1+\delta+2\nu\xi, \; 1+\mu+2\xi\lambda\right),$$

where $q = 2,\ 4,\ 6, \ldots,\ c > a$.

4.1.8　16th *General Formula*

$$N_{16} = \int_a^c dx(x-a)^\delta (c-x)^\mu \; \cos^m \left[b(x-a)^\nu (c-x)^\lambda \right]$$

$$= \frac{1}{2^{m-1}} \frac{1}{2i} \int_{-\beta+i\infty}^{-\beta-i\infty} d\xi \frac{b^{2\xi}}{\sin \pi \xi \; \Gamma(1+2\xi)} \; N'_q(\xi) \tag{4.9}$$

$$\times (c-a)^{1+\delta+2\nu\xi+\mu+2\lambda\xi}$$

$$\times B\left(1+\delta+2\nu\xi, \; 1+\mu+2\xi\lambda\right),$$

where $m = 1,\ 3,\ 5, \ldots,\ c > a$.

Now we obtain very specific main formulas for particular integrals.

4.1.9 17th *General Formula*

$$N_{17} = \int\limits_0^\infty \frac{dx}{\sqrt{x}} \, x^\mu (x+a)^{-\mu} (x+c)^{-\mu}$$

$$\times \sin^q \left[bx^\nu (x+a)^{-\nu} (x+c)^{-\nu} \right]$$

$$= \frac{\sqrt{\pi}}{2^{q-1}} \frac{1}{2i} \int\limits_{\alpha+i\infty}^{\alpha-i\infty} d\xi \frac{(2b)^{2\xi}}{\sin \pi\xi \, \Gamma(1+2\xi)} \, I_q(\xi)$$

$$\times \left(\sqrt{a} + \sqrt{c} \right)^{1-2(\mu+2\nu\xi)} \frac{\Gamma(\mu+2\nu\xi-\frac{1}{2})}{\Gamma(\mu+2\nu\xi)}.$$

(4.10)

4.1.10 18th *General Formula*

$$N_{18} = \int\limits_0^\infty \frac{dx}{\sqrt{x}} \, x^\mu (x+a)^{-\mu} (x+c)^{-\mu}$$

$$\times \sin^m \left[bx^\nu (x+a)^{-\nu} (x+c)^{-\nu} \right]$$

$$= \frac{\sqrt{\pi}}{2^{m-1}} \frac{1}{2i} \int\limits_{-\beta+i\infty}^{-\beta-i\infty} d\xi \frac{b^{2\xi+1}}{\sin \pi\xi \, \Gamma(2+2\xi)} \, N_m(\xi)$$

$$\times \left(\sqrt{a} + \sqrt{c} \right)^{1-2(\mu+\nu+2\nu\xi)} \frac{\Gamma(\mu+\nu+2\nu\xi-\frac{1}{2})}{\Gamma(\mu+\nu+2\nu\xi)}.$$

(4.11)

4.1.11 19th *General Formula*

$$N_{19} = \int\limits_0^\infty \frac{dx}{\sqrt{x}} \, x^\mu (x+a)^{-\mu} (x+c)^{-\mu}$$

$$\times \left\{ \cos^q \left[bx^\nu (x+a)^{-\nu} (x+c)^{-\nu} \right] - 1 \right\}$$

$$= \frac{\sqrt{\pi}}{2^{q-1}} \frac{1}{2i} \int\limits_{\alpha+i\infty}^{\alpha-i\infty} d\xi \frac{(2b)^{2\xi}}{\sin \pi\xi \, \Gamma(1+2\xi)} \, I'_q(\xi)$$

$$\times \left(\sqrt{a} + \sqrt{c} \right)^{1-2(\mu+2\nu\xi)} \frac{\Gamma(\mu+2\nu\xi-\frac{1}{2})}{\Gamma(\mu+2\nu\xi)}.$$

(4.12)

4.1.12　20th General Formula

$$N_{20} = \int\limits_{0}^{\infty} \frac{dx}{\sqrt{x}} \, x^{\mu}(x+a)^{-\mu}(x+c)^{-\mu}$$

$$\times \cos^{m}\left[bx^{\nu}(x+a)^{-\nu}(x+c)^{-\nu}\right]$$

$$= \frac{\sqrt{\pi}}{2^{m-1}} \frac{1}{2i} \int\limits_{-\beta+i\infty}^{-\beta-i\infty} d\xi \frac{b^{2\xi}}{\sin \pi\xi \, \Gamma(1+2\xi)} \, N'_{m}(\xi)$$

$$\times \left(\sqrt{a}+\sqrt{c}\right)^{1-2(\mu+2\nu\xi)} \frac{\Gamma(\mu+2\nu\xi-\frac{1}{2})}{\Gamma(\mu+2\nu\xi)}.$$

(4.13)

4.1.13　21st General Formula

$$N_{21} = \int\limits_{0}^{1} dx(1-x^{\sigma})^{\mu} \, \sin^{q}\left[b(1-x^{\sigma})^{\lambda}\right]$$

$$= \frac{1}{2^{q-1}} \frac{1}{2i\sigma} \int\limits_{\alpha+i\infty}^{\alpha-i\infty} d\xi \frac{(2b)^{2\xi}}{\sin \pi\xi \, \Gamma(1+2\xi)} \, I_{q}(\xi)$$

$$\times B\left(\frac{1}{\sigma}, \quad 1+\mu+2\lambda\xi\right).$$

(4.14)

4.1.14　22nd General Formula

$$N_{22} = \int\limits_{0}^{1} dx(1-x^{\sigma})^{\mu} \, \sin^{m}\left[b(1-x^{\sigma})^{\lambda}\right]$$

$$= \frac{1}{2^{m-1}} \frac{1}{2i\sigma} \int\limits_{-\beta+i\infty}^{-\beta-i\infty} d\xi \frac{b^{2\xi+1}}{\sin \pi\xi \, \Gamma(2+2\xi)} \, N_{m}(\xi)$$

$$\times B\left(\frac{1}{\sigma}, \quad 1+\mu+\lambda+2\lambda\xi\right).$$

(4.15)

4.1.15 23rd *General Formula*

$$N_{23} = \int\limits_0^1 dx(1 - x^\sigma)^\mu \left\{ \cos^q \left[b(1 - x^\sigma)^\lambda \right] - 1 \right\}$$

$$= \frac{1}{2^{q-1}} \frac{1}{2i\sigma} \int\limits_{\alpha+i\infty}^{\alpha-i\infty} d\xi \frac{b^{2\xi}}{\sin \pi\xi \; \Gamma(1 + 2\xi)} \; I_q'(\xi)$$

$$\times B\left(\frac{1}{\sigma}, \quad 1 + \mu + 2\lambda\xi \right).$$

(4.16)

4.1.16 24th *General Formula*

$$N_{24} = \int\limits_0^1 dx(1 - x^\sigma)^\mu \; \cos^m \left[b(1 - x^\sigma)^\lambda \right]$$

$$= \frac{1}{2^{m-1}} \frac{1}{2i\sigma} \int\limits_{-\beta+i\infty}^{-\beta-i\infty} d\xi \frac{b^{2\xi}}{\sin \pi\xi \; \Gamma(1 + 2\xi)} \; N_m'(\xi)$$

$$\times B\left(\frac{1}{\sigma}, \quad 1 + \mu + 2\lambda\xi \right).$$

(4.17)

In these four formulas (4.14)-(4.17), one can also put $\mu = -1/\nu$. For this substitution, all formulas are valid.

4.1.17 25th *General Formula*

$$N_{25} = \int\limits_0^1 dx(1 - \sqrt{x})^\mu \; \sin^q \left[b(1 - \sqrt{x})^\lambda \right] = \frac{1}{2^{q-1}} \frac{1}{i}$$

$$\times \int\limits_{\alpha+i\infty}^{\alpha-i\infty} d\xi \frac{(2b)^{2\xi} \; I_q(\xi)}{\sin \pi\xi \; \Gamma(1 + 2\xi)} \frac{1}{(1 + \mu + 2\lambda\xi)} \frac{1}{(2 + \mu + 2\lambda\xi)}.$$

(4.18)

4.1.18 26th *General Formula*

$$N_{26} = \int\limits_0^1 dx (1 - \sqrt{x})^\mu \, \sin^m \left[b(1 - \sqrt{x})^\lambda \right]$$

$$= \frac{1}{2^{m-1}} \frac{1}{i} \int\limits_{-\beta+i\infty}^{-\beta-i\infty} d\xi \frac{b^{2\xi+1} \, N_m(\xi)}{\sin \pi\xi \, \Gamma(2 + 2\xi)} \tag{4.19}$$

$$\times \frac{1}{(1 + \mu + \lambda + 2\lambda\xi)} \frac{1}{(2 + \mu + \lambda + 2\lambda\xi)}.$$

4.1.19 27th *General Formula*

$$N_{27}$$

$$= \int\limits_0^1 dx (1 - \sqrt{x})^\mu \left\{ \cos^q \left[b(1 - \sqrt{x})^\lambda \right] - 1 \right\} = \frac{1}{2^{q-1}} \frac{1}{i}$$

$$\times \int\limits_{\alpha+i\infty}^{\alpha-i\infty} d\xi \frac{b^{2\xi} \, I_q'(\xi)}{\sin \pi\xi \, \Gamma(1 + 2\xi)} \frac{1}{(1 + \mu + 2\lambda\xi)} \frac{1}{(2 + \mu + 2\lambda\xi)}. \tag{4.20}$$

4.1.20 28th *General Formula*

$$N_{28} = \int\limits_0^1 dx (1 - \sqrt{x})^\mu \, \cos^m \left[b(1 - \sqrt{x})^\lambda \right] = \frac{1}{2^{m-1}} \frac{1}{i}$$

$$\times \int\limits_{-\beta+i\infty}^{-\beta-i\infty} d\xi \frac{b^{2\xi} \, N_m'(\xi)}{\sin \pi\xi \, \Gamma(1 + 2\xi)} \frac{1}{(1 + \mu + 2\lambda\xi)} \frac{1}{(2 + \mu + 2\lambda\xi)}. \tag{4.21}$$

4.1.21 29th *General Formula*

$$N_{29} = \int_0^\infty dx\, x^\delta (1 + tx^\sigma)^{-\mu}\, \sin^q\left[bx^\nu(1 + tx^\sigma)^{-\lambda}\right] = \frac{1}{2^{q-1}}$$

$$\times \frac{1}{2i\sigma} \int_{\alpha+i\infty}^{\alpha-i\infty} d\xi\, \frac{(2b)^{2\xi}\, I_q(\xi)}{\sin \pi\xi\, \Gamma(1 + 2\xi)}\, t^{-\frac{\delta + 1 + 2\nu\xi}{\sigma}} \tag{4.22}$$

$$\times B\left(\frac{\delta + 1 + 2\nu\xi}{\sigma},\quad \mu + 2\lambda\xi - \frac{1 + \delta + 2\nu\xi}{\sigma}\right),$$

where $\sigma > 0$, $0 < \mathrm{Re}\,(1 + \delta + 2\nu\xi) < \sigma\, \mathrm{Re}\,(\mu + 2\lambda\xi)$.

4.1.22 30th *General Formula*

$$N_{30} = \int_0^\infty dx\, x^\delta (1 + tx^\sigma)^{-\mu}\, \sin^m\left[bx^\nu(1 + tx^\sigma)^{-\lambda}\right] = \frac{1}{2^{m-1}}\, \frac{1}{2i\sigma}$$

$$\times \int_{-\beta+i\infty}^{-\beta-i\infty} d\xi\, \frac{b^{2\xi+1}\, N_m(\xi)}{\sin \pi\xi\, \Gamma(2 + 2\xi)}\, t^{-\frac{\delta + 1 + \nu + 2\nu\xi}{\sigma}} \tag{4.23}$$

$$\times B\left(\frac{1 + \delta + \nu + +2\nu\xi}{\sigma},\quad \mu + \lambda + 2\lambda\xi - \frac{1 + \delta + \nu + 2\nu\xi}{\sigma}\right).$$

4.1.23 31st *General Formula*

$$N_{31} = \int_0^\infty dx\, x^\delta (1 + tx^\sigma)^{-\mu}\, \left\{\cos^q\left[bx^\nu(1 + tx^\sigma)^{-\lambda}\right] - 1\right\}$$

$$= \frac{1}{2^{q-1}}\, \frac{1}{2i\sigma} \int_{\alpha+i\infty}^{\alpha-i\infty} d\xi\, \frac{(2b)^{2\xi}\, I_q'(\xi)}{\sin \pi\xi\, \Gamma(1 + 2\xi)}\, t^{-\frac{\delta + 1 + 2\nu\xi}{\sigma}} \tag{4.24}$$

$$\times B\left(\frac{\delta + 1 + 2\nu\xi}{\sigma},\quad \mu + 2\lambda\xi - \frac{1 + \delta + 2\nu\xi}{\sigma}\right).$$

4.1.24 32nd *General Formula*

$$
N_{32} = \int\limits_{0}^{\infty} dx x^{\delta}(1 + tx^{\sigma})^{-\mu} \ \cos^{m}\left[bx^{\nu}(1 + tx^{\sigma})^{-\lambda}\right]
$$

$$
= \frac{1}{2^{m-1}} \frac{1}{2i\sigma} \int\limits_{-\beta+i\infty}^{-\beta-i\infty} d\xi \frac{b^{2\xi} \ N_m'(\xi)}{\sin \pi\xi \ \Gamma(1 + 2\xi)} \ t^{-\frac{\delta + 1 + 2\nu\xi}{\sigma}}
$$

$$
\times \ B\left(\frac{\delta + 1 + 2\nu\xi}{\sigma}, \ \ \mu + 2\lambda\xi - \frac{1 + \delta + 2\nu\xi}{\sigma}\right).
$$

(4.25)

4.2   Calculation of Particular Integrals

(1) The formula (4.1) with

$$
q = 2, \ \delta = 2, \ \sigma = 2, \ \nu = 1, \ \mu = -\frac{3}{2}, \ \lambda = -\frac{1}{2}
$$

reads:

$$
j_1 = \int\limits_{0}^{1} dx x^2 (1 - x^2)^{-\frac{3}{2}} \ \sin^2\left[bx(1 - x^2)^{-1/2}\right] = -\frac{1}{8i}
$$

$$
\times \int\limits_{\alpha+i\infty}^{\alpha-i\infty} d\xi \frac{(2b)^{2\xi}}{\sin \pi\xi \ \Gamma(1 + 2\xi)} \ \frac{\Gamma\left(\frac{3+2\xi}{2}\right) \ \Gamma\left(1 - \frac{3}{2} - \xi\right)}{\Gamma(1)},
$$

(4.26)

where

$$
\bullet \ \Gamma\left(\frac{3 + 2\xi}{2}\right) \Gamma\left(1 - \frac{3}{2} - \xi\right) = \frac{\pi}{\sin \pi\left(\frac{3}{2} + \xi\right)},
$$

$$
\bullet \ \sin \pi\left(\frac{3}{2} + \xi\right) = -\cos \pi\xi.
$$

After changing the integration variable $2\xi \to y$, we have

$$
j_1 = \frac{\pi}{8i} \int\limits_{\alpha+i\infty}^{\alpha-i\infty} dy \frac{(2b)^y}{\sin \pi\xi \ \Gamma(1 + y)} = \frac{\pi}{4}\left[e^{-2b} - 1\right].
$$

(4.27)

(2) The main formula (4.1) with

$$
q = 2, \ \delta = 1, \ \sigma = 4, \ \nu = 2, \ \mu = -\frac{1}{2}, \ \lambda = -\frac{1}{2}
$$

gives

$$j_2 = \int\limits_0^1 dx\, x(1 - x^4)^{-1/2}\, \sin^2\left[bx^2(1 - x^4)^{-\frac{1}{2}}\right]$$

$$= -\frac{1}{16i} \int\limits_{\alpha+i\infty}^{\alpha-i\infty} d\xi \frac{(2b)^{2\xi}}{\sin \pi\xi\, \Gamma(1 + 2\xi)} \frac{\Gamma\left(\frac{1}{2} + \xi\right)\, \Gamma\left(\frac{1}{2} - \xi\right)}{\Gamma(1)}$$

$$= -\frac{\pi}{16i} \int\limits_{\alpha+i\infty}^{\alpha-i\infty} d\xi \frac{(2b)^{2\xi}}{\frac{1}{2}\sin 2\pi\xi\, \Gamma(1 + 2\xi)}.$$

Substitution $2\xi \to x$ gives

$$j_2 = -\frac{\pi}{8} \sum_{n=1}^{\infty} \frac{(-1)^n (2b)^n}{n!} = \frac{\pi}{8}\left(1 - e^{-2b}\right). \tag{4.28}$$

(3) Assuming

$$\delta = 7,\ \sigma = 8,\ \nu = -4,\ \mu = -1,\ m = 1,\ \lambda = \frac{1}{2}$$

in the formula (4.3), one gets

$$j_3 = \int\limits_0^1 dx\, x^7 (1 - x^8)^{-1}\, \sin\left[bx^{-4}\left(1 - x^8\right)^{1/2}\right]$$

$$= \frac{1}{16i} \int\limits_{-\beta+i\infty}^{-\beta-i\infty} d\xi \frac{b^{2\xi+1}}{\sin \pi\xi\, \Gamma(2 + 2\xi)} \frac{\Gamma\left(\frac{1}{2} - \xi\right)\, \Gamma\left(\frac{1}{2} + \xi\right)}{\Gamma(1)}$$

$$= \frac{\pi b}{8i} \int\limits_{-\beta+i\infty}^{-\beta-i\infty} d\xi \frac{b^{2\xi}}{\sin 2\pi\xi\, \Gamma(2 + 2\xi)}.$$

Going to the integration variable

$$\xi \to x - \frac{1}{2},\ \beta \to \beta',\ \frac{1}{2} < \beta' < 1$$

and taking residue at the points $x = n$, one obtains:

$$j_3 = -\frac{b\pi}{16i} \int\limits_{-\beta'+i\infty}^{-\beta'-i\infty} dx \frac{b^{2x-1}}{\sin \pi x\, \cos \pi x\, \Gamma(1 + 2x)}$$

$$= -\frac{\pi}{8} \sum_{n=1}^{\infty} \frac{b^{2n}}{(2n)!} = \frac{\pi}{8}(1 - \cosh b), \tag{4.29}$$

where

$$\cosh b = \frac{1}{2}\left(e^{-b} + e^b\right)$$

is the hyperbolic cosine function.

(4) The formula (4.3) with parameters

$$\delta = -1, \ \sigma = 6, \ \nu = -3, \ \mu = 0, \ m = 1, \ \lambda = \frac{1}{2}$$

gives the following integral

$$j_4 = \int\limits_0^1 dx x^{-1} \ \sin\left[bx^{-3}(1 - x^6)^{1/2}\right]$$

$$= \frac{1}{12i} \int\limits_{-\beta+i\infty}^{-\beta-i\infty} d\xi \frac{b^{2\xi+1}}{\sin \pi\xi \ \Gamma(2 + 2\xi)} \frac{\Gamma\left(\frac{3}{2} + \xi\right) \Gamma\left(1 - \frac{3}{2} - \xi\right)}{\Gamma(1)}$$

$$= \frac{b\pi}{12i} \int\limits_{-\beta+i\infty}^{-\beta-i\infty} d\xi \frac{b^{2\xi}}{\sin \pi\xi \ \Gamma(2 + 2\xi)} \frac{1}{\sin\left(\frac{3}{2} + \xi\right)\pi}. \tag{4.30}$$

Going to the variable $\xi \to x - \frac{1}{2}$, one gets

$$j_4 = \frac{\pi}{6i} \int\limits_{-\beta+i\infty}^{-\beta-i\infty} dx \frac{b^{2x}}{\sin 2\pi x \ \Gamma(1 + 2x)}.$$

Shifting the integration contour to the right and calculating residue at the points $x = n$, we obtain

$$j_4 = \frac{\pi}{6} \sum_{n=0}^{\infty} \frac{b^{2n}}{(2n)!} = \frac{\pi}{6} \cosh b. \tag{4.31}$$

(5) If instead of the above parameters in (4.30), we choose other parameters:

$$\delta = 11, \ \sigma = 6, \ \nu = -3, \ \mu = 0, \ m = 1, \ \lambda = \frac{1}{2},$$

then

$$j_5 = \int\limits_0^1 dx \ x^{11} \ \sin\left[bx^{-3}(1 - x^6)^{1/2}\right]$$

$$= \frac{1}{12i} \int\limits_{-\beta+i\infty}^{-\beta-i\infty} d\xi \frac{b^{2\xi+1}}{\sin \pi\xi \ \Gamma(2 + 2\xi)} \frac{\Gamma\left(\frac{3}{2} + \xi\right) \Gamma\left(3 - \frac{3}{2} - \xi\right)}{\Gamma(3)},$$

where as in Chapter 3, we have

$$\Gamma\left(\frac{3}{2} - \xi\right) = -\left(\frac{1}{4} - \xi^2\right)\Gamma\left(1 - \frac{3}{2} - \xi\right).$$

So that

$$j_5 = \frac{\pi b}{6} \frac{1}{2i} \int\limits_{-\beta+i\infty}^{-\beta-i\infty} d\xi \frac{b^{2\xi}}{\sin 2\pi\xi \ \Gamma(2 + 2\xi)} \left(\frac{1}{4} - \xi^2\right).$$

Going to the integration variable

$$2\xi \to y - 1, \quad d\xi = \frac{1}{2}dy, \quad \sin\pi(y - 1) = -\sin\pi y$$

and the substitution (3.27) in Chapter 3, one gets

$$j_5 = \frac{\pi b}{48}(1 + b)e^{-b}. \tag{4.32}$$

(6) The main formula (4.4) with

$$q = 2, \ \delta = 1, \ \sigma = 4, \ \nu = 2, \ \mu = \frac{3}{2}, \ \lambda = -\frac{1}{2}$$

reads

$$j_6 = \int\limits_0^1 dx\, x(1 - x^4)^{\frac{3}{2}}\left\{\cos^2\left[bx^2\left(1 - x^4\right)^{-1/2}\right] - 1\right\} = \frac{1}{8}$$

$$\times \frac{1}{2i}\int\limits_{\alpha+i\infty}^{\alpha-i\infty} d\xi \frac{(2b)^{2\xi}}{\sin\pi\xi\,\Gamma(1 + 2\xi)}\, \frac{\Gamma\left(\frac{1}{2} + \xi\right)\,\Gamma\left(\frac{5}{2} - \xi\right)}{\Gamma(3)}, \tag{4.33}$$

where

$$\Gamma\left(\frac{5}{2} - \xi\right) = \left(\frac{3}{4} - 2\xi + \xi^2\right)\Gamma\left(\frac{1}{2} - \xi\right).$$

The Mellin representation (4.33) is similar to the one of (3.25) in Chapter 3. Thus, after some calculations, we have

$$j_6 = \frac{\pi}{16}\left\{\frac{3}{4} - \frac{2}{2}(-2b) + \frac{1}{4}\left[(2b)^2 - (2b)\right]\right\}(e^{-2b} - 1)$$

$$= \frac{\pi}{64}\left[3 + 6b + 4b^2\right](e^{-2b} - 1). \tag{4.34}$$

(7) We put

$$\delta = -1, \ \sigma = 6, \ \nu = -3, \ \mu = 0, \ \lambda = 1, \ q = 2$$

in (4.4) and obtain

$$j_7 = \int\limits_0^1 dx\, x^{-1}\left\{\cos^2\left[bx^{-3}(1 - x^6)\right] - 1\right\} = \frac{1}{12}\frac{1}{2i}$$

$$\times \int\limits_{\alpha+i\infty}^{\alpha-i\infty} d\xi \frac{(2b)^{2\xi}}{\sin\pi\xi\,\Gamma(1 + 2\xi)}\, \frac{\Gamma(-\xi)\,\Gamma(1 + 2\xi)}{\Gamma(1 + \xi)}, \tag{4.35}$$

where

$$\Gamma(-\xi) = -\frac{\pi}{\sin\pi\xi\,\Gamma(1 + \xi)}.$$

Thus,

$$j_7 = -\frac{\pi}{12}\frac{1}{2i}\int\limits_{\alpha+i\infty}^{\alpha-i\infty} d\xi \frac{(2b)^{2\xi}}{\sin^2 \pi\xi \, \Gamma^2(1+\xi)}.$$

Using the formula (1.10) in Chapter 1 for calculation of residue at poles of second order, one gets

$$
\begin{aligned}
j_7 &= -\frac{1}{12}\sum_{n=1}^{\infty}\frac{(2b)^{2n}}{\Gamma^2(1+n)}\left[\ln(4b^2) - 2\Psi(1+n)\right] \\
&= \frac{1}{12}\sum_{n=1}^{\infty}\frac{(2b)^{2n}}{(n!)^2}\left[2\Psi(1+n) - \ln(4b^2)\right] \\
&= \frac{1}{6}\left[K_0(z) - \Psi(1) + \ln(2b)\right],
\end{aligned}
\tag{4.36}
$$

where $z = 4b$.

(8) The main formula (4.5) with parameters

$$m = 1, \ \delta = 3, \ \sigma = 8, \ \nu = -4, \ \mu = -\frac{1}{2}, \ \lambda = \frac{1}{2}$$

gives an integral

$$
\begin{aligned}
j_8 &= \int\limits_0^1 dx x^3 (1 - x^8)^{-1/2} \cos\left(bx^{-4}(1 - x^8)^{1/2}\right) = \frac{1}{8}\frac{1}{2i} \\
&\times \int\limits_{-\beta+i\infty}^{-\beta-i\infty} d\xi \frac{b^{2\xi}}{\sin \pi\xi \, \Gamma(1 + 2\xi)}\frac{\Gamma\left(\frac{1}{2}+\xi\right)\Gamma\left(\frac{1}{2}-\xi\right)}{\Gamma(1)}.
\end{aligned}
\tag{4.37}
$$

Changing the integration variable $\xi = x + \frac{1}{2}$ and calculating residue at the points $x = n$, one gets

$$j_8 = -\frac{\pi}{8}\sinh b,\tag{4.38}$$

where

$$\sinh b = \frac{1}{2}\left(e^x - e^{-x}\right)$$

is the hyperbolic sine function.

(9) Let

$$m = 1, \ \delta = 3, \ \sigma = 8, \ \nu = 4, \ \mu = \frac{1}{2}, \ \lambda = -\frac{1}{2}$$

in the formula (4.5). Then we have

$$
\begin{aligned}
j_9 &= \int\limits_0^1 dx x^3 (1 - x^8)^{1/2} \cos\left[bx^4(1 - x^8)^{-1/2}\right] = \frac{1}{8}\frac{1}{2i} \\
&\times \int\limits_{-\beta+i\infty}^{-\beta-i\infty} d\xi \frac{b^{2\xi}}{\sin \pi\xi \, \Gamma(1 + 2\xi)}\Gamma\left(\frac{1}{2}+\xi\right)\Gamma\left(\frac{3}{2}-\xi\right),
\end{aligned}
\tag{4.39}
$$

where

$$\bullet \; \Gamma\left(\frac{3}{2} - \xi\right) = \Gamma\left(\frac{1}{2} - \xi\right)\left(\frac{1}{2} - \xi\right),$$

$$\bullet \; \Gamma\left(\frac{1}{2} + \xi\right)\Gamma\left(\frac{1}{2} - \xi\right) = \frac{\pi}{\cos\pi\xi}.$$

After some calculations, we have

$$j_9 = \frac{\pi}{16}(1 + b)e^{-b}. \tag{4.40}$$

(10) The case $m = 3$ in the previous integral (4.39) reads

$$j_{10} = \int\limits_0^1 dx x^3 (1 - x^8)^{1/2} \; \cos^3[bx^4(1 - x^8)^{-1/2}]$$

$$= \frac{\pi}{32} \frac{1}{2i} \int\limits_{-\beta+i\infty}^{-\beta-i\infty} d\xi \frac{b^{2\xi}}{\sin\pi\xi \; \cos\pi\xi \; \Gamma(1 + 2\xi)}$$

$$\times \left(\frac{1}{2} - \xi\right)(3 + 3^{2\xi}). \tag{4.41}$$

Simple calculation results

$$j_{10} = \frac{\pi}{64}\left[3(1 + b)\, e^{-b} + (1 + 3b)\, e^{-3b}\right]. \tag{4.42}$$

(11) In the general formula (4.6), we put

$$\delta = \frac{1}{2}, \; \nu = \frac{1}{2}, \; \mu = -\frac{1}{2}, \; q = 2, \; \lambda = -\frac{1}{2}, \; I_2(\xi) = -1$$

and obtain

$$j_{11} = \int\limits_a^c dx (x - a)^{1/2}(c - x)^{-1/2} \; \sin^2\left[b(x - a)^{1/2}(c - x)^{-1/2}\right]$$

$$= -\frac{1}{2}\frac{1}{2i}\int\limits_{\alpha+i\infty}^{\alpha-i\infty} d\xi \frac{(2b)^{2\xi}(c - a)}{\sin\pi\xi \; \Gamma(1 + 2\xi)} \; B\left(\frac{3}{2} + \xi, \; \frac{1}{2} - \xi\right)$$

$$= -\frac{(c - a)\pi}{2}\frac{1}{2i}\int\limits_{\alpha+i\infty}^{\alpha-i\infty} d\xi \frac{(2b)^{2\xi}\left(\frac{1}{2} + \xi\right)}{\sin\pi\xi \; \cos\pi\xi \; \Gamma(1 + 2\xi)}. \tag{4.43}$$

Changing variable $2\xi \to y$ and shifting the integral to right, one gets

$$j_{11} = -\frac{(c - a)\pi}{4}\left\{(e^{-2b} - 1) - 2b\, e^{-2b}\right\}$$

$$= \frac{\pi}{4}(c - a)\left[1 - (1 - 2b)\, e^{-2b}\right]. \tag{4.44}$$

(12) The formula (4.6) with parameters:

$$q = 4, \ \delta = -\frac{1}{2}, \ \nu = -\frac{1}{2}, \ \mu = -\frac{1}{2}, \ \lambda = \frac{1}{2}, \ I_4(\xi) = 2^{2\xi} - 4$$

reads

$$j_{12} = \int\limits_a^c dx (x-a)^{-1/2}(c-x)^{-1/2} \sin^4 \left[b(x-a)^{-1/2}(c-x)^{1/2} \right]$$

or

$$j_{12} = \frac{1}{8} \frac{1}{2i} \int\limits_{\alpha+i\infty}^{\alpha-i\infty} d\xi \frac{(2b)^{2\xi}(2^{2\xi}-4)}{\sin \pi\xi \ \Gamma(1+2\xi)} \frac{\Gamma\left(\frac{1}{2}+\xi\right) \Gamma\left(\frac{1}{2}-\xi\right)}{\Gamma(1)}.$$

The changing the integration variable $2\xi \to x$ leads to

$$j_{12} = \frac{\pi}{8} \left[e^{-4b} - 1 - 4 \left(e^{-2b} - 1 \right) \right]$$

$$= \frac{\pi}{8} \left[3 + e^{-4b} - 4 \ e^{-2b} \right]. \tag{4.45}$$

(13) The formula (4.7) with parameters

$$\delta = -1, \ \nu = \frac{1}{2}, \ \mu = 0, \ \lambda = -\frac{1}{2}, \ m = 5,$$

$$N_5(\xi) = 5^{2\xi+1} - 5 \cdot 3^{2\xi+1} + 10$$

gives an integral

$$j_{13} = \int\limits_a^c dx (x-a)^{-1} \sin^5 \left[b(x-a)^{1/2}(c-x)^{-1/2} \right] = \frac{1}{16} \frac{1}{2i}$$

$$\times \int\limits_{-\beta+i\infty}^{-\beta-i\infty} d\xi \frac{b^{2\xi+1} \left[5^{2\xi+1} - 5 \cdot 3^{2\xi+1} + 10 \right]}{\sin \pi\xi \ \Gamma(2+2\xi)} \Gamma\left(\frac{1}{2}-\xi\right) \Gamma\left(\frac{1}{2}+\xi\right)$$

$$= \frac{\pi b}{8} \frac{1}{2i} \int\limits_{-\beta+i\infty}^{-\beta-i\infty} d\xi \frac{b^{2\xi}}{\sin 2\pi\xi \ \Gamma(2+2\xi)} \left(5^{2\xi+1} - 5 \cdot 3^{2\xi+1} + 10 \right).$$

Changing the integration variable $\xi \to x - \frac{1}{2}$ and the contour of integration $\beta \to \beta'$, $1/2 < \beta' < 1$ and calculating residue at the points $x = n$, one gets

$$j_{13} = -\frac{\pi}{16} \frac{1}{2i} \int\limits_{-\beta'+i\infty}^{-\beta'-\infty} dx \frac{b^{2x} \left(5^{2x} - 5 \cdot 3^{2x} + 10 \right)}{\sin \pi x \ \cos \pi x \ \Gamma(1+2x)}$$

$$= -\frac{\pi}{16} \left\{ \cosh(5b) - 1 - 5(\cos(3b) - 1) + 10(\cos b - 1) \right\}$$

$$= \frac{\pi}{16} \left[6 - 10 \cosh b - \cosh(5b) + 5 \cosh(3b) \right]. \tag{4.46}$$

(14) The formula (4.7) with

$$\delta = 0, \ \nu = \frac{1}{2}, \ \mu = -1, \ \lambda = -\frac{1}{2}, \ m = 7,$$

$$N_7(\xi) = -7^{2\xi+1} + 7 \cdot 5^{2\xi+1} - 21 \cdot 3^{2\xi+1} + 35$$

gives the following integral

$$j_{14} = \int_a^c dx (c - x)^{-1} \sin^7 \left[b(x - a)^{1/2} (c - x)^{-1/2} \right]$$

$$= \frac{1}{64} \frac{1}{2i} \int_{-\beta+i\infty}^{-\beta-i\infty} d\xi \frac{b^{2\xi+1}}{\sin \pi \xi \, \Gamma(2 + 2\xi)}$$

$$\times \left(-7^{2\xi+1} + 7 \cdot 5^{2\xi+1} - 21 \cdot 3^{2\xi+1} + 35 \right) \Gamma\left(\frac{3}{2} + \xi \right)$$

$$\times \Gamma\left(1 - \frac{3}{2} - \xi \right). \tag{4.47}$$

The change of the integration variable $\xi \to x - \frac{1}{2}$ leads to

$$j_{14} = \frac{\pi}{32} \frac{1}{2i} \int_{-\beta+i\infty}^{-\beta-i\infty} dx \frac{b^{2x} \left(-7^{2x} + 7 \cdot 5^{2x} - 21 \cdot 3^{2x} + 35 \right)}{\sin 2\pi x \, \Gamma(1 + 2x)}.$$

Displacing the integration contour to the right and taking residue at the points $x = n$, one gets

$$j_{14} = \frac{\pi}{64} \left[35 \cosh b + 7 \cosh(5b) - \cosh(7b) - 21 \cosh(3b) \right]. \tag{4.48}$$

(15) The formula (4.8) with parameters:

$$\delta = \frac{1}{2}, \ \nu = \frac{1}{2}, \ \mu = \frac{1}{2}, \ \lambda = -\frac{1}{2}, \ q = 4,$$

$$I_4'(\xi) = 2^{2\xi} + 4$$

leads to an integral

$$j_{15} = \int_a^c dx (x - a)^{1/2} (c - x)^{1/2} \left\{ \cos^4 \left[b(x - a)^{1/2}(c - x)^{-1/2} \right] - 1 \right\}$$

$$= \frac{1}{8} \frac{1}{2i} (c - a)^2 \int_{\alpha+i\infty}^{\alpha-i\infty} d\xi \frac{(2b)^{2\xi} \left(2^{2\xi} + 4 \right)}{\sin \pi \xi \, \Gamma(1 + 2\xi)} \frac{\Gamma\left(\frac{3}{2} + \xi \right) \Gamma\left(\frac{3}{2} - \xi \right)}{\Gamma(3)},$$

where

$$\bullet \ \Gamma\left(\frac{3}{2} - \xi \right) = -\left(\frac{1}{4} - \xi^2 \right) \Gamma\left(1 - \frac{3}{2} - \xi \right),$$

$$\bullet \ \Gamma\left(\frac{3}{2} + \xi \right) \Gamma\left(1 - \frac{3}{2} - \xi \right) = \frac{\pi}{\sin \pi \left(\frac{3}{2} - \xi \right)} = -\frac{\pi}{\cos \pi \xi}.$$

Thus,

$$j_{15} = \frac{\pi}{8} \frac{(c-a)^2}{2i} \int\limits_{a+i\infty}^{a-i\infty} d\xi \frac{(2b)^{2\xi} \, (2^{2\xi}+4)}{\sin 2\pi\xi \, \Gamma(2+2\xi)} \left(\frac{1}{4} - \xi^2\right)(1+2\xi), \quad (4.49)$$

where

$$\Lambda_1 = \left(2^{2\xi} + 4\right)\left(\frac{1}{4} - \xi^2\right)(1+2\xi)$$

$$= 4\left(\frac{1}{4} + \frac{\xi}{2} - \xi^2 - 2\xi^3\right) + 2^{2\xi}\left(\frac{1}{4} + \frac{\xi}{2} - \xi^2 - 2\xi^3\right).$$

In the integral (4.49), we do a change of variable

$$2\xi \to y-1, \ d\xi = \frac{1}{2}dy, \ \sin\pi(y-1) = -\sin\pi y$$

and

$$\Lambda_1 \to 2y^2 - y^3 + \frac{2^y}{8}(2y^2 - y^3).$$

Then

$$j_{15} = -\frac{\pi}{16} \frac{(c-a)^2}{2i} \int\limits_{a+i\infty}^{a-i\infty} dy \frac{(2b)^{y-1}}{\sin\pi y} \frac{1}{\Gamma(1+y)}$$

$$\times \left[2y^2 - y^3 + \frac{2^y}{8}(2y^2 - y^3)\right]. \quad (4.50)$$

Here we use the substitution (3.27) in Chapter 3 and calculate the residue at the poles $y = n$. The result reads

$$j_{15} = -\frac{\pi}{16}\left\{\frac{5}{4} + 3b - 8b^2 + (4b^2 - 2b - 1)\, e^{-2b}\right.$$

$$\left. + \left(4b^2 - b - \frac{1}{4}\right)\, e^{-4b}\right\}(c-a)^2. \quad (4.51)$$

(16) The choice of parameters

$$\delta = -\frac{1}{2}, \ \nu = \frac{1}{2}, \ \mu = -\frac{1}{2}, \ \lambda = -\frac{1}{2}, \ m = 3,$$

$$N_3'(\xi) = 3 + 3^{2\xi}$$

in the formula (4.9) gives an integral:

$$j_{16} = \int\limits_a^c dx(x-a)^{-1/2}(c-x)^{-1/2}\cos^3\left[b(x-a)^{1/2}(c-x)^{-1/2}\right]$$

$$= \frac{1}{4}\frac{1}{2i} \int\limits_{-\beta+i\infty}^{-\beta-i\infty} d\xi \frac{b^{2\xi}\, (3 + 3^{2\xi})}{\sin\pi\xi\, \Gamma(2+2\xi)}$$

$$\times \frac{\Gamma\left(\frac{1}{2} - \xi\right)\Gamma\left(\frac{1}{2} + \xi\right)}{\Gamma(1)}(1+2\xi), \quad (4.52)$$

where

$$\Lambda_2 = (3 + 3^{2\xi})(1 + 2\xi) = 3 + 3^{2\xi} + 2\xi(3 + 3^{2\xi}).$$

We carry out the same procedure as in the previous case in formula (4.49):

$$2\xi \to y - 1, \quad \sin \pi(y - 1) = -\sin \pi y$$

at which

$$\Lambda_2 \to 3 + 3^{y-1} + (y - 1)(3 + 3^{y-1}) = y(3 + 3^{y-1}).$$

After some simple calculations, we have

$$j_{16} = -\frac{\pi}{4} \frac{1}{2i} \int\limits_{-\beta+i\infty}^{-\beta-i\infty} dy \frac{b^{y-1}}{\sin \pi y} \frac{y}{\Gamma(1+y)} (3 + 3^{y-1})$$

$$= -\frac{\pi}{4b} \left\{ -3b \, e^{-b} - \frac{1}{3} \, 3b \, e^{-3b} \right\}$$

$$= \frac{\pi}{4} \left(3 \, e^{-b} + e^{-3b} \right). \tag{4.53}$$

(17) We put

$$\mu = \frac{1}{2}, \quad \nu = -\frac{1}{2}, \quad q = 2, \quad I_2(\xi) = -1$$

in the general formula (4.10) and obtain

$$j_{17} = \int\limits_0^\infty dx \left[\frac{1}{(x+a)(x+c)} \right]^{1/2} \sin^2 \left[b \left(\frac{(x+a)(x+c)}{x} \right)^{1/2} \right]$$

$$= \frac{\sqrt{\pi}}{2} \frac{-1}{2i} \int\limits_{\alpha+i\infty}^{\alpha-i\infty} d\xi \frac{(2bA)^{2\xi}}{\sin \pi\xi \, \Gamma(1+2\xi)} \frac{\Gamma(-\xi)}{\Gamma\left(\frac{1}{2} - \xi\right)}, \tag{4.54}$$

where

- $A = \sqrt{a} + \sqrt{c}, \quad \Gamma(-\xi) = -\dfrac{\pi}{\Gamma(1+\xi) \, \sin \pi\xi},$

- $\Gamma\left(\dfrac{1}{2} - \xi\right) = \sqrt{\pi} \, 2^{2\xi+1} \dfrac{\Gamma(-2\xi)}{\Gamma(-\xi)}$

and

$$\Gamma(1 + 2\xi) \, \Gamma(-2\xi) = -\frac{\pi}{\sin 2\pi\xi}.$$

Thus,

$$j_{17} = \frac{\pi}{2} \frac{1}{2i} \int\limits_{\alpha+i\infty}^{\alpha-i\infty} d\xi \frac{(bA)^{2\xi}}{\sin^2 \pi\xi} \frac{\cos \pi\xi}{\Gamma^2(1+\xi)}. \tag{4.55}$$

Displacing the contour of integration to the right and calculating residue at the points $\xi = n$, one gets

$$j_{17} = \sum_{n=1}^{\infty} \frac{(-1)^n (bA)^{2n}}{(n!)^2} \left[\ln(bA) - \Psi(1+n) \right]$$

$$= \sum_{n=1}^{\infty} \frac{(ibA)^{2n}}{(n!)^2} \left[\ln i^{-1} + \ln(ibA) - \Psi(1+n) \right]$$

$$= \ln i^{-1} \left[J_0(2bA) - 1 \right] - K_0(2ibA) + \Psi(1) - \ln(ibA), \qquad (4.56)$$

where

$$\ln i^{-1} = \ln(e^{-i\pi/2}) = -i\frac{\pi}{2}.$$

(18) The formula (4.10) with

$$q = 2, \ \mu = \frac{3}{2}, \ \nu = 1$$

leads to an integral

$$j_{18} = \int_0^{\infty} dx\, x \left[(x+a)(x+c) \right]^{-3/2} \sin^2 \left[b\frac{x}{(x+a)(x+c)} \right]$$

$$= \frac{\sqrt{\pi}}{2} (\sqrt{a} + \sqrt{c})^{-2} \frac{-1}{2i} \int_{\alpha+i\infty}^{\alpha-i\infty} d\xi \frac{\left[2b/(\sqrt{a} + \sqrt{c})^2 \right]^{2\xi}}{\sin \pi\xi \ \Gamma\left(\frac{3}{2} + 2\xi\right)}$$

$$= -\frac{\sqrt{\pi}}{2} \frac{1}{(\sqrt{a} + \sqrt{c})^2} \sum_{n=1}^{\infty} \frac{(-1)^n \left[2b/(\sqrt{a} + \sqrt{c})^2 \right]^{2n}}{\Gamma\left(\frac{3}{2} + 2n\right)}. \qquad (4.57)$$

(19) Assuming $\mu = \nu = \frac{1}{2}$, $m = 1$ in (4.11), one gets

$$j_{19} = \int_0^{\infty} dx \frac{1}{\sqrt{x+a}} \frac{1}{\sqrt{x+c}} \sin \left[b\sqrt{x} \frac{1}{\sqrt{(x+a)(x+c)}} \right]$$

$$= 2\pi \frac{1}{2i} \int_{-\beta+i\infty}^{-\beta-i\infty} d\xi \frac{[b/(2A)]^{2\xi+1}}{\sin \pi\xi \ \Gamma^2(1+\xi)(1+2\xi)}$$

$$= 2\pi I \left(\frac{b}{2A} \right). \qquad (4.58)$$

Then

$$\frac{\partial}{\partial Y} I \left(\frac{b}{2A} \right) = \frac{1}{2i} \int_{-\beta+i\infty}^{-\beta-i\infty} d\xi \frac{(b/2A)^{2\xi}}{\sin \pi\xi \ \Gamma^2(1+\xi)}$$

$$= J_0 \left(\frac{b}{A} \right), \quad Y = \frac{b}{2A}. \qquad (4.59)$$

Therefore, from (4.58), it follows that

$$j_{19} = 2\pi \int dY J_0 \left(\frac{b}{A}\right) = \pi \int dx J_0(x)$$

$$= 2\pi \sum_{k=0}^{\infty} J_{1+2k} \left(\frac{b}{A}\right). \tag{4.60}$$

(20) The choice of parameters

$$\nu = -\frac{1}{2}, \ \mu = 1, \ m = 1$$

in (4.11) leads to an integral

$$j_{20} = \int_0^{\infty} dx \sqrt{x} (x+a)^{-1} (x+c)^{-1}$$

$$\times \sin\left[b\frac{1}{\sqrt{x}} \sqrt{(x+a)(x+c)}\right]$$

$$= -\pi \frac{b}{2i} \int_{\beta+i\infty}^{-\beta-i\infty} d\xi \frac{(bA/2)^{2\xi} \cos \pi \xi}{\sin^2 \pi \xi \ \Gamma^2(1+\xi)(1+2\xi)}. \tag{4.61}$$

Here we have obtained a similar integral as in (4.55). After the same calculation, we have

$$j_{20} = -2b \sum_{n=0}^{\infty} \frac{(ibA/2)^{2n}}{(n!)^2} \left(\ln\frac{ibA}{2} - \Psi(1+n)\right)$$

$$= 2bK_0(ibA), \tag{4.62}$$

where $A = \sqrt{a} + \sqrt{c}$.

(21) The formula (4.12) with

$$\mu = \frac{3}{2}, \ \nu = 1, \ q = 2$$

gives

$$j_{21} = \int_0^{\infty} dx x \left[(x+a)(x+c)\right]^{-\frac{3}{2}} \left\{\cos^2\left[bx\frac{1}{(x+a)(x+c)}\right] - 1\right\}$$

$$= \frac{\sqrt{\pi}}{2} \frac{A^{-2}}{2i} \int_{\alpha+i\infty}^{\alpha-i\infty} d\xi \frac{(2b/A^2)^{2\xi}}{\sin \pi \xi \ \Gamma\left(\frac{3}{2} + 2\xi\right)}$$

$$= \frac{\sqrt{\pi}}{2A^2} \sum_{n=1}^{\infty} \frac{(-1)^n}{\Gamma\left(\frac{3}{2} + 2n\right)} \left(\frac{2b}{A^2}\right)^{2n}. \tag{4.63}$$

(22) The case of

$$\mu = \frac{5}{2}, \ \nu = 1, \ q = 2$$

in (4.12) leads to an integral

$$j_{22} = \int\limits_0^\infty \frac{dx}{\sqrt{x}} \left[\frac{x}{(x+a)(x+c)} \right]^{\frac{5}{2}} \left\{ \cos^2 \left[b \frac{x}{(x+a)(x+c)} \right] - 1 \right\}$$

$$= 2\sqrt{\pi} \frac{1}{A^4} \frac{1}{2i} \int\limits_{\alpha+i\infty}^{\alpha-i\infty} d\xi \frac{(2b/A^2)^{2\xi} \ (1+2\xi)}{\sin \pi\xi (3+4\xi)(1+4\xi) \ \Gamma\left(\frac{1}{2}+2\xi\right)}$$

$$= 2\sqrt{\pi} \frac{1}{A^4} \sum_{n=1}^\infty \frac{(-1)^n \ (2b/A^2)^{2n} \ (1+2n)}{(3+2n) \ (1+2n) \ \Gamma\left(\frac{1}{2}+2n\right)}. \tag{4.64}$$

(23) Similar calculations for the formula (4.13) can be carried out. For example, if

$$\mu = \frac{1}{2}, \ \nu = -\frac{1}{2}, \ m = 1,$$

then one gets

$$j_{23} = \int\limits_0^\infty dx (x+a)^{-1/2} (x+c)^{-1/2} \cos\left[b\sqrt{x} (x+a)^{1/2} (x+c)^{1/2} \right]$$

$$= -\pi \frac{1}{2i} \int\limits_{-\beta+i\infty}^{-\beta-i\infty} d\xi \frac{(bA/2)^{2\xi} \ \cos \pi\xi}{\sin^2 \pi\xi \ \Gamma^2(1+\xi)}$$

$$= -\sum_{n=0}^\infty \frac{(-1)^n \ (bA/2)^{2n}}{(n!)^2} \left[2\ln\left(\frac{bA}{2}\right) - 2\Psi(1+n) \right],$$

where

$$(-1)^n \left(\frac{bA}{2}\right)^{2n} = \left(\frac{bA}{2} i\right)^{2n}$$

and

$$\ln\left(\frac{bA}{2}\right) = \ln\left(i\frac{bA}{2i}\right) = \ln i^{-1} + \ln\left(i\frac{bA}{2}\right).$$

So that

$$j_{23} = -2\ln i^{-1} \ J_0(bA) + 2K_0(ibA). \tag{4.65}$$

(24) In the previous case, if we assume

$$\mu = \frac{1}{2}, \ \nu = -\frac{1}{2}, \ m = 3, \ N'_\xi = 3 + 3^{2\xi},$$

then from (4.13) we have

$$j_{24} = \int\limits_0^\infty dx (x+a)^{-1/2} (x+c)^{-1/2} \cos^3 \left[b\sqrt{x}(x+a)^{1/2}(x+c)^{1/2} \right]$$

$$= -\frac{\pi}{4} \frac{1}{2i} \int\limits_{-\beta+i\infty}^{-\beta-i\infty} d\xi \frac{(bA/2)^{2\xi}}{\sin^2 \pi\xi} \frac{\cos \pi\xi}{\Gamma^2(1+\xi)} (3 + 3^{2\xi})$$

$$= \frac{1}{2} \left\{ 3 \left[-J_0(bA) \ln i^{-1} + 2K_0(ibA) \right] - J_0(3bA) \ln i^{-1} \right.$$

$$+ 2K_0(3ibA) \right\} = \frac{1}{2} \left\{ -\ln i^{-1} \left[3J_0(bA) + J_0(3bA) \right] \right.$$

$$+ 6K_0(ibA) + 2K_0(3ibA) \right\}. \tag{4.66}$$

(25) The formula (4.14) with parameters

$$\lambda = 1, \ \mu = 0, \ \sigma = 1, \ q = 2, \ I_2(\xi) = -1$$

reads

$$j_{25} = \int\limits_0^1 dx \sin^2 (b(1-x)) = \frac{1}{2} \frac{-1}{2i}$$

$$\times \int\limits_{\alpha+i\infty}^{\alpha-i\infty} d\xi \frac{(2b)^{2\xi}}{\sin \pi\xi \ \Gamma(1+2\xi)} \frac{\Gamma(1+2\xi)}{\Gamma(2+2\xi)}$$

$$= \frac{1}{4b} [2b - \sin(2b)]. \tag{4.67}$$

(26) Let

$$\mu = -1, \ \lambda = -\frac{1}{2}, \ \sigma = 2, \ q = 2$$

be in the formula (4.14). Then

$$j_{26} = \int\limits_0^1 dx (1-x^2)^{-1} \sin^2 \left[b(1-x^2)^{-1/2} \right]$$

$$= \frac{\pi}{4} \frac{1}{2i} \int\limits_{\alpha+i\infty}^{\alpha-i\infty} d\xi \frac{b^{2\xi}}{\sin^2 \pi\xi} \frac{\cos \pi\xi}{\Gamma^2(1+\xi)}$$

$$= 2 \left\{ \frac{i\pi}{2} [1 - J_0(2b)] + \Psi(1) - \ln(ib) - K_0(2ib) \right\}. \tag{4.68}$$

(27) We put

$$\mu = \lambda = -\frac{1}{2}, \ \sigma = 2, \ m = 1$$

in (4.15) and obtain

$$j_{27} = \int\limits_0^1 dx(1-x^2)^{-1/2} \, \sin\left[b(1-x^2)^{-1/2}\right] \tag{4.69}$$

$$= -\frac{\pi}{2b}\frac{1}{2i} \int\limits_{-\beta+i\infty}^{-\beta-i\infty} d\xi \frac{(b/2)^{2\xi} \, \cos\pi\xi}{\sin^2\pi\xi \, \Gamma^2(1+\xi)(1+2\xi)}$$

$$= \frac{1}{b}\sum_{n=0}^{\infty} \frac{(-1)^n \, (b/2)^{2n}}{(n!)^2(1+2n)}\left[-\ln\frac{b}{2} + \Psi(1+n) + \frac{1}{1+2n}\right].$$

(28) We put

$$\sigma = 2, \ \mu = -\frac{3}{2}, \ \lambda = -\frac{1}{2}, \ q = 2$$

in (4.16) and get

$$j_{28} = \int\limits_0^1 dx(1-x^2)^{-3/2}\left\{\cos^2\left[b(1-x^2)^{-1/2}\right]-1\right\}$$

$$= \frac{\pi}{4}\frac{1}{2i} \int\limits_{\alpha+i\infty}^{\alpha-i\infty} d\xi \frac{(b/2)^{2\xi}}{\cos\pi\xi \, \Gamma\left(\frac{1}{2}+\xi\right) \, \Gamma\left(\frac{3}{2}+\xi\right)}$$

$$= -\frac{\pi}{4}\sum_{m=1}^{\infty} \frac{(b/2)^{2m+1}}{(m!)^2(1+m)}. \tag{4.70}$$

(29) Let

$$\sigma = 2, \ \mu = -\frac{1}{2}, \ \lambda = -1, \ m = 7,$$

$$N_7' = 7^{2\xi} + 7\cdot 5^{2\xi} + 21\cdot 3^{2\xi} + 35$$

be in the general formula (4.17), then

$$j_{29} = \int\limits_0^1 dx(1-x^2)^{-1/2}\cos^7\left[b(1-x^2)^{-1}\right] = \frac{\sqrt{\pi}}{2^7}\frac{1}{2i}$$

$$\times \int\limits_{-\beta+i\infty}^{-\beta-i\infty} d\xi \frac{b^{2\xi}\left[7^{2\xi} + 7\cdot 5^{2\xi} + 21\cdot 3^{2\xi} + 35\right]}{\xi\,\Gamma\left(\frac{1}{2}+2\xi\right)}\frac{\cos\pi\xi}{\cos 2\pi\xi}.$$

Going to the integration variable $2\xi = x$, one gets

$$j_{29} = \frac{\sqrt{\pi}}{2^7}\frac{1}{2i} \int\limits_{-\beta+i\infty}^{-\beta-i\infty} dx \frac{b^x\left(7^x + 7\cdot 5^x + 21\cdot 3^x + 35\right)\cos\frac{\pi}{2}x}{x\,\Gamma\left(\frac{1}{2}+x\right)\,\cos\pi x} = \frac{\pi}{2}$$

$$- \frac{1}{2^6}\sqrt{\frac{\pi b}{2}}\sum_{m=1}^{\infty} \frac{(-1)^m\left(\sqrt{7}\,7^m + 7\sqrt{5}\,5^m + 21\sqrt{3}\,3^m + 35\right)}{(1+2m)m!}$$

$$\times \left(\cos\frac{\pi}{2}m - \sin\frac{\pi}{2}m\right)b^m. \tag{4.71}$$

(30) The choice of parameters

$$\mu = 0, \ \lambda = -\frac{1}{2}, \ q = 2, \ I_2(\xi) = -1$$

in the formula (4.18) leads to an integral

$$j_{30} = \int_0^1 dx \, \sin^2\left[b(1 - \sqrt{x})^{-1/2}\right]$$

$$= -\frac{1}{2i} \int_{\alpha+i\infty}^{\alpha-i\infty} d\xi \frac{(2b)^{2\xi}}{\sin \pi\xi \, \Gamma(1 + 2\xi)} \frac{1}{(1 - \xi)} \frac{1}{(2 - \xi)}.$$

Given the function has double poles at the points $\xi = 1$, $\xi = 2$. After some calculations, we have:

$$j_{30} = 2b^2 \left(2\Psi(3) - 2\ln(2b) - 1\right) - \frac{(2b)^4}{\Gamma(5)} \left(\ln(4b^2) - 2\Psi(5) - 1\right)$$

$$- \sum_{n=3}^{\infty} (-1)^n \frac{(2b)^{2n}}{\Gamma(1 + 2n)} \frac{1}{(1 - n)(2 - n)}$$

$$= 2b^2 \left(2\Psi(3) - 2\ln(2b) - 1\right) + \frac{2}{3}b^4 \left(2\Psi(5) - 2\ln(2b) + 1\right)$$

$$- \sum_{n=3}^{\infty} (-1)^n \frac{(2b)^{2n}}{(2n)!(1 - n)(2 - n)}. \tag{4.72}$$

(31) Another choice $\mu = \lambda = -\frac{1}{2}$, $m = 1$ in the main formula (4.19) gives

$$j_{31} = \int_0^1 dx (1 - \sqrt{x})^{-1/2} \sin\left[b(1 - \sqrt{x})^{-1/2}\right]$$

$$= \frac{1}{i} \int_{-\beta+i\infty}^{-\beta-i\infty} d\xi \frac{b^{2\xi+1}}{\sin \pi\xi \, \Gamma(2 + 2\xi)} \frac{1}{-\xi} \frac{1}{(1 - \xi)}.$$

We see that this function has poles of the second order at the points $\xi = 0$, $\xi = 1$. Calculation of residue at these points and $\xi = n$, $n = 2, 3, \ldots$, gives

$$j_{31} = -2b \left[\ln b^2 - 2\Psi(2) + 1\right] - \frac{1}{3}b^3 \left[\ln b^2 - 2\Psi(4) - 1\right]$$

$$- 2 \sum_{n=2}^{\infty} (-1)^n \frac{b^{2n+1}}{(1 + 2n)! \, n \, (1 - n)}. \tag{4.73}$$

(32) Let

$$\mu = -\frac{1}{2}, \ \lambda = -\frac{1}{2}, \ q = 2$$

be in (4.20), then

$$j_{32} = \int\limits_0^1 dx (1 - \sqrt{x})^{-1/2} \left\{ \cos^2 \left[b(1 - \sqrt{x})^{-1/2} \right] - 1 \right\}$$

$$= \frac{1}{2i} \int\limits_{\alpha+i\infty}^{\alpha-i\infty} d\xi \frac{b^{2\xi}}{\sin \pi \xi \; \Gamma(1 + 2\xi)} \frac{1}{\left(\frac{1}{2} - \xi \right)} \frac{1}{\left(\frac{3}{2} - \xi \right)}.$$

Now we calculate necessary residues at the points

$$\xi = n, \; n = 1, \, 2, \, 3, \ldots, \; \xi = \frac{1}{2}, \; \xi = \frac{3}{2}$$

and derive

$$j_{32} = \pi b + \frac{\pi b^3}{6} + 4 \sum_{n=1}^{\infty} \frac{(-1)^n \, b^{2n}}{(2n)! \, (1 - 2n) \, (3 - 2n)}. \tag{4.74}$$

(33) The formula (4.21) with

$$\mu = 0, \; \lambda = 1, \; m = 3, \; N_3'(\xi) = 3 + 3^{2\xi}$$

reads

$$j_{33} = \int\limits_0^1 dx \cos^3 \left[b(1 - \sqrt{x}) \right],$$

$$j_{33} = \frac{1}{4i} \int\limits_{-\beta+i\infty}^{-\beta-i\infty} d\xi \frac{b^{2\xi} \left(3 + 3^{2\xi} \right)}{\sin \pi \xi \; \Gamma(1 + 2\xi)} \frac{1}{1 + 2\xi}$$

$$\times \frac{1}{2 + 2\xi} = \frac{1}{2} \sum_{n=0}^{\infty} \frac{(-1)^n}{(2 + 2n)!} \left(3 + 3^{2n} \right) b^{2n}. \tag{4.75}$$

(34) Let

$$\delta = 1, \; \sigma = 2, \; \mu = 2, \; \nu = 2, \; \lambda = 1, \; q = 2$$

be in the formula (4.22), then

$$j_{34} = \int\limits_0^{\infty} dx x (1 + t x^2)^{-2} \; \sin^2 \left[b x^2 (1 + t x^2)^{-1} \right]$$

$$= -\frac{1}{4t} \frac{1}{2i} \int\limits_{\alpha+i\infty}^{\alpha-i\infty} d\xi \frac{(2b/t)^{2\xi}}{\sin \pi \xi \; \Gamma(1 + 2\xi)} \frac{\Gamma(1 + 2\xi)}{\Gamma(2 + 2\xi)}$$

$$= -\frac{1}{4t} \sum_{n=1}^{\infty} (-1)^n \frac{(2b/t)^{2n}}{(1 + 2n)!} = \frac{1}{8b} \left(\frac{2b}{t} - \sin \frac{2b}{t} \right). \tag{4.76}$$

(35) The formula (4.23) with parameters

$$\delta = -\frac{1}{2}, \ \sigma = 1, \ \mu = \frac{3}{2}, \ \nu = -\frac{1}{2}, \ \lambda = \frac{1}{2}, \ m = 1$$

gives

$$j_{35} = \int\limits_0^\infty dx \frac{(1+tx)^{-3/2}}{\sqrt{x}} \ \sin\left[bx^{-1/2}(1+tx)^{-1/2}\right] \qquad (4.77)$$

$$= \frac{1}{2i} \int\limits_{-\beta+i\infty}^{-\beta-i\infty} d\xi \frac{b^{2\xi+1} \ (\sqrt{t})^{2\xi}}{\sin\pi\xi \ \Gamma(2+2\xi)} \ \frac{\Gamma(-\xi) \ \Gamma(2+2\xi)}{\Gamma(2+\xi)}$$

$$= -\frac{\pi}{2i} \int\limits_{-\beta+i\infty}^{-\beta-i\infty} d\xi \frac{b^{2\xi+1} \ (\sqrt{t})^{2\xi}}{\sin^2\pi\xi \ \Gamma^2(1+\xi) \ (1+\xi)}$$

$$= -b \sum_{n=0}^\infty \frac{(b\sqrt{t})^{2n}}{(n!)^2(1+n)} \left[\ln(b^2 t) - \Psi(1+n) - \frac{1}{1+n}\right].$$

(36) Let

$$\delta = -\frac{1}{2}, \ \sigma = \frac{1}{2}, \ \nu = \frac{1}{2}, \ \mu = 2, \ \lambda = 1, \ q = 2$$

be in (4.24). Then

$$j_{36} = \int\limits_0^\infty dx \frac{(1+t\sqrt{x})^{-2}}{\sqrt{x}} \ \left\{\cos^2\left[b\sqrt{x}(1+t\sqrt{x})\right] - 1\right\}$$

$$= \frac{1}{2it} \int\limits_{\alpha+i\infty}^{\alpha-i\infty} d\xi \frac{(2b/t)^{2\xi}}{\sin\pi\xi \ \Gamma(1+2\xi)} \ \frac{\Gamma(1+2\xi)}{\Gamma(2+2\xi)}.$$

After some elementary calculations, we have

$$j_{36} = \frac{1}{t} \sum_{n=1}^\infty \frac{(-1)^n}{(1+2n)!} \left(\frac{2b}{t}\right)^{2n} = \frac{1}{2b} \left[\sin\frac{2b}{t} - \frac{2b}{t}\right]. \qquad (4.78)$$

(37) The choice of parameters

$$\delta = -1, \ \sigma = 4, \ \mu = 1, \ \nu = -2, \ \lambda = \frac{1}{2}, \ m = 1$$

in (4.25) reads

$$
j_{37} = \int\limits_0^\infty dx \frac{(1 + tx^4)^{-1}}{x} \; \cos \left[bx^{-2}(1 + tx^4)^{-1/2} \right]
$$

$$
= \frac{1}{4} \frac{1}{2i} \int\limits_{-\beta+i\infty}^{-\beta-i\infty} d\xi \frac{(b\sqrt{t})^{2\xi}}{\sin \pi\xi \; \Gamma(1 + 2\xi)} \; \frac{\Gamma(-\xi) \; \Gamma(1 + 2\xi)}{\Gamma(1 + \xi)}
$$

$$
= -\frac{\pi}{4} \frac{1}{2i} \int\limits_{-\beta+i\infty}^{-\beta-i\infty} d\xi \frac{(b\sqrt{t})^{2\xi}}{\sin^2 \pi\xi \; \Gamma(1 + \xi)}
$$

$$
= -\frac{1}{4} \sum_{n=0}^\infty \frac{(b\sqrt{t})^{2n}}{(n!)^2} \left[\ln(b^2 t) - 2\Psi(1 + n) \right]. \tag{4.79}
$$

(38) Finally, we consider the case:

$$
\delta = 1, \; \sigma = 4, \; \nu = -2, \; \lambda = -\frac{1}{2}, \; \mu = 1, \; m = 1
$$

in (4.25) which gives

$$
j_{38} = \int\limits_0^\infty dx\, x(1 + tx^4)^{-1} \; \cos \left[bx^{-2}(1 + tx^4)^{1/2} \right]
$$

$$
= \frac{1}{\sqrt{t}} \frac{\pi}{4} \frac{1}{2i} \int\limits_{-\beta+i\infty}^{-\beta-i\infty} d\xi \frac{(b\sqrt{t}/2)^{2\xi}}{\xi \cos \pi\xi \; \Gamma^2 \left(\frac{1}{2} + \xi \right)}
$$

$$
= \frac{\pi}{4} \frac{1}{\sqrt{t}} + \frac{\pi b}{4} \sum_{n=1}^\infty \frac{(-1)^n \; (b\sqrt{t}/2)^{2n}}{(n!)^2(2n + 1)}. \tag{4.80}
$$

Chapter 5

Integrals Involving x^γ, $\dfrac{1}{(p + tx^\rho)^\lambda}$, e^{-ax^ν} and Trigonometric Functions

5.1 Universal Formulas for Integrals Involving Exponential Functions

5.1.1 33rd General Formula

$$N_{33} = \int_0^\infty dx e^{-bx^\nu} = \frac{1}{2i} \int_{-\beta+i\infty}^{-\beta-i\infty} d\xi \frac{b^\xi}{\sin \pi\xi} \frac{1}{\Gamma(1+\xi)}$$

$$\times \lim_{\varepsilon \to 0} \int_\varepsilon^\infty dx x^{\nu\xi} = -\lim_{\varepsilon \to 0} \frac{1}{2i} \int_{-\beta+i\infty}^{-\beta-i\infty} d\xi \frac{b^\xi \, \varepsilon^{1+\nu\xi}}{\sin \pi\xi \, \Gamma(1+\xi)(1+\nu\xi)}$$

$$= \frac{1}{\nu} \Gamma\left(\frac{1}{\nu}\right) b^{-\frac{1}{\nu}}. \tag{5.1}$$

Thus,

$$N_{33} = \int_0^\infty dx e^{-bx^\nu} = \frac{1}{\nu} \Gamma\left(\frac{1}{\nu}\right) b^{-\frac{1}{\nu}}. \tag{5.2}$$

(1) Let $\nu = 1$, then

$$e_1 = \int_0^\infty dx e^{-bx} = \frac{1}{b}. \tag{5.3}$$

(2) Assuming $\nu = 2$, one gets the famous Gaussian integral

$$e_2 = \int_0^\infty dx x e^{-bx^2} = \frac{1}{2} \sqrt{\frac{\pi}{b}}, \tag{5.4}$$

where

$$G_2 = \int_{-\infty}^\infty dx e^{-x^2} = 2 \int_0^\infty dx e^{-x^2} = \sqrt{\pi}. \tag{5.5}$$

(3) Let $\nu = -2$, then one gets

$$e_3 = \int_0^\infty dx\, e^{-bx^{-2}} = -\frac{1}{2}\,\Gamma\left(-\frac{1}{2}\right)\,b^{\frac{1}{2}} = \sqrt{\pi b}.$$

(5.6)

(4) Let $\nu = \frac{1}{2}$, then one obtains

$$e_4 = \int_0^\infty dx\, e^{-b\sqrt{x}} = 2\,\Gamma(2)\,b^{-2} = \frac{2}{b^2}.$$

(5.7)

(5) Making use of $\nu = \frac{4}{3}$, one gets

$$e_5 = \int_0^\infty dx\, e^{-b\sqrt[3]{x^4}} = \frac{3}{4}\,\Gamma\left(\frac{3}{4}\right)\,b^{-\frac{3}{4}} = \frac{3}{4}\sqrt{2}\,\pi\frac{1}{\Gamma\left(\frac{1}{4}\right)}\,b^{-\frac{3}{4}}.$$

(5.8)

5.1.2   34th *General Formula*

$$N_{34} = \int_0^1 dx\, x^\delta (1 - x^\sigma)^\mu\, \exp\left[-bx^\nu (1 - x^\sigma)^\lambda\right] = \frac{1}{2i\sigma}$$

$$\times \int_{-\beta+i\infty}^{-\beta-i\infty} d\xi \frac{b^\xi}{\sin\pi\xi\,\Gamma(1+\xi)}\, B\left(\frac{1 + \delta + \nu\xi}{\sigma},\quad 1 + \mu + \lambda\xi\right).$$

(5.9)

5.1.3   35th *General Formula*

$$N_{35} = \int_a^c dx\,(x - a)^\delta (c - x)^\mu\, \exp\left[-b(x - a)^\nu (c - x)^\lambda\right]$$

$$= \frac{1}{2i} \int_{-\beta+i\infty}^{-\beta-i\infty} d\xi \frac{b^\xi\,(c - a)^{1+\delta+\mu+\nu\xi+\lambda\xi}}{\sin\pi\xi\,\Gamma(1+\xi)}$$

$$\times B(1 + \delta + \nu\xi,\quad 1 + \mu + \lambda\xi).$$

(5.10)

5.1.4 36th *General Formula*

$$
\begin{aligned}
N_{36} &= \int\limits_{0}^{\infty} dx \frac{x^\mu \ (x+a)^{-\mu}(x+c)^{-\mu}}{\sqrt{x}} \\
&\quad \times \exp\left[-bx^\nu(x+a)^{-\nu}(x+c)^{-\nu}\right] \\
&= \sqrt{\pi}\ \frac{1}{2i} \int\limits_{-\beta+i\infty}^{-\beta-i\infty} d\xi \frac{b^\xi\ (\sqrt{a}+\sqrt{c})^{1-2(\mu+\nu\xi)}}{\sin\pi\xi\ \Gamma(1+\xi)}\ \frac{\Gamma\left(\mu+\nu\xi-\frac{1}{2}\right)}{\Gamma(\mu+\nu\xi)}.
\end{aligned}
$$

(5.11)

5.1.5 37th *General Formula*

$$
\begin{aligned}
N_{37} &= \int\limits_{0}^{1} dx(1-x^\sigma)^\mu\ \exp\left[-b(1-x^\sigma)^\lambda\right] = \frac{1}{\sigma} \\
&\quad \times \frac{1}{2i} \int\limits_{-\beta+i\infty}^{-\beta-i\infty} d\xi \frac{b^\xi}{\sin\pi\xi\ \Gamma(1+\xi)}\ B\left(\frac{1}{\sigma},\ 1+\mu+\lambda\xi\right).
\end{aligned}
$$

(5.12)

5.1.6 38th *General Formula*

$$
\begin{aligned}
N_{38} &= \int\limits_{0}^{1} dx(1-\sqrt{x})^\mu\ \exp\left[-b(1-\sqrt{x})^\lambda\right] = \frac{1}{i} \\
&\quad \times \int\limits_{-\beta+i\infty}^{-\beta-i\infty} d\xi \frac{b^\xi}{\sin\pi\xi\ \Gamma(1+\xi)}\ \frac{1}{(1+\mu+\lambda\xi)}\ \frac{1}{(2+\mu+\lambda\xi)}.
\end{aligned}
$$

(5.13)

5.1.7   39th *General Formula*

$$N_{39} = \int\limits_0^\infty dx\, x^\delta (1 + tx^\sigma)^{-\mu}\, \exp\left[-bx^\nu (1+tx^\sigma)^{-\lambda}\right]$$

$$= \frac{1}{\sigma}\, \frac{1}{2i} \int\limits_{-\beta+i\infty}^{-\beta-i\infty} d\xi \frac{b^\xi\, t^{-\frac{\delta+1+\nu\xi}{\sigma}}}{\sin \pi\xi\, \Gamma(1+\xi)}$$

$$\times B\left(\frac{1+\delta+\nu\xi}{\sigma},\quad \mu + \lambda\xi - \frac{1+\delta+\nu\xi}{\sigma}\right). \tag{5.14}$$

5.1.8   40th *General Formula*

$$N_{40} = \int\limits_0^\infty dx \frac{x^\gamma}{[p + tx^\rho]^\lambda}\, \exp[-bx^\nu]$$

$$= \frac{1}{\rho}\, \frac{1}{\Gamma(\lambda)}\, \frac{1}{p^\lambda}\, \left(\frac{p}{t}\right)^{\frac{\gamma+1}{\rho}}\, \frac{1}{2i} \int\limits_{-\beta+i\infty}^{-\beta-i\infty} d\xi \frac{b^\xi}{\sin \pi\xi\, \Gamma(1+\xi)}$$

$$\times \left(\frac{p}{t}\right)^{\frac{\nu\xi}{\rho}}\, \Gamma\left(\frac{1+\gamma}{\rho} + \frac{\nu\xi}{\rho}\right) \Gamma\left(\lambda - \frac{\gamma+1+\nu\xi}{\rho}\right). \tag{5.15}$$

5.2   Calculation of Concrete Integrals

(6) Assuming

$$\delta = 1,\ \nu = 2,\ \mu = -1,\ \lambda = -1,\ \sigma = 2$$

in (5.9), one gets

$$e_6 = \int_0^1 dx \, x(1-x^2)^{-1} \exp\left[-bx^2(1-x^2)^{-1}\right]$$

$$= \frac{1}{2}\frac{1}{2i} \int_{-\beta+i\infty}^{-\beta-i\infty} d\xi \frac{b^\xi}{\sin \pi\xi \, \Gamma(1+\xi)} \frac{\Gamma(1+\xi)\,\Gamma(-\xi)}{\Gamma(1)}$$

$$= -\frac{\pi}{2}\frac{1}{2i} \int_{-\beta+i\infty}^{-\beta-i\infty} d\xi \frac{b^\xi}{\sin^2 \pi\xi \, \Gamma(1+\xi)} = -\frac{1}{2}\sum_{n=0}^{\infty} \frac{b^n}{n!}[\ln b - \Psi(1+n)]$$

$$= -\frac{1}{2}\ln b \, e^b + \frac{1}{2}\sum_{n=0}^{\infty} \frac{b^n}{n!}\Psi(1+n). \tag{5.16}$$

(7) Let

$$\delta = -\frac{1}{2}, \ \mu = -\frac{1}{2}, \ \nu = -1, \ \lambda = 1$$

be in the main formula (5.10), then we have

$$e_7 = \int_a^c dx (x-a)^{-1/2}(c-x)^{-1/2} \exp\left[-b(x-a)^{-1}(c-x)\right]$$

$$= \frac{1}{2i} \int_{-\beta+i\infty}^{-\beta-i\infty} d\xi \frac{b^\xi}{\sin \pi\xi \, \Gamma(1+\xi)} \frac{\Gamma\left(\frac{1}{2}-\xi\right)\Gamma\left(\frac{1}{2}+\xi\right)}{\Gamma(1)}$$

$$= \frac{\pi}{i} \int_{-\beta+i\infty}^{-\beta-i\infty} d\xi \frac{b^\xi}{\sin 2\pi\xi \, \Gamma(1+\xi)} = \pi \sum_{n=0}^{\infty} \frac{(-1)^n(\sqrt{b})^n}{\Gamma\left(1+\frac{n}{2}\right)}, \tag{5.17}$$

where we have used the change of the integration variable: $2\xi = x$, $b > 0$. Now we separate off even and odd numbers of n in the summation (5.17) and obtain

$$e_7 = \pi e^b - 2\sqrt{\pi b} \sum_{k=0}^{\infty} \frac{(2b)^k}{(2k+1)!!},$$

where

$$(2k+1)!! = 1 \cdot 3 \cdot \ldots \cdot (2k+1).$$

(8) Assuming $\delta = 0$, $\mu = -1$, $\nu = 1$, $\lambda = 1$ in (5.10), one obtains

$$e_8 = \int_a^c dx (c-x)^{-1} \exp\left[-b(x-a)(c-x)\right]$$

$$= \frac{1}{2i} \int_{-\beta+i\infty}^{-\beta-i\infty} d\xi \frac{b^\xi (c-a)^{2\xi}}{\sin \pi\xi \, \Gamma(1+\xi)} \frac{\Gamma(1+\xi)\,\Gamma(\xi)}{\Gamma(1+2\xi)},$$

where

$$\Gamma(1 + 2\xi) = 2\xi \, \frac{2^{2\xi-1}}{\sqrt{\pi}} \, \Gamma(\xi) \, \Gamma\left(\frac{1}{2} + \xi\right), \quad b > 0.$$

So that

$$e_8 = \frac{\sqrt{\pi}}{2i} \int\limits_{-\beta+i\infty}^{-\beta-i\infty} d\xi \frac{\left(\sqrt{b} \, \frac{(c-a)}{2}\right)^{2\xi}}{\xi \, \sin \pi\xi \, \Gamma\left(\frac{1}{2} + \xi\right)} = 2\ln\left[\frac{\sqrt{b}}{2}(c - a)\right]$$

$$- \Psi\left(\frac{1}{2}\right) + \sum_{n=1}^{\infty} \frac{(-1)^n}{n} \frac{\left(\sqrt{\frac{b}{2}}(c - a)\right)^{2n}}{(2n - 1)!!}. \tag{5.18}$$

(9) In the formula (5.11), we put $\mu = \frac{3}{2}$, $\nu = 1$ and get

$$e_9 = \int\limits_0^\infty \frac{dx}{\sqrt{x}} \left[\frac{x}{(x + a)(x + c)}\right]^{\frac{3}{2}} \exp\left[- b \, \frac{x}{(x + a)(x + c)}\right]$$

$$= \frac{\sqrt{\pi}}{A^2} \frac{1}{2i} \int\limits_{-\beta+i\infty}^{-\beta-i\infty} d\xi \frac{(b/A^2)^\xi}{\sin \pi\xi \, \Gamma(1 + \xi)} \frac{\Gamma(1 + \xi)}{\Gamma\left(\frac{3}{2} + \xi\right)}$$

$$= \frac{2}{A^2} \sum_{n=0}^{\infty} (-1)^n \frac{(\sqrt{2b}/A)^{2n}}{(1 + 2n)(2n - 1)!!}, \tag{5.19}$$

where

$$A = \sqrt{a} + \sqrt{c}.$$

(10) Let

$$\sigma = -2, \quad \mu = -\frac{1}{2}, \quad \lambda = -1,$$

be in (5.12), then

$$e_{10} = \int\limits_0^1 dx (1 - x^{-2})^{-1/2} \exp\left[- b(1 - x^{-2})^{-1}\right]$$

$$= -\frac{1}{2} \frac{1}{2i} \int\limits_{-\beta+i\infty}^{-\beta-i\infty} d\xi \frac{b^\xi}{\sin \pi\xi \, \Gamma(1 + \xi)} \frac{\Gamma\left(-\frac{1}{2}\right) \Gamma\left(\frac{1}{2} - \xi\right)}{\Gamma(-\xi)}$$

$$= -\sqrt{\pi} \frac{1}{2i} \int\limits_{-\beta+i\infty}^{-\beta-i\infty} d\xi \frac{b^\xi}{\cos \pi\xi \, \Gamma\left(\frac{1}{2} + \xi\right)} = \sqrt{\pi b} \, e^{-b}. \tag{5.20}$$

(11) Assuming $\mu = -1$, $\lambda = 1$ in (5.13), one gets

$$e_{11} = \int_0^1 dx\, (1 - \sqrt{x})^{-1} \exp\left[-b(1 - \sqrt{x})\right]$$

$$= \frac{1}{i} \int_{-\beta+i\infty}^{-\beta-i\infty} d\xi \frac{b^\xi}{\sin \pi\xi\, \Gamma(1+\xi)} \frac{1}{\xi(1+\xi)} = \Omega(b). \tag{5.21}$$

From this integral (5.21), it follows

$$\frac{\partial^2}{\partial b^2}\left[b\Omega(b)\right] = \frac{1}{b}\frac{1}{i} \int_{-\beta+i\infty}^{-\beta-i\infty} d\xi \frac{b^\xi}{\sin \pi\xi\, \Gamma(1+\xi)} = \frac{2}{b} e^{-b}. \tag{5.22}$$

Therefore,

$$\frac{\partial}{\partial b}\left[b\Omega(b)\right] = 2E_i(-b).$$

So that

$$e_{11} = \Omega(b) = \frac{2}{b}\int db\, E_i(-b). \tag{5.23}$$

(12) We put

$$\delta = 1,\ \nu = -4,\ \sigma = 4,\ \mu = 0,\ \lambda = -1$$

in (5.14) and obtain

$$e_{12} = \int_0^\infty dx\, x\, \exp\left[-bx^{-4}(1 + tx^4)\right]$$

$$= \frac{1}{\sqrt{t}}\frac{1}{4}\frac{1}{2i} \int_{-\beta+i\infty}^{-\beta-i\infty} d\xi \frac{(bt)^\xi}{\sin \pi\xi\, \Gamma(1+\xi)} \frac{\Gamma\left(\frac{1}{2} - \xi\right)\Gamma\left(-\frac{1}{2}\right)}{\Gamma(-\xi)}.$$

After some transformation of gamma-functions, we have

$$e_{12} = \frac{\sqrt{\pi}}{2\sqrt{t}}\frac{1}{2i} \int_{-\beta+i\infty}^{-\beta-i\infty} d\xi \frac{(bt)^\xi}{\cos \pi\xi\, \Gamma\left(\frac{1}{2} + \xi\right)}. \tag{5.24}$$

Calculation of residues at the points $\xi = n + \frac{1}{2}$ gives

$$e_{12} = -\frac{\sqrt{\pi b}}{2}\sum_{n=0}^\infty \frac{(-1)^n\,(bt)^n}{n!} = -\frac{\sqrt{\pi b}}{2} e^{-bt}. \tag{5.25}$$

(13) Let

$$\rho = 2, \ \lambda = 1, \ \gamma = 0, \ \nu = 2, \ t = 1, \ p = \beta^2$$

be in (5.15), then

$$e_{13} = \int_0^\infty dx \frac{e^{-bx^2}}{[\beta^2 + x^2]} = \frac{\pi}{2\beta} \frac{1}{2i} \int_{-\beta+i\infty}^{-\beta-i\infty} d\xi \frac{(b\beta^2)^\xi}{\sin \pi\xi \ \cos \pi\xi \ \Gamma(1+\xi)},$$

where we have used the relation

$$\Gamma\left(\frac{1}{2} - \xi\right) \Gamma\left(\frac{1}{2} + \xi\right) = \frac{\pi}{\cos \pi\xi}.$$

Now we calculate residues at the points $\xi = n$, $\xi = n + \frac{1}{2}$, and obtain

$$e_{13} = \frac{\pi}{2\beta} \left\{ \sum_{n=0}^\infty \frac{(b\beta^2)^n}{n!} - \sum_{n=0}^\infty \frac{(b\beta^2)^{n+1/2}}{\Gamma\left(\frac{3}{2} + n\right)} \right\}.$$

Using the formula

$$\Gamma\left(\frac{3}{2} + n\right) = \frac{\sqrt{\pi}}{2^{n+1}}(2n+1)!!,$$

one gets

$$e_{13} = \frac{\pi}{2\beta} e^{b\beta^2} \left[1 - \Phi(\sqrt{b}\ \beta)\right], \tag{5.26}$$

where

$$\Phi(x) = \frac{2}{\sqrt{\pi}} \sum_{n=0}^\infty \frac{2^n \ x^{2n+1}}{(2n+1)!!} e^{-x^2}$$

is called the probability integral.

5.3 Simple Formulas for Integrals Involving Exponential and Polynomial Functions

5.3.1 41st *General Formula*

$$N_{41} = \int_0^\infty dx x^\gamma e^{-bx^\nu} = \frac{1}{\nu} b^{-\frac{\gamma+1}{\nu}} \Gamma\left(\frac{\gamma+1}{\nu}\right). \tag{5.27}$$

5.3.2   42nd *General Formula*

For any natural n-numbers

$$
\begin{aligned}
N_{42} &= \int\limits_0^\infty dx\, e^{-bx^\nu} \left[p + tx^\rho \right]^n \\
&= \frac{1}{\nu} \sum_{k=0}^n C_n^k\, p^{n-k}\, t^k\, b^{-\frac{\rho k+1}{\nu}}\, \Gamma\left(\frac{\rho k + 1}{\nu} \right),
\end{aligned}
\tag{5.28}
$$

where the binomial coefficients C_n^k are positive and integers n, k are defined by the well-known formula

$$
C_n^k = \begin{cases} \frac{n!}{k!(n-k)!} & \text{for} \quad 0 \le k \le n \\ 0 & \text{for} \quad 0 \le n < k. \end{cases}
\tag{5.29}
$$

This definition may be extended for any real numbers a and integers $k \ge 0$:

$$
C_k^a = \binom{a}{k} = \begin{cases} \frac{a(a-1)(a-2)\ldots(a-k+1)}{k!} & \text{for} \quad k > 0 \\ 1 & \text{for} \quad k = 0. \end{cases}
\tag{5.30}
$$

For example

$$
\binom{5}{3} = \frac{5!}{3!(5-3)!} = 10, \quad \binom{-2}{3} = \frac{-2(-2-1)(-2-2)}{3!} = -4,
$$

$$
\binom{2}{5} = 0, \quad \binom{\sqrt{2}}{4} = \frac{\sqrt{2}(\sqrt{2}-1)(\sqrt{2}-2)(\sqrt{2}-3)}{4!} = \frac{13 - 9\sqrt{2}}{12}.
$$

(14) Assuming

$$
n = 1, \ \nu = 2, \ t = 2\beta, \ p = 1, \ \rho = 2
$$

in (5.28) and using the formula (5.27), one gets

$$
e_{14} = \int\limits_0^\infty dx(1 + 2\beta x^2)\, e^{-bx^2} = \frac{b+\beta}{2} \sqrt{\frac{\pi}{b^3}}.
\tag{5.31}
$$

(15) Due to the formula (5.27), the following integral is easily calculated:

$$
e_{15} = \int\limits_0^\infty dx \frac{e^{-bx} - e^{-ax}}{x} = \ln\frac{a}{b},
\tag{5.32}
$$

where we have used the L'Hôpital rule (1.5) in Chapter 1, that gives

$$
\lim_{\varepsilon \to 0} \Gamma(\varepsilon)\left(b^{-\varepsilon} - a^{-\varepsilon} \right) = \frac{\pi}{\Gamma(1)} \lim_{\varepsilon \to 0} \frac{b^{-\varepsilon} - a^{-\varepsilon}}{\sin \pi\xi} = \ln\frac{a}{b}.
$$

(16) From the formula (5.27) with $\gamma = -2$, $\nu = 2$, it follows directly

$$e_{16} = \int_0^\infty dx \frac{e^{-bx^2} - e^{-ax^2}}{x^2} = \sqrt{\pi}\left(\sqrt{a} - \sqrt{b}\right),$$ (5.33)

where we have used the equality $\Gamma(-\frac{1}{2}) = -2\sqrt{\pi}$.

(17) Now by using the formula (5.15), we derive an integral which requires a more complicated procedure of calculation:

$$e_{17} = \int_0^\infty dx \frac{e^{-bx}}{[x^2 + u^2]^{1-\sigma}},$$

where

$$\lambda = 1 - \sigma, \ \gamma = 0, \ p = u^2, \ t = 1, \ \rho = 2, \ \nu = 1.$$

Thus,

$$e_{17} = \frac{1}{2}\frac{1}{\Gamma(1-\sigma)}\frac{1}{(u^2)^{1/2-\sigma}}\frac{1}{2i}\int_{-\beta+i\infty}^{-\beta-i\infty} d\xi \frac{(bu)^\xi}{\sin \pi\xi\, \Gamma(1+\xi)}$$

$$\times \Gamma\left(\frac{1}{2} + \frac{\xi}{2}\right)\Gamma\left(\frac{1}{2} - \sigma - \frac{\xi}{2}\right).$$ (5.34)

The change of the integration variable $\xi \to 2x$ leads to the following integral

$$e_{17} = \frac{1}{\Gamma(1-\sigma)}\frac{1}{(u^2)^{1/2-\sigma}}\frac{1}{2i}\int_{-\beta+i\infty}^{-\beta-i\infty} dx \frac{(bu)^{2x}\,\Gamma\left(\frac{1}{2}+x\right)\Gamma\left(\frac{1}{2}-\sigma-x\right)}{\sin 2\pi x\, \Gamma(1+2x)},$$

where

$$\bullet\ \Gamma\left(\frac{1}{2}+x\right)\Gamma(x)\frac{2^{2x-1}}{\sqrt{\pi}} = \Gamma(2x),$$

$$\bullet\ \Gamma\left(\frac{1}{2}-\sigma-x\right) = \frac{\pi}{\Gamma\left(\frac{1}{2}+\sigma+x\right)\cos\pi(\sigma+x)}.$$

It turns out that in our integral, there are three poles at the points

$$x = n, \ x = n+\frac{1}{2}, \ \text{and } \sigma + x = n+\frac{1}{2}.$$

First, we calculate the residue due to the points $x \to n+\frac{1}{2}$, and obtain

$$e_{17} = \frac{\pi\sqrt{\pi}}{(u^2)^{1/2-\sigma}}\frac{1}{\Gamma(1-\sigma)}\frac{1}{2i}\int_{-\beta+i\infty}^{-\beta-i\infty} dx \frac{(bu/2)^{2x}}{2\sin\pi x\, \cos\pi x}$$

$$\times \frac{1}{\Gamma(1+x)}\frac{1}{\cos\pi(\sigma+x)\,\Gamma\left(\frac{1}{2}+\sigma+x\right)},$$ (5.35)

where

- $(\cos \pi x)' = -\pi \sin \pi x,$

- $\sin \pi \left(n + \dfrac{1}{2} \right) = \sin \pi n \, \cos \dfrac{\pi}{2} + \sin \dfrac{\pi}{2} \, \cos \pi n = (-1)^n,$

- $\cos \pi \left(\sigma + n + \dfrac{1}{2} \right) = \cos \pi \left(\sigma + \dfrac{1}{2} \right) \cos \pi n$

$$- \sin \pi \left(\sigma + \dfrac{1}{2} \right) \sin \pi n = -(-1)^n \, \sin \pi \sigma.$$

Thus,

$$e_{17}^1 = \frac{\sqrt{\pi}}{2} \left(\frac{2u}{b} \right)^{\sigma - \frac{1}{2}} \Gamma(\sigma) \, \mathbf{H}_{\sigma - \frac{1}{2}}(ub), \tag{5.36}$$

where

$$\mathbf{H}_\sigma(z) = \sum_{n=0}^{\infty} (-1)^n \frac{\left(\frac{z}{2} \right)^{2n+\sigma+1}}{\Gamma \left(n + \frac{3}{2} \right) \Gamma \left(\sigma + n + \frac{3}{2} \right)}$$

is called the Struve function.

Calculation of residues at the points $\sigma + x = n + \frac{1}{2}$ and $x = n$, where

- $\left[\cos \pi(\sigma + x) \right]' = -\pi \sin \pi(\sigma + x),$

- $\sin 2\pi x = \sin \pi(2n + 1 - 2\sigma) = \sin 2\pi\sigma,$

- $\Gamma(1 + x) \to \Gamma \left(\dfrac{3}{2} - \sigma + n \right),$

- $\Gamma \left(\dfrac{1}{2} + \sigma + x \right) \to \Gamma(1 + n),$

- $\cos \pi(\sigma + n) = (-1)^n \, \cos \pi\sigma$

leads to

$$e_{17}^2 = -\frac{\sqrt{\pi}}{2} \left(\frac{2u}{b} \right)^{\sigma - \frac{1}{2}} \Gamma(\sigma) \, N_{\sigma - \frac{1}{2}}(bu), \tag{5.37}$$

where

$$N_\sigma(z) = \frac{1}{\sin \pi\sigma} \left\{ \cos \pi\sigma \left(\frac{z}{2} \right)^\sigma \sum_{k=0}^{\infty} (-1)^k \frac{z^{2k}}{2^{2k} \, k! \, \Gamma(\sigma + k + 1)} \right.$$

$$\left. - \left(\frac{z}{2} \right)^{-\sigma} \sum_{k=0}^{\infty} (-1)^k \frac{z^{2k}}{2^{2k} \, k! \, \Gamma(k - \sigma + 1)} \right\}$$

is called the Bessel function of the second kind or the Neumann function, also denoted by $Y_\sigma(z)$. Finally, collecting two results we have

$$e_{17} = e_{17}^1 + e_{17}^2 = \frac{\sqrt{\pi}}{2} \left(\frac{2u}{b} \right)^{\sigma - \frac{1}{2}} \Gamma(\sigma)$$

$$\times \left[\mathbf{H}_{\sigma - \frac{1}{2}}(ub) - N_{\sigma - \frac{1}{2}}(bu) \right]. \tag{5.38}$$

Notice that our derived general formula (5.15) is also verified by this integral.

(18^a) The formula (5.15) with

$$\gamma = -\frac{1}{2}, \ \nu = 1, \ p \to a, \ t = 1, \ \rho = 1, \ \lambda = 1/2$$

gives the integral

$$e_{18}^a = \int_0^\infty dx \frac{e^{-bx}}{\sqrt{x}\,\sqrt{x+a}} = -\frac{\pi}{\sqrt{\pi}}\frac{1}{2i}\int_{-\beta+i\infty}^{-\beta-i\infty} d\xi \frac{(ba)^\xi \, \Gamma\left(\frac{1}{2}+\xi\right)}{\sin^2 \pi\xi \, \Gamma^2(1+\xi)}$$

$$= -\frac{1}{\sqrt{\pi}}\sum_{n=0}^\infty \frac{(ba)^n}{(n!)^2}\,\Gamma\left(\frac{1}{2}+n\right)$$

$$\times \left[\ln(ab) - 2\Psi(1+n) + \Psi\left(\frac{1}{2}+n\right)\right], \tag{5.39}$$

where

$$\Gamma\left(\frac{1}{2}+n\right) = \frac{\sqrt{\pi}}{2^n}(2n-1)!!$$

and we have used the formulas

$$\bullet \ \Psi\left(\frac{1}{2}+n\right) = -C + 2\left[\sum_{k=1}^n \frac{1}{2k-1} - \ln 2\right],$$

$$\bullet \ \Psi(n+1) = -C + \sum_{k=1}^n \frac{1}{k},$$

where C is the Euler number. By definition of the modified Bessel function of the second kind $K_0(z)$, we have from (5.39)

$$e_{18}^a = e^{ab/2}\, K_0\left(\frac{ab}{2}\right).$$

(18^b) A similar integral with respect to (5.39) is derived by the formula (5.15) with parameters

$$\gamma = 0, \ \rho = 1, \ t = 1, \ \nu = 1, \ \lambda = 1,$$

$$e_{18}^b = \int_0^\infty dx \frac{e^{-bx}}{x+p} = -\frac{\pi}{2i}\int_{-\beta+i\infty}^{-\beta-i\infty} d\xi \frac{(bp)^\xi}{\sin^2 \pi\xi \, \Gamma(1+\xi)}$$

$$= -\sum_{n=0}^\infty \frac{(bp)^n}{n!}\left[\ln(bp) - \Psi(n+1)\right] = -e^{bp}\, E_i(-bp),$$

where

$$\Psi(n+1) = -C + \sum_{k=1}^n \frac{1}{k},$$

$C = -\Psi(1)$ is Euler's number, and

$$E_i(x) = C + \ln(-x) + \sum_{k=1}^\infty \frac{x^k}{k\,k!} \qquad \text{if} \quad x < 0,$$

$$E_i(x) = C + \ln x + \sum_{k=1}^\infty \frac{x^k}{k\,k!} \qquad \text{if} \quad x > 0$$

is the integral exponential function.

5.4 Unified Formulas for Integrals Containing Exponential and Trigonometric Functions

5.4.1 43rd *General Formula*

$$
N_{43} = \int\limits_0^\infty dx\, x^\gamma e^{-ax^\mu} \sin^q(bx^\nu) = \frac{1}{\mu} a^{-\frac{\gamma+1}{\mu}} \frac{1}{2^{q-1}}
$$

$$
\times \frac{1}{2i} \int\limits_{\alpha+i\infty}^{\alpha-i\infty} d\xi \frac{\left(\dfrac{2b}{a^{\nu/\mu}}\right)^{2\xi}}{\sin \pi\xi\, \Gamma(1+2\xi)} I_q(\xi)\, \Gamma\left(\frac{\gamma+1+2\nu\xi}{\mu}\right)
$$

(5.40)

or

$$
N_{43} = \frac{\sqrt{\pi}}{2\nu} b^{-\frac{\gamma+1}{\nu}} \frac{1}{2^{q-1}} \frac{1}{2i} \int\limits_{-\beta+i\infty}^{-\beta-i\infty} d\eta \frac{\left(\dfrac{a}{b^{\mu/\nu}}\right)^\eta}{\sin \pi\eta\, \Gamma(1+\eta)}
$$

$$
\times I_q\left(\xi = -\frac{\gamma+1+\mu\eta}{2\nu}\right) \frac{\Gamma\left(\dfrac{1+\gamma+\mu\eta}{2\nu}\right)}{\Gamma\left(\dfrac{1}{2}-\dfrac{1+\gamma+\mu\eta}{2\nu}\right)}.
$$

(5.41)

Here there exist two equivalent representations for the integral (5.40), where $q = 2, 4, 6, \dots$ and $I_q(\xi)$ is given by (1.32) in Chapter 1.

5.4.2 44th *General Formula*

$$
N_{44} = \int\limits_0^\infty dx\, x^\gamma\, e^{-ax^\mu} \sin^m(bx^\nu)
$$

$$
= \frac{1}{\mu} a^{-\frac{\gamma+1+\nu}{\mu}} \frac{1}{2^{m-1}} \frac{1}{2i} \int\limits_{-\beta+i\infty}^{-\beta-i\infty} d\xi \frac{b\left(\dfrac{b}{a^{\nu/\mu}}\right)^{2\xi}}{\sin \pi\xi\, \Gamma(2+2\xi)}
$$

$$
\times N_m(\xi)\Gamma\left(\frac{\gamma+1+\nu+2\nu\xi}{\mu}\right)
$$

(5.42)

or

$$N_{44} = \frac{\sqrt{\pi}}{2\nu} \left(\frac{2}{b}\right)^{\frac{\gamma+1}{\nu}} \frac{1}{2^{m-1}} \frac{1}{2i} \int\limits_{-\beta+i\infty}^{-\beta-i\infty} d\eta \frac{\left[a\,(2/b)^{\mu/\nu}\right]^{\eta}}{\sin \pi\eta \; \Gamma(1+\eta)}$$

$$\times N_m \left(\xi = -\frac{1}{2}\left[1 + \frac{\gamma+1+\mu\eta}{\nu}\right]\right) \frac{\Gamma\left(\frac{1}{2}\left[1 + \frac{\gamma+1+\mu\eta}{\nu}\right]\right)}{\Gamma\left(1 - \frac{\gamma+1+\mu\eta}{2\nu}\right)}, \qquad (5.43)$$

$$m = 1,\ 3,\ 5,\ 7, \ldots.$$

5.4.3   45th General Formula

$$N_{45} = \int\limits_{0}^{\infty} dx\, x^{\gamma}\, e^{-ax^{\mu}} \cos^m(bx^{\nu}) = \frac{1}{\mu}\, a^{-\frac{\gamma+1}{\mu}}\, \frac{1}{2^{m-1}}$$

$$\times \frac{1}{2i} \int\limits_{-\beta+i\infty}^{-\beta-i\infty} d\xi \frac{(b/a^{\nu/\mu})^{2\xi}\, N_m'(\xi)}{\sin \pi\xi \; \Gamma(1+2\xi)} \Gamma\left(\frac{\gamma+1+2\nu\xi}{\mu}\right) \qquad (5.44)$$

or

$$N_{45} = \frac{\sqrt{\pi}}{2\nu} \left(\frac{2}{b}\right)^{\frac{\gamma+1}{\nu}} \frac{1}{2^{m-1}} \frac{1}{2i} \int\limits_{-\beta'+i\infty}^{-\beta'-i\infty} d\eta \frac{\left[a\,(2/b)^{\mu/\nu}\right]^{\eta}}{\sin \pi\eta \; \Gamma(1+\eta)}$$

$$\times N_m' \left(\xi = -\frac{\gamma+1+\mu\eta}{\nu}\right) \frac{\Gamma\left(\frac{\gamma+1+\mu\eta}{2\nu}\right)}{\Gamma\left(\frac{1}{2} - \frac{\gamma+1+\mu\eta}{2\nu}\right)}. \qquad (5.45)$$

5.4.4   46th General Formula

$$N_{46} = \int\limits_{0}^{\infty} dx\, x^{\gamma}\, e^{-ax^{\mu}} \left[\cos^q(bx^{\nu}) - 1\right] = \frac{1}{\mu}\, a^{-\frac{\gamma+1}{\mu}}\, \frac{1}{2^{q-1}}$$

$$\times \frac{1}{2i} \int\limits_{\alpha+i\infty}^{\alpha-i\infty} d\xi \frac{(2b/a^{\nu/\mu})^{2\xi}\, I_q'(\xi)}{\sin \pi\xi \; \Gamma(1+2\xi)} \Gamma\left(\frac{\gamma+1+2\nu\xi}{\mu}\right) \qquad (5.46)$$

or

$$N_{46} = \frac{\sqrt{\pi}}{2\nu} b^{-\frac{\gamma+1}{\nu}} \frac{1}{2^{q-1}} \frac{1}{2i} \int\limits_{-\beta'+i\infty}^{-\beta'-i\infty} d\eta \frac{\left(a/b^{\mu/\nu}\right)^\eta}{\sin \pi\eta \, \Gamma(1+\eta)}$$

$$\times I_q'\left(\xi = -\frac{\gamma+1+\mu\eta}{2\nu}\right) \frac{\Gamma\left(\dfrac{1+\gamma+\mu\eta}{2\nu}\right)}{\Gamma\left(\dfrac{1}{2} - \dfrac{1+\gamma+\mu\eta}{2\nu}\right)}, \tag{5.47}$$

$$q = 2, \, 4, \, 6, \ldots .$$

In these six formulas, expressions $N_m(\xi)$, $N_m'(\xi)$ and $I_q'(\xi)$ are defined by (1.34), (1.36) and (1.38) in Chapter 1, respectively.

Notice that in accordance with the formulas (2.3), (2.5), (2.6), (2.7) and (5.27), these four general formulas written in two variants are derived automatically. Such types of the chain rule are also valid for complicated integrals, in particular, those containing special functions, which are determined by double or triple Mellin representations (also see Chapter 6).

5.5   Calculation of Particular Integrals Arising from the General Formulas in Section 5.4

(19) Assuming

$$\gamma = 1, \, a \to p^2, \, \mu = 2, \, m = 1, \, \nu = 1$$

in (5.42), one gets

$$e_{19} = \int\limits_0^\infty dx\, x\, e^{-p^2 x^2} \, \sin bx$$

$$= \frac{b}{2p^3} \sum_{n=0}^\infty \frac{(-1)^n}{(2n+1)!} \left(\frac{b^2}{p^2}\right)^n \Gamma\left(\frac{3}{2}+n\right). \tag{5.48}$$

In this expression, we use the following relations:

$$\Gamma\left(\frac{3}{2}+n\right) = \Gamma\left(\frac{1}{2}+1+n\right) = \frac{\sqrt{\pi}}{2^{n+1}}(2n+1)!!$$

and

$$\frac{(2n+1)!!}{(2n+1)!} = \frac{1}{2^n}\frac{1}{n!}.$$

So that

$$e_{19} = \frac{b\sqrt{\pi}}{4p^3} \sum_{n=0}^\infty (-1)^n \frac{\left(b^2/4p^2\right)^n}{n!} = \frac{b\sqrt{\pi}}{4p^3} \exp\left[-\frac{b^2}{4p^2}\right].$$

Here we see that by using our general formula (5.42), such types of integrals are easily calculated.

(20) We consider the following integral

$$e_{20} = \int\limits_0^\infty dx x^2 \, e^{-p^2 x^2} \, \cos bx$$

which is obtained by the substitution of the parameters

$$\gamma = 2, \ a \to p^2, \ \mu = 2, \ \nu = 1, \ m = 1$$

in the formula (5.44). Thus.

$$e_{20} = \frac{1}{2p^3} \frac{1}{2i} \int\limits_{-\beta+i\infty}^{-\beta-i\infty} d\xi \frac{(b/p)^{2\xi} \, \Gamma\left(\frac{3}{2} + \xi\right)}{\sin \pi\xi \, \Gamma(1 + 2\xi)}$$

$$= \frac{\sqrt{\pi}}{4p^3} \sum_{n=0}^\infty \frac{(b^2/p^2)^n}{(2n)!} \frac{(2n+1)!!}{2^n}, \qquad (5.49)$$

where

$$\frac{(2n+1)!!}{(2n)!} = \frac{(2n+1)}{(2n+1)} \frac{1}{(2n)!} (2n+1)!!$$

$$= \frac{(2n+1)!!}{(2n+1)!} + 2n\frac{(2n+1)!!}{(2n+1)!}.$$

So that

$$e_{20} = \frac{\sqrt{\pi}}{4p^3} \left\{ e^{-\frac{b^2}{4p^2}} + \frac{b^2}{p^2} \, 2 \, \frac{\partial}{\partial z} \, e^{-\frac{1}{4}z} \right\},$$

where $z = b^2/p^2$, and finally we have

$$e_{20} = \frac{\sqrt{\pi}}{8p^5} \left(2p^2 - b^2 \right) \exp\left[-\frac{b^2}{4p^2} \right]. \qquad (5.50)$$

(21) The case

$$\gamma = -1, \ a \to p^2, \ \mu = 2, \ m = 1, \ \nu = 1$$

in (5.42) leads to an integral

$$e_{21} = \int\limits_0^\infty dx \frac{e^{-p^2 x^2}}{x} \, \sin bx$$

$$= \frac{1}{2} \frac{b}{p} \frac{1}{2i} \int\limits_{-\beta+i\infty}^{-\beta-i\infty} d\xi \frac{(b/p)^{2\xi}}{\sin \pi\xi \, \Gamma(2 + 2\xi)} \, \Gamma\left(\frac{1}{2} + \xi\right). \qquad (5.51)$$

Using the relations

$$\frac{(2n+1)!!}{(2n+1)!} = \frac{1}{2^n} \frac{1}{n!}, \qquad \Gamma\left(\frac{1}{2} + n\right) = \frac{\sqrt{\pi}}{2^n} (2n-1)!!$$

for integers n, we have

$$e_{21} = \frac{b\sqrt{\pi}}{2p} \sum_{n=0}^\infty \frac{(-1)^n}{n!(2n+1)} \left(\frac{b}{2p} \right)^{2n} \qquad (5.52)$$

or

$$e_{21} = \sqrt{\pi} \int dz \, e^{-z^2}, \qquad z = \frac{b}{2p}.$$

(22) Let

$$\gamma = \rho - 1, \ \mu = 2, \ m = 1, \ \nu = 1$$

be in (5.42), where $a > 0$, $\rho > -1$, then

$$e_{22} = \int_0^\infty dx x^{\rho-1} \, e^{-ax^2} \, \sin bx = \frac{1}{2} \, a^{-\frac{1+\rho}{2}}$$

$$\times \frac{1}{2i} \int_{-\beta+i\infty}^{-\beta-i\infty} d\xi \frac{b(b/\sqrt{a})^{2\xi}}{\sin \pi\xi \, \Gamma(2+2\xi)} \, \Gamma\left(\frac{\rho+1}{2}+\xi\right), \tag{5.53}$$

where

$$\Gamma(2+2\xi) = \frac{1}{\sqrt{\pi}} \, 2^{2(1+\xi)-1} \, \Gamma(1+\xi) \, \Gamma\left(\frac{3}{2}+\xi\right)$$

and

$$\Gamma\left(\frac{1}{2}+n+\frac{\rho}{2}\right) = \frac{\left[2\left(n+\frac{\rho}{2}\right)-1\right]!!}{2^{(n+\rho/2)}} \sqrt{\pi}$$

for integers n. So that by definition of the degenerating hypergeometric function

$$\Phi(\alpha, \beta, z) = {}_1F_1(\alpha, \beta, z) = 1 + \frac{\alpha}{\beta} \frac{z}{1!}$$

$$+ \frac{\alpha(\alpha+1)}{\beta(\beta+1)} \frac{z^2}{2!} + \frac{\alpha(\alpha+1)(\alpha+2)}{\beta(\beta+1)(\beta+2)} \frac{z^3}{3!} + \cdots, \tag{5.54}$$

we have

$$e_{22} = \frac{\pi}{4} \frac{b}{\sqrt{a}} \left(\frac{1}{2a}\right)^{\rho/2} \sum_{n=0}^\infty \frac{(-1)^n \left(\frac{b^2}{4a}\right)^n}{n! \, \Gamma\left(\frac{3}{2}+n\right)} \left[2\left(n+\frac{\rho}{2}\right)-1\right]!!$$

$$= \frac{b \, e^{-b^2/4a}}{2 \, a^{(\rho+1)/2}} \, \Gamma\left(\frac{1+\rho}{2}\right) \, {}_1F_1\left(1-\frac{\rho}{2}; \frac{3}{2}; \frac{b^2}{4a}\right). \tag{5.55}$$

(23) The formula (5.42) with

$$\gamma = 3, \ \nu = -4, \ \mu = 8, \ m = 1$$

gives

$$e_{23} = \int_0^\infty dx x^3 \, e^{-ax^8} \, \sin(bx^{-4})$$

$$= -\frac{\pi\sqrt{\pi}}{16} \frac{1}{2i} \int_{-\beta+i\infty}^{-\beta-i\infty} d\xi \frac{(b\sqrt{a})^{2\xi}}{\sin^2 \pi\xi \, \Gamma^2(1+\xi) \, \Gamma\left(\frac{3}{2}+\xi\right)}$$

$$= -\frac{\sqrt{\pi}}{16} \sum_{n=0}^\infty \frac{(b\sqrt{a})^{2n}}{(n!)^2 \, \Gamma\left(\frac{3}{2}+n\right)} \left[2\ln(b\sqrt{a}) - 2\Psi(1+n)\right.$$

$$\left. - \Psi\left(\frac{3}{2}+n\right)\right], \tag{5.56}$$

$a > 0$, $b > 0$, where

$$\bullet\ \Gamma\left(\frac{3}{2} + n\right) = \frac{\sqrt{\pi}}{2^{n+1}}(2n+1)!!,$$

$$\bullet\ \Psi\left(\frac{3}{2} + n\right) = \Psi\left(\frac{1}{2} + n\right) + \frac{2}{1+2n}$$

and

$$\bullet\ \Psi(n+1) = -C + \sum_{k=1}^{n} \frac{1}{k},$$

$$\bullet\ \Psi\left(\frac{1}{2} + n\right) = -C + 2\left[\sum_{k=1}^{n} \frac{1}{2k-1} - \ln 2\right].$$

(24) Assuming

$$\gamma = \rho - 1,\ \mu = 1,\ m = 1,\ \nu = 1$$

in the formula (5.42), one gets

$$e_{24} = \int_{0}^{\infty} dx\, x^{\rho-1}\, e^{-ax}\, \sin bx$$

$$= \frac{b}{a^{1+\rho}}\, \frac{1}{2i} \int_{-\beta+i\infty}^{-\beta-i\infty} d\xi \frac{(b/a)^{2\xi}}{\sin \pi\xi\, \Gamma(2+2\xi)}\, \Gamma(1+\rho+2\xi)$$

$$= \frac{1}{a^{\rho}} \sum_{n=0}^{\infty} \frac{(-1)^{n}\, (b/a)^{2n+1}}{(2n+1)!}\, \Gamma(1+\rho+2n). \tag{5.57}$$

So that we have the relation

$$e_{24} = \frac{\Gamma(\rho)}{\left[a^{2}+b^{2}\right]^{\rho/2}}\, \sin\left(\rho \arctan \frac{b}{a}\right). \tag{5.58}$$

(25) Similarly, we put

$$\gamma = \rho - 1,\ \mu = 1,\ m = 1,\ \nu = 1$$

in the formula (5.44) and get

$$e_{25} = \int_{0}^{\infty} dx\, x^{\rho-1}\, e^{-ax}\, \cos bx$$

$$= a^{-\rho}\, \frac{1}{2i} \int_{-\beta+i\infty}^{-\beta-i\infty} d\xi \frac{(b/a)^{2\xi}}{\sin \pi\xi\, \Gamma(1+2\xi)}\, \Gamma(\rho+2\xi)$$

$$= a^{-\rho} \sum_{n=0}^{\infty} \frac{(-1)^{n}}{(2n)!} \left(\frac{b}{a}\right)^{2n}\, \Gamma(\rho+2n)$$

$$= \frac{\Gamma(\rho)}{\left(a^{2}+b^{2}\right)^{\rho/2}}\, \cos\left(\rho \arctan \frac{b}{a}\right). \tag{5.59}$$

(26) If we change the notation $a \to a\cos t$ and $b \to a\sin t$, where $|t| < \frac{\pi}{2}$, Re $\rho > -1$, $a > 0$ in the integral (5.57), then we get automatically

$$e_{26} = \int\limits_0^\infty dx\, x^{\rho-1}\, e^{-a\cos t\, x}\, \sin[a\sin t\, x]$$

$$= \Gamma(\mu)\, a^{-\rho}\, \sin(\rho t).\tag{5.60}$$

(27) The limit $\rho \to 0$ in the integral (5.57) leads to

$$e_{27} = \int\limits_0^\infty dx\, x^{-1}\, e^{-ax}\, \sin(bx)$$

$$= \sum_{n=0}^\infty \frac{(-1)^n}{(1+2n)!} \left(\frac{b}{a}\right)^{1+2n} \Gamma(1+2n)$$

$$= \sum_{n=0}^\infty \frac{(-1)^n\,(b/a)^{2n+1}}{(2n+1)} = \arctan\left(\frac{b}{a}\right),\tag{5.61}$$

where $b/a \le 1$.

(28) Assuming

$$\gamma = 0,\ \mu = 2,\ \nu = -1,\ m = 1$$

in (5.44), one gets

$$e_{28} = \int\limits_0^\infty dx\, e^{-ax^2}\, \cos\left(bx^{-1}\right)$$

$$= \frac{1}{2\sqrt{a}}\, \frac{1}{2i} \int\limits_{-\beta+i\infty}^{-\beta-i\infty} d\xi \frac{(b\sqrt{a})^{2\xi}}{\sin\pi\xi\,\Gamma(1+2\xi)}\, \Gamma\left(\frac{1}{2}-\xi\right)$$

$$= \frac{\sqrt{\pi}}{2\sqrt{a}}\, \frac{1}{2i} \int\limits_{-\beta+i\infty}^{-\beta-i\infty} dx \frac{(2b\sqrt{a})^x\,\Gamma\left(1+\frac{x}{2}\right)}{\sin\pi x\,\Gamma^2(1+x)},\tag{5.62}$$

where we have used the relation

$$\Gamma(1+\xi) = \frac{2^{2\xi}}{\sqrt{\pi}}\, \Gamma\left(\frac{1+\xi}{2}\right) \Gamma\left(1+\frac{\xi}{2}\right)$$

and changed the integration variable $2\xi = x$. The result reads

$$e_{28} = \frac{\sqrt{\pi}}{2\sqrt{a}} \sum_{n=0}^\infty \frac{(-1)^n\,(2b\sqrt{a})^n}{(n!)^2}\, \Gamma\left(1+\frac{n}{2}\right),\tag{5.63}$$

where $a > 0$.

(29) If we choose

$$\gamma = -1, \ \mu = 1, \ m = 1, \ \nu = 1$$

in (5.42) and assume $a/b \leq 1$, then we should carry out our calculation by using the main formula (5.43). Thus

$$e_{29} = \int_0^\infty dx x^{-1} \, e^{-ax} \, \sin(bx)$$

$$= \frac{\sqrt{\pi}}{2} \frac{1}{2i} \int_{-\beta'+i\infty}^{-\beta'-i\infty} d\eta \frac{\left(\frac{2a}{b}\right)^\eta \, \Gamma\left(\frac{1}{2}(1+\eta)\right)}{\sin \pi\eta \, \Gamma(1+\eta) \, \Gamma\left(1-\frac{1}{2}\eta\right)}, \qquad (5.64)$$

where

$$\Gamma(1+\eta) = \frac{2^\eta}{\sqrt{\pi}} \, \Gamma\left(\frac{1+\eta}{2}\right) \, \Gamma\left(1+\frac{\eta}{2}\right).$$

Then

$$e_{29} = \frac{\pi}{4i} \int_{-\beta'+i\infty}^{-\beta'-i\infty} d\eta \frac{(a/b)^\eta \, \Gamma(1+\eta)}{\sin \pi\eta \, \Gamma(1+\eta) \, \Gamma\left(1-\frac{\eta}{2}\right) \, \Gamma\left(1+\frac{\eta}{2}\right)}. \qquad (5.65)$$

The change of the integration variable

$$\eta \to 2x + 1$$

reads

$$e_{29} = -\frac{1}{2i} \int_{-\beta''+i\infty}^{-\beta''-i\infty} dx \frac{(a/b)^{2x+1}}{(2x+1) \, \sin \pi x},$$

where $-1 < \beta'' < -\frac{1}{2}$.

Displacement of the integration contour to the right gives

$$e_{29} = \frac{\pi}{2} - \sum_{n=0}^\infty \frac{(-1)^n}{(2n+1) \, (b/a)^{2n+1}}, \qquad (5.66)$$

where $b/a \geq 1$.

Thus

$$e_{29} = \arctan\left(\frac{b}{a}\right)$$

as it should be.

(30) According to the general formula (5.43), the previous integral (5.64) with

$$m = 3, \qquad N_3 = 3^{1+2\xi} - 3,$$

$$m = 5, \qquad N_5 = 5^{1+2\xi} - 5 \cdot 3^{1+2\xi} + 10,$$

$$m = 7, \qquad N_7 = -7^{1+2\xi} + 7 \cdot 5^{1+2\xi} - 21 \cdot 3^{1+2\xi} + 35$$

is easy to calculate. The results are:

$$e_{30} = \int_0^\infty dx \, x^{-1} \, e^{-ax} \, \sin^3(bx)$$

$$= \arctan\left(\frac{b}{3a}\right) - 3 \arctan\left(\frac{b}{a}\right), \tag{5.67}$$

(31)

$$e_{31} = \int_0^\infty dx \, x^{-1} \, e^{-ax} \, \sin^5(bx)$$

$$= \arctan\left(\frac{b}{5a}\right) - 5 \arctan\left(\frac{b}{3a}\right) + 10 \arctan\left(\frac{b}{a}\right), \tag{5.68}$$

(32)

$$e_{32} = \int_0^\infty dx \, x^{-1} \, e^{-ax} \, \sin^7(bx)$$

$$= -\arctan\left(\frac{b}{7a}\right) + 7 \arctan\left(\frac{b}{5a}\right)$$

$$- 21 \arctan\left(\frac{b}{3a}\right) + 35 \arctan\left(\frac{b}{a}\right). \tag{5.69}$$

(33) Let

$$\gamma = 7, \ \mu = 8, \ q = 2, \ \nu = 4$$

be in (5.40), then

$$e_{33} = \int_0^\infty dx \, x^7 \, e^{-ax^8} \, \sin^2(bx^4)$$

$$= \frac{1}{16a} \frac{-1}{2i} \int_{\alpha+i\infty}^{\alpha-i\infty} d\xi \frac{(2b/\sqrt{a})^{2\xi}}{\sin \pi\xi \, \Gamma(1+2\xi)} \, \Gamma(1+\xi)$$

$$= -\frac{\sqrt{\pi}}{16a} \sum_{n=1}^\infty (-1)^n \frac{(b/\sqrt{a})^{2n}}{\Gamma\left(\frac{1}{2}+n\right)}, \tag{5.70}$$

where

$$\Gamma\left(\frac{1}{2}+n\right) = \frac{\sqrt{\pi}}{2^n}(2n+1)!!.$$

Thus,

$$e_{33} = -\frac{1}{16a} \sum_{n=1}^\infty (-1)^n \frac{\left(b\sqrt{2/a}\right)^{2n}}{(2n-1)!!}. \tag{5.71}$$

(34) Assuming

$$\gamma = 7, \ \mu = 8, \ q = 2, \ \nu = 8$$

in (5.40), one gets

$$e_{34} = \int_0^\infty dx x^7 \ e^{-ax^8} \ \sin^2(bx^8)$$

$$= -\frac{1}{16a} \frac{1}{2i} \int_{\alpha+i\infty}^{\alpha-i\infty} d\xi \frac{(2b/a)^{2\xi}}{\sin \pi\xi \ \Gamma(1+2\xi)} \ \Gamma(1+2\xi)$$

$$= -\frac{1}{16a} \sum_{n=1}^\infty (-1)^n \left(\frac{4b^2}{a^2}\right)^n = \frac{1}{16a}\left[1 - \frac{a^2}{a^2 + 4b^2}\right]. \qquad (5.72)$$

(35) The choice of the parameters

$$\gamma = -1, \ \mu = 2, \ \nu = 2, \ q = 2$$

in the formula (5.40), reads

$$e_{35} = \int_0^\infty dx \frac{e^{-ax^2} \ \sin^2(bx^2)}{x}$$

$$= -\frac{1}{8} \frac{1}{2i} \int_{\alpha+i\infty}^{\alpha-i\infty} d\xi \frac{(2b/a)^{2\xi}}{\sin \pi\xi \ \xi} = -\frac{1}{8} \sum_{n=1}^\infty (-1)^n \frac{(4b^2/a^2)^n}{n}$$

$$= \frac{1}{8} \ln\left(1 + \frac{4b^2}{a^2}\right). \qquad (5.73)$$

(36) The formula (5.40) with parameters

$$\gamma = -3, \ \mu = 2, \ \nu = 2, \ q = 2$$

leads to an integral

$$e_{36} = \int_0^\infty dx x^{-3} \ e^{-ax^2} \ \sin^2(bx^2)$$

$$= -\frac{a}{8} \frac{1}{2i} \int_{\alpha+i\infty}^{\alpha-i\infty} d\xi \frac{(2b/a)^{2\xi}}{\sin \pi\xi \ \xi(2\xi-1)}. \qquad (5.74)$$

After some elementary calculations, we have

$$e_{36} = -\frac{a}{8} \sum_{n=1}^\infty (-1)^n \frac{(4b^2/a^2)^n}{n(2n-1)}$$

$$= -\frac{a}{8}\left\{\ln\left(1 + \frac{4b^2}{a^2}\right) - \frac{4b}{a} \ \arctan\left(\frac{2b}{a}\right)\right\}. \qquad (5.75)$$

(37) Notice that our general formulas allow us to give many sets of families of equivalent integrals.

For example, if we choose parameters

$$\gamma = 0,\ \mu = 1,\ \nu = 1,\ q = 2$$

in the formula (5.40), then we obtain

$$e_{37} = \int_0^\infty dx\ e^{-ax}\ \sin^2(bx)$$

$$= -\frac{1}{2a}\frac{1}{2i} \int_{\alpha+i\infty}^{\alpha-i\infty} d\xi \frac{(2b/a)^{2\xi}}{\sin \pi\xi} = 8\ e_{34}. \tag{5.76}$$

(38) We consider yet one integral of this family, with parameters

$$\gamma = 2,\ \mu = 3,\ \nu = 3,\ q = 2$$

in (5.40)

$$e_{38} = \int_0^\infty dx x^2\ e^{-ax^3}\ \sin^2(bx^3) = \frac{8}{3}\ e_{34} \tag{5.77}$$

etc.

Similar families of equivalent integrals can be obtained from other general formulas of (5.42), (5.44) and (5.46).

(39) The integral (5.42) with

$$\gamma = 2,\ \mu = 6,\ \nu = 3,\ m = 1$$

gives

$$e_{39} = \int_0^\infty dx x^2\ e^{-ax^6}\ \sin(bx^3)$$

$$= \frac{1}{6a}\frac{1}{2i} \int_{-\beta+i\infty}^{-\beta-i\infty} d\xi \frac{b(b/\sqrt{a})^{2\xi}}{\sin \pi\xi\ \Gamma(2+2\xi)}\ \Gamma(1+\xi)$$

$$= \frac{b}{6a} \sum_{n=0}^\infty (-1)^n \frac{(b/\sqrt{2a})^{2n}}{(2n+1)!!}, \tag{5.78}$$

where $a > 0$.

(40) Let

$$\mu = 6,\ \nu = 6,\ \gamma = 5,\ m = 1$$

be in (5.42), then

$$e_{40} = \int\limits_0^\infty dx x^5 \, e^{-ax^6} \sin(bx^6)$$

$$= \frac{b}{6a^2} \frac{1}{2i} \int\limits_{-\beta+i\infty}^{-\beta-i\infty} d\xi \frac{(b/a)^{2\xi}}{\sin \pi \xi \, \Gamma(2+2\xi)} \Gamma(2+2\xi)$$

$$= \frac{b}{6} \frac{1}{a^2+b^2}, \tag{5.79}$$

where $b/a \le 1$.

(41) The main formula (5.42) with

$$\gamma = -1, \ \mu = 9, \ \nu = 9, \ m = 1$$

reads

$$e_{41} = \int\limits_0^\infty dx x^{-1} \, e^{-ax^9} \sin(bx^9)$$

$$= \frac{1}{9a} \frac{1}{2i} \int\limits_{-\beta+i\infty}^{-\beta-i\infty} d\xi \frac{b(b/a)^{2\xi}}{\sin \pi \xi \, \Gamma(2+2\xi)} \Gamma(1+2\xi)$$

and therefore

$$e_{41} = \frac{1}{9} \sum_{n=0}^\infty \frac{(-1)^n \, (b/a)^{2n+1}}{2n+1} = \frac{1}{9} \arctan\left(\frac{b}{a}\right), \tag{5.80}$$

where $b^2/a^2 \le 1$.

(42) If

$$\gamma = -4, \ \nu = 3, \ \mu = 3, \ m = 1$$

in (5.42), then we derive

$$e_{42} = \int\limits_0^\infty dx x^{-4} \, e^{-ax^3} \sin(bx^3)$$

$$= \frac{b}{3} \frac{1}{2i} \int\limits_{-\beta+i\infty}^{-\beta-i\infty} d\xi \frac{(b/a)^{2\xi}}{\sin \pi \xi \, \Gamma(2+2\xi)} \Gamma(2\xi)$$

$$= \frac{b}{3}\left[\ln\left(\frac{b}{a}\right) - 1\right] + \frac{b}{6} \sum_{n=1}^\infty \frac{(-1)^n \, (b/a)^{2n}}{n(2n+1)}. \tag{5.81}$$

Since

$$\frac{1}{n(2n+1)} = \frac{1}{n} - \frac{2}{2n+1}.$$

Then, we have

$$e_{42} = -\frac{b}{6}\ln\left(1 + \frac{b^2}{a^2}\right) - \frac{a}{3}\arctan\left(\frac{b}{a}\right) + \frac{b}{3}\ln\left(\frac{b}{a}\right), \qquad (5.82)$$

where $b^2/a^2 \leq 1$.

Similar procedure of calculating integrals in (5.44) and (5.46) holds.

(43) The integral with

$$\gamma = 7, \ \mu = 8, \ m = 3, \ \nu = 4$$

in (5.44) is given by the expression

$$e_{43} = \int_0^\infty dx x^7 \, e^{-ax^8} \, \cos^3(bx^4)$$

$$= \frac{1}{32a}\frac{1}{2i}\int_{-\beta+i\infty}^{-\beta-i\infty} d\xi \frac{(b/\sqrt{a})^{2\xi} \, (3 + 3^{2\xi})}{\sin\pi\xi \, \Gamma(1+2\xi)} \, \Gamma(1+\xi)$$

$$= \frac{1}{32a}\sum_{n=0}^\infty \frac{(-1)^n}{(2n-1)!!}\left[3\left(\frac{b}{\sqrt{2a}}\right)^{2n} + \left(3\frac{b}{\sqrt{2a}}\right)^{2n}\right]. \qquad (5.83)$$

(44) The choice of the parameters $\gamma = 7, \ \mu = 8, \ m = 3, \ \nu = 8$ in (5.44) gives

$$e_{44} = \int_0^\infty dx x^7 \, e^{-ax^8} \, \cos^3(bx^8)$$

$$= \frac{1}{32a}\frac{1}{2i}\int_{-\beta+i\infty}^{-\beta-i\infty} d\xi \frac{(b/a)^{2\xi} \, (3 + 3^{3\xi})}{\sin\pi\xi}$$

$$= \frac{1}{32a}\left[\frac{3}{1 + \frac{b^2}{a^2}} + \frac{1}{1 + \frac{9b^2}{a^2}}\right]. \qquad (5.84)$$

(45) The integral with

$$\gamma = -1, \ \mu = 2, \ \nu = 2, \ m = 3$$

in (5.44) is given by

$$e_{45} = \int_0^\infty dx \frac{e^{-ax^2} \, \cos^3(bx^2)}{x} \qquad (5.85)$$

$$= \frac{1}{16}\frac{1}{2i}\int_{-\beta+i\infty}^{-\beta-i\infty} d\xi \frac{(b/a)^{2\xi}}{\xi \, \sin\pi\xi} \, (3 + 3^{2\xi})$$

$$= \frac{1}{8}\left(\ln\frac{b}{a} + \ln 3\right) + \frac{1}{16}\sum_{n=1}^\infty \frac{(-1)^n}{n}\left[3\left(\frac{b}{a}\right)^{2n} + \left(\frac{3b}{a}\right)^{2n}\right]$$

$$= \frac{1}{8}\left[\ln\left(\frac{b}{a}\right) + \ln 3\right] - \frac{3}{16}\ln\left(1 + \frac{b^2}{a^2}\right) - \frac{1}{16}\ln\left(1 + \frac{9b^2}{a^2}\right),$$

where $9b^2/a^2 \leq 1$.

(46) The main formula (5.44) with

$$\gamma = -3, \ \mu = 2, \ \nu = 2, \ m = 3$$

reads

$$e_{46} = \int\limits_0^\infty dx x^{-3} \ e^{-ax^2} \ \cos^3(bx^2) \tag{5.86}$$

$$= \frac{a}{16} \frac{1}{2i} \int\limits_{-\beta+i\infty}^{-\beta-i\infty} d\xi \frac{(b/a)^{2\xi} \ (3 + 3^{2\xi})}{\sin \pi \xi \ \xi(2\xi - 1)}$$

$$= \frac{a}{8} \left[\ln\left(\frac{b}{a}\right) + \ln 3 - 2 \right] + \frac{a}{16} \sum\limits_{n=1}^\infty \frac{(-1)^n \ (b/a)^{2n} \ (3 + 3^{2n})}{n(2n - 1)}$$

$$= \frac{a}{8} \left[\ln\left(\frac{b}{a}\right) + \ln 3 - 2 \right] + \frac{a}{16} \left\{ \left[3 \ln\left(1 + \frac{b^2}{a^2}\right) + \ln\left(1 + \frac{9b^2}{a^2}\right) \right] \right.$$

$$\left. - \frac{6b}{a} \arctan\left(\frac{b}{a}\right) - \frac{6b}{a} \arctan\left(\frac{3b}{a}\right) \right\}.$$

(47) The formula (5.46) with

$$\gamma = 7, \ \mu = 8, \ q = 4, \ \nu = 4,$$

where

$$I_4'(\xi) = 2^{2\xi} + 4$$

gives

$$e_{47} = \int\limits_0^\infty dx x^7 \ e^{-ax^8} \left[\cos^4(bx^4) - 1 \right]$$

$$= \frac{1}{32a} \frac{1}{2i} \int\limits_{\alpha+i\infty}^{\alpha-i\infty} d\xi \frac{(2b/\sqrt{a})^{2\xi} \ (4 + 2^{2\xi})}{\sin \pi \xi \ \Gamma(1 + 2\xi)} \Gamma(1 + \xi)$$

$$= \frac{\sqrt{\pi}}{32a} \sum\limits_{n=1}^\infty \frac{(-1)^n}{\Gamma\left(\frac{1}{2} + n\right)} \left[4 \left(\frac{b}{\sqrt{a}}\right)^{2n} + \left(\frac{2b}{\sqrt{a}}\right)^{2n} \right]. \tag{5.87}$$

Here

$$\Gamma\left(\frac{1}{2} + n\right) = \frac{\sqrt{\pi}}{2^n}(2n - 1)!!.$$

(48) We put

$$\gamma = 7, \ \mu = 8, \ q = 4, \ \nu = 8$$

in (5.46) and obtain

$$e_{48} = \int_0^\infty dx x^7 \, e^{-ax^8} \left[\cos^4(bx^8) - 1 \right]$$

$$= \frac{1}{32a} \frac{1}{2i} \int_{\alpha+i\infty}^{\alpha-i\infty} d\xi \frac{(2b/a)^{2\xi} \, (4 + 2^{2\xi})}{\sin \pi \xi}$$

$$= \frac{1}{32a} \left[4 \left(\frac{1}{1 + \frac{4b^2}{a^2}} - 1 \right) + \frac{1}{1 + \frac{16b^2}{a^2}} - 1 \right]$$

$$= \frac{1}{32a} \left\{ \frac{4}{1 + \frac{4b^2}{a^2}} + \frac{1}{1 + \frac{16b^2}{a^2}} - 5 \right\}, \tag{5.88}$$

where $4b^2/a^2 \le 1$.

(49) Parameters

$$\gamma = -1, \ \mu = 2, \ \nu = 2, \ q = 4$$

in (5.46) give

$$e_{49} = \int_0^\infty \frac{dx}{x} \, e^{-ax^2} \left[\cos^4(bx^2) - 1 \right]$$

$$= \frac{1}{16} \frac{1}{2i} \int_{\alpha+i\infty}^{\alpha-i\infty} d\xi \frac{(2b/a)^{2\xi}}{\sin \pi \xi \, \xi} (4 + 2^{2\xi})$$

$$= -\frac{1}{16} \left\{ 4 \ln \left(1 + \frac{4b^2}{a^2} \right) + \ln \left(1 + \frac{16b^2}{a^2} \right) \right\}. \tag{5.89}$$

(50) We put

$$\gamma = -3, \ \mu = 2, \ \nu = 2, \ q = 4$$

in the formula (5.46) and derive

$$e_{50} = \int_0^\infty dx x^{-3} \, e^{-ax^2} \left[\cos^4(bx^2) - 1 \right]$$

$$= \frac{a}{16} \frac{1}{2i} \int_{\alpha+i\infty}^{\alpha-i\infty} d\xi \frac{(2b/a)^{2\xi} \, (4 + 2^{2\xi})}{\sin \pi \xi \, \xi(2\xi - 1)}. \tag{5.90}$$

By carrying out some necessary calculations, one gets

$$e_{50} = \frac{a}{16} \left\{ 4 \left[\ln \left(1 + \frac{4b^2}{a^2} \right) - \frac{4b}{a} \arctan \left(\frac{2b}{a} \right) \right] \right.$$

$$\left. + \ln \left(1 + \frac{16b^2}{a^2} \right) - \frac{8b}{a} \arctan \left(\frac{4b}{a} \right) \right\}. \tag{5.91}$$

5.6   Universal Formulas for Integrals Involving Exponential, Trigonometric and $x^\gamma \left[p + tx^\rho \right]^{-\lambda}$ -Functions

5.6.1 47th *General Formula*

$$
N_{47} = \int\limits_0^\infty dx \frac{x^\gamma \, e^{-ax^\mu}}{\left[p + tx^\rho \right]^\lambda} \, \sin^q(bx^\nu) = \frac{1}{\rho} \, \frac{1}{\Gamma(\lambda)} \, \frac{1}{p^\lambda} \left(\frac{p}{t} \right)^{\frac{\gamma+1}{\rho}}
$$

$$
\times \frac{1}{2^{q-1}} \frac{1}{2i} \int\limits_{\alpha+i\infty}^{\alpha-i\infty} d\xi \frac{\left[2b \left(\frac{p}{t} \right)^{\nu/\rho} \right]^{2\xi}}{\sin \pi\xi \, \Gamma(1 + 2\xi)} \, I_q(\xi)
$$

$$
\times \frac{1}{2i} \int\limits_{-\beta'+i\infty}^{-\beta'-i\infty} d\eta \frac{\left[a \left(\frac{p}{t} \right)^{\mu/\rho} \right]^\eta}{\sin \pi\eta \, \Gamma(1 + \eta)} \, \Gamma \left(\frac{1 + \gamma + 2\nu\xi}{\rho} + \frac{\mu\eta}{\rho} \right)
$$

$$
\times \Gamma \left(\lambda - \frac{\gamma + 1 + 2\nu\xi + \mu\eta}{\rho} \right)
$$

(5.92)

or

$$
N_{47} = \frac{1}{\rho} \, \frac{1}{\Gamma(\lambda)} \, \frac{1}{p^\lambda} \left(\frac{p}{t} \right)^{\frac{\gamma+1}{\rho}} \frac{1}{2i} \int\limits_{-\beta'+i\infty}^{-\beta'-i\infty} d\eta \frac{\left[a \left(\frac{p}{t} \right)^{\mu/\rho} \right]^\eta}{\sin \pi\eta \, \Gamma(1 + \eta)} \frac{1}{2^{q-1}}
$$

$$
\times \frac{1}{2i} \int\limits_{\alpha+i\infty}^{\alpha-i\infty} d\xi \frac{\left[2b \left(\frac{p}{t} \right)^{\nu/\rho} \right]^{2\xi}}{\sin \pi\xi \, \Gamma(1 + 2\xi)} \, I_q(\xi) \, \Gamma \left(\frac{\gamma + 1 + \mu\eta + 2\nu\xi}{\rho} \right)
$$

(5.93)

$$
\times \Gamma \left(\lambda - \frac{\gamma + 1 + \mu\eta + 2\nu\xi}{\rho} \right),
$$

where $q = 2, \ 4, \ 6, \ldots,$ $I_q(\xi)$ is given by (1.32) in Chapter 1, and $0 < \alpha < 1$, $-1 < \beta' < 0$.

5.6.2   48th *General Formula*

$$N_{48} = \int_0^\infty dx \frac{x^\gamma \, e^{-ax^\mu}}{\left[p + tx^\rho\right]^\lambda} \, \sin^m(bx^\nu) = \frac{1}{\rho} \, \frac{1}{\Gamma(\lambda)} \, \frac{1}{p^\lambda} \, \left(\frac{p}{t}\right)^{\frac{\gamma+1+\nu}{\rho}}$$

$$\times \frac{1}{2^{m-1}} \, \frac{1}{2i} \int_{-\beta+i\infty}^{-\beta-i\infty} d\xi \frac{b \left[b \left(\frac{p}{t}\right)^{\nu/\rho}\right]^{2\xi}}{\sin \pi\xi \, \Gamma(2+2\xi)} N_m(\xi) \qquad (5.94)$$

$$\times \frac{1}{2i} \int_{-\beta'+i\infty}^{-\beta'-i\infty} d\eta \frac{\left[a \left(\frac{p}{t}\right)^{\mu/\rho}\right]^\eta}{\sin \pi\eta \, \Gamma(1+\eta)} \, \Gamma\left(\frac{1+\gamma+\nu+2\nu\xi+\mu\eta}{\rho}\right)$$

$$\times \Gamma\left(\lambda - \frac{1+\gamma+\nu+2\nu\xi+\mu\eta}{\rho}\right)$$

or

$$N_{48} = \frac{1}{\rho} \, \frac{1}{\Gamma(\lambda)} \, \frac{1}{p^\lambda} \, \left(\frac{p}{t}\right)^{\frac{\gamma+1+\nu}{\rho}} \frac{1}{2i} \int_{-\beta'+i\infty}^{-\beta'-i\infty} d\eta \frac{\left[a \left(\frac{p}{t}\right)^{\mu/\rho}\right]^\eta}{\sin \pi\eta \, \Gamma(1+\eta)} \, \frac{1}{2^{m-1}}$$

$$\times \frac{1}{2i} \int_{-\beta+i\infty}^{-\beta-i\infty} d\xi \frac{b \left[b \left(\frac{p}{t}\right)^{\nu/\rho}\right]^{2\xi}}{\sin \pi\xi \, \Gamma(2+2\xi)} \, \Gamma\left(\frac{1+\gamma+\nu+2\nu\xi+\mu\eta}{\rho}\right) \qquad (5.95)$$

$$\times \Gamma\left(\lambda - \frac{1+\gamma+\nu+2\nu\xi+\mu\eta}{\rho}\right) N_m(\xi),$$

where $m = 1, 3, 5, 7, \ldots$, $-1 < \beta, \beta' < 0$ and $N_m(\xi)$ is defined by (1.34) in Chapter 1.

5.6.3 49ᵗʰ *General Formula*

$$N_{49} = \int\limits_0^\infty dx\, x^\gamma \frac{e^{-ax^\mu}}{\left[p + tx^\rho\right]^\lambda} \cos^m(bx^\nu) = \frac{1}{\rho}\, \frac{1}{\Gamma(\lambda)}\, \frac{1}{p^\lambda}\, \left(\frac{p}{t}\right)^{\frac{\gamma+1}{\rho}}$$

$$\times \frac{1}{2^{m-1}}\, \frac{1}{2i} \int\limits_{-\beta+i\infty}^{-\beta-i\infty} d\xi \frac{\left[b\left(\frac{p}{t}\right)^{\nu/\rho}\right]^{2\xi}}{\sin\pi\xi\,\Gamma(1+2\xi)} N_m'(\xi)$$

$$\times \frac{1}{2i} \int\limits_{-\beta'+i\infty}^{-\beta'-i\infty} d\eta \frac{\left[a\left(\frac{p}{t}\right)^{\mu/\rho}\right]^{\eta}}{\sin\pi\eta\,\Gamma(1+\eta)}\, \Gamma\left(\frac{1+\gamma+2\nu\xi+\mu\eta}{\rho}\right)$$

$$\times \Gamma\left(\lambda - \frac{\gamma+1+2\nu\xi+\mu\eta}{\rho}\right) \tag{5.96}$$

or

$$N_{49} = \frac{1}{\rho}\, \frac{1}{\Gamma(\lambda)}\, \frac{1}{p^\lambda}\, \left(\frac{p}{t}\right)^{\frac{\gamma+1}{\rho}} \frac{1}{2i} \int\limits_{-\beta'+i\infty}^{-\beta'-i\infty} d\eta \frac{\left[a\left(\frac{p}{t}\right)^{\mu/\rho}\right]^{\eta}}{\sin\pi\eta\,\Gamma(1+\eta)}\, \frac{1}{2^{m-1}}$$

$$\times \frac{1}{2i} \int\limits_{-\beta+i\infty}^{-\beta-i\infty} d\xi \frac{\left[b\left(\frac{p}{t}\right)^{\nu/\rho}\right]^{2\xi}}{\sin\pi\xi\,\Gamma(1+2\xi)} N_m'(\xi)\, \Gamma\left(\frac{1+\gamma+2\nu\xi+\mu\eta}{\rho}\right) \tag{5.97}$$

$$\times \Gamma\left(\lambda - \frac{\gamma+1+2\nu\xi+\mu\eta}{\rho}\right),$$

where $m = 1, 3, 5, 7, \ldots$ and $N_m'(\xi)$ is given by (1.36) in Chapter 1.

5.6.4   50th *General Formula*

$$N_{50} = \int\limits_0^\infty dx \frac{x^\gamma \, e^{-ax^\mu}}{\left[p + tx^\rho\right]^\lambda} \left[\cos^q(bx^\nu) - 1\right] = \frac{1}{\rho} \, \frac{1}{\Gamma(\lambda)} \, \frac{1}{p^\lambda} \left(\frac{p}{t}\right)^{\frac{\gamma+1}{\rho}}$$

$$\times \frac{1}{2^{q-1}} \frac{1}{2i} \int\limits_{\alpha+i\infty}^{\alpha-i\infty} d\xi \frac{\left[2b\left(\frac{p}{t}\right)^{\nu/\rho}\right]^{2\xi}}{\sin \pi\xi \, \Gamma(1+2\xi)} \, I'_q(\xi)$$

$$\times \frac{1}{2i} \int\limits_{-\beta'+i\infty}^{-\beta'-i\infty} d\eta \frac{\left[a\left(\frac{p}{t}\right)^{\mu/\rho}\right]^\eta}{\sin \pi\eta \, \Gamma(1+\eta)} \, \Gamma\left(\frac{1+\gamma+2\nu\xi+\mu\eta}{\rho}\right)$$

$$\times \Gamma\left(\lambda - \frac{\gamma+1+2\nu\xi+\mu\eta}{\rho}\right)$$

(5.98)

or

$$N_{50} = \frac{1}{\rho} \, \frac{1}{\Gamma(\lambda)} \, \frac{1}{p^\lambda} \left(\frac{p}{t}\right)^{\frac{\gamma+1}{\rho}} \frac{1}{2i} \int\limits_{-\beta'+i\infty}^{-\beta'-i\infty} d\eta \frac{\left[a\left(\frac{p}{t}\right)^{\mu/\rho}\right]^\eta}{\sin \pi\eta \, \Gamma(1+\eta)} \, \frac{1}{2^{q-1}}$$

$$\times \frac{1}{2i} \int\limits_{\alpha+i\infty}^{\alpha-i\infty} d\xi \frac{\left[2b\left(\frac{p}{t}\right)^{\nu/\rho}\right]^{2\xi}}{\sin \pi\xi \, \Gamma(1+2\xi)} \, I'_q(\xi) \, \Gamma\left(\frac{1+\gamma+2\nu\xi+\mu\eta}{\rho}\right)$$

$$\times \Gamma\left(\lambda - \frac{\gamma+1+2\nu\xi+\mu\eta}{\rho}\right).$$

(5.99)

5.6.5   51st *General Formula*

$$N_{51} = \int\limits_0^\infty dx \frac{x^\gamma \ \sin^m(bx^\nu)}{\left[p + tx^\rho\right]^\lambda} \ \exp\left[-a\left(p + tx^\rho\right)^\sigma\right] = \frac{1}{\rho}\frac{1}{p^\lambda}$$

$$\times \left(\frac{p}{t}\right)^{\frac{\gamma+1+\nu}{\rho}} \frac{1}{2^{m-1}}\frac{1}{2i} \int\limits_{-\beta+i\infty}^{-\beta-i\infty} d\xi \frac{\left[b\left(\frac{p}{t}\right)^{\nu/\rho}\right]^{2\xi}}{\sin\pi\xi \ \Gamma(2 + 2\xi)} \ N_m(\xi)$$

$$\times \Gamma\left(\frac{\gamma + 1 + \nu + 2\nu\xi}{\rho}\right) \tag{5.100}$$

$$\times \frac{1}{2i} \int\limits_{-\beta'+i\infty}^{-\beta'-i\infty} d\eta \frac{\left[ap^\sigma\right]^\eta}{\sin\pi\eta \ \Gamma(1 + \eta)} \frac{\Gamma\left(\lambda - \sigma\eta - \dfrac{\gamma + 1 + \nu + 2\xi\nu}{\rho}\right)}{\Gamma(\lambda - \sigma\eta)},$$

$$m = 1, \ 3, \ 5, \ 7, \ldots.$$

5.6.6   52nd *General Formula*

$$N_{52} = \int\limits_0^\infty dx \frac{x^\gamma \ \cos^m(bx^\nu)}{\left[p + tx^\rho\right]^\lambda} \ \exp\left[-a\left(p + tx^\rho\right)^\sigma\right]$$

$$= \frac{1}{\rho}\frac{1}{p^\lambda}\left(\frac{p}{t}\right)^{\frac{\gamma+1}{\rho}} \frac{1}{2^{m-1}}$$

$$\times \frac{1}{2i} \int\limits_{-\beta+i\infty}^{-\beta-i\infty} d\xi \frac{\left[b\left(\frac{p}{t}\right)^{\nu/\rho}\right]^{2\xi}}{\sin\pi\xi \ \Gamma(1 + 2\xi)} N_m'(\xi) \ \Gamma\left(\frac{\gamma + 1 + 2\nu\xi}{\rho}\right) \tag{5.101}$$

$$\times \frac{1}{2i} \int\limits_{-\beta'+i\infty}^{-\beta'-i\infty} d\eta \frac{\left[ap^\sigma\right]^\eta}{\sin\pi\eta \ \Gamma(1 + \eta)} \frac{\Gamma\left(\lambda - \sigma\eta - \dfrac{\gamma + 1 + 2\nu\xi}{\rho}\right)}{\Gamma(\lambda - \sigma\eta)},$$

$$m = 1, \ 3, \ 5, \ 7, \ldots.$$

Similar general formulas can be obtained by using functions $\sin^q(bx^\nu)$ and $\left[\cos^q(bx^\nu) - 1\right]$ as in (5.100) and (5.101).

5.7   Some Consequences of the General Formulas Obtained in Section 5.6

(51) The formula (5.94) with

$$\gamma = 1, \ \mu = 2, \ \nu = 1, \ p = \omega^2, \ t = 1, \ \rho = 2, \ \lambda = 1, \ m = 1$$

gives

$$e_{51} = \int\limits_0^\infty dx \frac{x\, e^{-ax^2}}{[\omega^2 + x^2]} \, \sin(bx)$$

$$= \frac{1}{2} \frac{1}{\Gamma(1)} \frac{1}{\omega^2} (\omega^2)^{3/2} \frac{1}{2i} \int\limits_{-\beta+i\infty}^{-\beta-i\infty} d\xi \frac{b[b\,\omega]^{2\xi}}{\sin \pi\xi \, \Gamma(2+2\xi)}$$

$$\times \frac{1}{2i} \int\limits_{-\beta'+i\infty}^{-\beta'-i\infty} d\eta \frac{[a\,\omega^2]^\eta}{\sin \pi\eta \, \Gamma(1+\eta)}$$

$$\times \Gamma\left(\frac{3}{2} + \xi + \eta\right) \Gamma\left(-\frac{1}{2} - \xi - \eta\right), \tag{5.102}$$

where

$$\bullet \ \Gamma\left(\frac{3}{2} + \xi + \eta\right) = \left(\frac{1}{2} + \xi + \eta\right) \Gamma\left(\frac{1}{2} + \xi + \eta\right),$$

$$\bullet \ \Gamma\left(-\frac{1}{2} - \xi - \eta\right) = \frac{\left(-\frac{1}{2} - \xi - \eta\right)}{\left(-\frac{1}{2} - \xi - \eta\right)} \Gamma\left(-\frac{1}{2} - \xi - \eta\right)$$

$$= \frac{1}{-\frac{1}{2} - \xi - \eta} \Gamma\left(\frac{1}{2} - \xi - \eta\right),$$

$$\bullet \ \Gamma\left(\frac{1}{2} + \xi + \eta\right) \Gamma\left(\frac{1}{2} - \xi - \eta\right) = \frac{\pi}{\cos(\xi + \eta)}.$$

Displacing contours of the integration variables to the right and after some calculations, we have

$$e_{51} = -\frac{\pi}{4} e^{a\omega^2} \left[2\sinh(b\omega) + \Phi\left(\omega\sqrt{a} - \frac{b}{2\sqrt{a}}\right) \right.$$

$$\left. - \Phi\left(\omega\sqrt{a} + \frac{b}{2\sqrt{a}}\right) \right], \tag{5.103}$$

where $a > 0$, Re $\omega > 0$.

(52) Similar calculation from the formula (5.96) with parameters

$$\gamma = 0, \ m = 1, \ \nu = 1, \ p \to \omega^2, \ t = 1, \ \rho = 2, \ \lambda = 1, \ \mu = 2$$

reads

$$e_{52} = \int\limits_0^\infty \frac{dx}{\omega^2 + x^2} \, e^{-ax^2} \, \cos(bx)$$

$$= \frac{1}{2} \frac{1}{\Gamma(1)} \frac{1}{\omega^2} (\omega^2)^{1/2} \frac{1}{2i} \int\limits_{-\beta+i\infty}^{-\beta-\infty} d\xi \frac{[b\,\omega]^{2\xi}}{\sin \pi\xi \, \Gamma(1 + 2\xi)}$$

$$\times \frac{1}{2i} \int\limits_{-\beta'+i\infty}^{-\beta'-i\infty} d\eta \frac{[a\omega^2]^\eta}{\sin \pi\eta \, \Gamma(1 + \eta)}$$

$$\times \Gamma\left(\frac{1}{2} + \xi + \eta\right) \Gamma\left(\frac{1}{2} - \xi - \eta\right), \tag{5.104}$$

where

$$\Gamma\left(\frac{1}{2} + \xi + \eta\right) \Gamma\left(\frac{1}{2} - \xi - \eta\right) = \frac{\pi}{\cos \pi(\xi + \eta)}.$$

Displacement of contours in turn for the integration variables ξ and η to the right and calculation of their residues results in

$$e_{52} = \frac{\pi}{4\omega} e^{a\omega^2} \left[2\cosh(b\omega) - \Phi\left(\omega\sqrt{a} - \frac{b}{2\sqrt{a}}\right) \right.$$

$$\left. - \Phi\left(\omega\sqrt{a} + \frac{b}{2\sqrt{a}}\right) \right]. \tag{5.105}$$

(53) If we put

$$\gamma = 1, \ \nu = 1, \ t = 1, \ \rho = 2, \ p \to \omega^2, \ \sigma = \frac{1}{2}, \ \lambda = \frac{1}{2}, \ m = 1$$

in the formula (5.100), then we have

$$e_{53} = \int\limits_0^\infty dx \frac{x}{\sqrt{\omega^2 + x^2}} \, \sin(bx) \, \exp\left[-a(\omega^2 + x^2)^{1/2} \right]$$

$$= \frac{1}{2} \frac{1}{\omega} (\omega^2)^{3/2} \frac{1}{2i} \int\limits_{-\beta+i\infty}^{-\beta-i\infty} d\xi \frac{b[b\,\omega]^{2\xi}}{\sin \pi\xi \, \Gamma(2 + 2\xi)} \Gamma\left(\frac{3}{2} + \xi\right)$$

$$\times \frac{1}{2i} \int\limits_{-\beta'+i\infty}^{-\beta'-i\infty} d\eta \frac{[a\,\omega]^\eta}{\sin \pi\eta \, \Gamma(1 + \eta)} \frac{\Gamma\left(\frac{1}{2} - \frac{1}{2}\eta - \frac{3}{2} - \xi\right)}{\Gamma\left(\frac{1}{2} - \frac{1}{2}\eta\right)}. \tag{5.106}$$

After some elementary calculations, we get

$$e_{53} = \frac{b\,\omega}{\sqrt{b^2 + a^2}} \, K_1\left(\omega\sqrt{b^2 + a^2}\right). \tag{5.107}$$

(54) The formula (5.101) with

$$\gamma = 0, \ \sigma = \frac{1}{2}, \ \lambda = \frac{1}{2}, \ \nu = 1, \ m = 1, \ p \to \omega^2, \ \rho = 2$$

gives

$$e_{54} = \int\limits_0^\infty dx \frac{\cos(bx)}{\sqrt{\omega^2 + x^2}} \exp\left[-a(\omega^2 + x^2)^{1/2}\right]$$

$$= \frac{1}{2}\frac{1}{\omega}(\omega^2)^{1/2}\frac{1}{2i} \int\limits_{-\beta+i\infty}^{-\beta-i\infty} d\xi \frac{[b\,\omega]^{2\xi}}{\sin \pi\xi \ \Gamma(1+2\xi)} \ \Gamma\left(\frac{1}{2}+\xi\right)$$

$$\times \frac{1}{2i} \int\limits_{-\beta'+i\infty}^{-\beta'-i\infty} d\eta \frac{[a\,\omega]^\eta}{\sin \pi\eta \ \Gamma(1+\eta)} \ \frac{\Gamma\left(\frac{1}{2} - \frac{1}{2}\eta - \frac{1}{2} - \xi\right)}{\Gamma\left(\frac{1}{2} - \frac{1}{2}\eta\right)}. \tag{5.108}$$

Similar calculations read

$$e_{54} = K_0\left(\omega\sqrt{a^2 + b^2}\right). \tag{5.109}$$

Chapter 6

Integrals Containing Bessel Functions

6.1 Integrals Involving $J_\mu(x)$, x^γ and $[p + tx^\rho]^{-\lambda}$

6.1.1 53rd *General Formula*

$$N_{53} = \int\limits_0^\infty dx\, x^\gamma\, J_\sigma(bx^\nu)$$

$$= \frac{1}{2i} \int\limits_{-\beta+i\infty}^{-\beta-i\infty} d\xi \frac{(b/2)^{2\xi+\sigma}}{\sin \pi\xi\, \Gamma(1+\xi)\, \Gamma(\sigma+1+\xi)}$$

$$\times \lim_{\varepsilon \to 0} \int\limits_\varepsilon^\infty dx\, x^{\gamma+2\nu\xi+\nu\sigma}. \tag{6.1}$$

By calculating the integral over x, taking residue at the point $\xi = -\frac{\gamma+1+\nu\sigma}{2\nu}$ and going to the limit $\varepsilon \to 0$, one gets

$$N_{53} = \frac{1}{2\nu} \left(\frac{b}{2}\right)^{-\frac{\gamma+1}{\nu}} \frac{\Gamma\left(\dfrac{\gamma+1+\nu\sigma}{2\nu}\right)}{\Gamma\left(1+\sigma - \dfrac{\gamma+1+\nu\sigma}{2\nu}\right)}, \tag{6.2}$$

$$b > 0.$$

6.1.2   54th *General Formula*

$$N_{54} = \int\limits_0^1 dx x^\delta \left(1 - x^æ\right)^\mu J_\sigma \left[bx^\nu (1 - x^æ)^\lambda\right]$$

$$= \frac{1}{2iæ} \int\limits_{-\beta+i\infty}^{-\beta-i\infty} d\xi \frac{(b/2)^{2\xi+\sigma}}{\sin \pi\xi \, \Gamma(1+\xi) \, \Gamma(\sigma+\xi+1)} \tag{6.3}$$

$$\times B\left(\frac{1 + \delta + \nu\sigma + 2\nu\xi}{æ}, \quad \mu + 1 + \lambda\sigma + 2\lambda\xi\right).$$

Ranges (or upper and lower bounds) of parameters γ, σ, and ν are established from original integrals (6.1) and (6.3).

For example, if $\nu = 1$, $\gamma > 0$ in (6.1), then $\text{Re}(\gamma + \sigma) > -1$, $\text{Re } \gamma < \frac{1}{2}$. If $\nu = 1$, $\gamma < 0$ in (6.1) then $\text{Re}(1 + \sigma) > \text{Re } \gamma > -\frac{1}{2}$, etc.

6.1.3   55th *General Formula*

$$N_{55} = \int\limits_a^c dx (x-a)^\delta (c-x)^\mu J_\sigma \left[b(x-a)^\nu (c-x)^\lambda\right]$$

$$= \frac{1}{2i} \int\limits_{-\beta+i\infty}^{-\beta-i\infty} d\xi \frac{(b/2)^{\sigma+2\xi} (c-a)^{1+\delta+\mu+\sigma(\nu+\lambda)+2\xi(\nu+\lambda)}}{\sin \pi\xi \, \Gamma(1+\xi) \, \Gamma(1+\sigma+\xi)} \tag{6.4}$$

$$\times B(1 + \delta + \nu\sigma + 2\nu\xi, \quad 1 + \mu + \lambda\sigma + 2\lambda\xi).$$

6.1.4   56th *General Formula*

$$N_{56} = \int\limits_0^\infty \frac{dx}{\sqrt{x}} x^\mu (x+a)^{-\mu} (x+c)^{-\mu} J_\sigma \left[bx^\nu (x+a)^{-\nu} (x+c)^{-\nu}\right]$$

$$= \frac{\sqrt{\pi}}{2i} \int\limits_{-\beta+i\infty}^{-\beta-i\infty} d\xi \frac{(b/2)^{\sigma+2\xi} (\sqrt{a}+\sqrt{c})^{1-2(\mu+\nu(\sigma+2\xi))}}{\sin \pi\xi \, \Gamma(1+\xi) \, \Gamma(1+\sigma+\xi)} \tag{6.5}$$

$$\times \frac{\Gamma\left(\mu + \nu(\sigma + 2\xi) - \dfrac{1}{2}\right)}{\Gamma\left(\mu + \nu(\sigma + 2\xi)\right)}.$$

6.1.5 57th *General Formula*

$$N_{57} = \int\limits_0^1 dx(1-x^{\text{æ}})^\mu \, J_\sigma\left[b(1-x^{\text{æ}})^\lambda\right] = \frac{1}{\text{æ}} \frac{1}{2i} \times$$

$$\int\limits_{-\beta+i\infty}^{-\beta-i\infty} d\xi \frac{(b/2)^{\sigma+2\xi}}{\sin \pi\xi \, \Gamma(1+\xi) \, \Gamma(1+\sigma+\xi)} \, B\left(\frac{1}{\text{æ}}, 1+\mu+\lambda(\sigma+2\xi)\right).$$

(6.6)

6.1.6 58th *General Formula*

$$N_{58} = \int\limits_0^1 dx(1-\sqrt{x})^\mu \, J_\sigma\left[b(1-\sqrt{x})^\lambda\right]$$

$$= \frac{1}{i} \int\limits_{-\beta+i\infty}^{-\beta-i\infty} d\xi \frac{(b/2)^{\sigma+2\xi}}{\sin \pi\xi \, \Gamma(1+\xi) \, \Gamma(1+\sigma+\xi)}$$

$$\times \frac{1}{1+\mu+\lambda(\sigma+2\xi)} \frac{1}{2+\mu+\lambda(\sigma+2\xi)}.$$

(6.7)

6.1.7 59th *General Formula*

$$N_{59} = \int\limits_0^\infty dx\, x^\delta (1+tx^{\text{æ}})^{-\mu} \, J_\sigma\left[bx^\nu\,(1+tx^{\text{æ}})^{-\lambda}\right]$$

$$= \frac{1}{\text{æ}} \frac{1}{2i} \int\limits_{-\beta+i\infty}^{-\beta-i\infty} d\xi \frac{(b/2)^{\sigma+2\xi}}{\sin \pi\xi \, \Gamma(1+\xi) \, \Gamma(1+\sigma+\xi)} \, t^{-\frac{1+\delta+\nu(\sigma+2\xi)}{\text{æ}}}$$

$$\times B\left(\frac{1+\delta+\nu(\sigma+2\xi)}{\text{æ}}, \quad \mu+\lambda(\sigma+2\xi) - \frac{1+\delta+\nu(\sigma+2\xi)}{\text{æ}}\right).$$

(6.8)

6.1.8 60th *General Formula*

$$N_{60} = \int\limits_0^\infty dx \frac{x^\gamma}{\left[p + tx^æ\right]^\lambda} \, J_\sigma(bx^\nu) = \frac{1}{æ} \, \frac{1}{\Gamma(\lambda)} \, \frac{1}{p^\lambda} \left(\frac{p}{t}\right)^{\frac{\gamma+1+\nu\sigma}{æ}}$$

$$\times \frac{1}{2i} \int\limits_{-\beta+i\infty}^{-\beta-\infty} d\xi \frac{(b/2)^{\sigma+2\xi} \left(\frac{p}{t}\right)^{2\nu\xi/æ}}{\sin \pi\xi \; \Gamma(1+\xi) \; \Gamma(1+\sigma+\xi)} \tag{6.9}$$

$$\times \Gamma\left(\frac{1+\gamma+\nu(\sigma+2\xi)}{æ}\right) \, \Gamma\left(\lambda - \frac{1+\gamma+\nu(\sigma+2\xi)}{æ}\right).$$

6.1.9 61st *General Formula*

$$N_{61} = \int\limits_0^\infty dx x^\gamma \, J_{\sigma_1}(ax^\nu) \, J_{\sigma_2}(bx^\nu) = \frac{1}{\Gamma(1+\sigma_2)}$$

$$\times \left(\frac{a}{2}\right)^{-\frac{\gamma+1+\nu\sigma_2}{\nu}} \frac{1}{2\nu} \left(\frac{b}{2}\right)^{\sigma_2} \frac{\Gamma\left(\dfrac{\gamma+1+\nu(\sigma_1+\sigma_2)}{2\nu}\right)}{\Gamma\left(1+\sigma_1 - \dfrac{\gamma+1+\nu(\sigma_1+\sigma_2)}{2\nu}\right)} \tag{6.10}$$

$$\times F\left(\frac{\gamma+1+\nu(\sigma_1+\sigma_2)}{2\nu}, -\sigma_1 + \frac{\gamma+1+\nu(\sigma_1+\sigma_2)}{2\nu}; 1+\sigma_2, \frac{b^2}{a^2}\right),$$

$$b/a \leq 1.$$

6.1.10 62nd *General Formula*

$$N_{62} = \int\limits_0^\infty dx x^\gamma \, J_{\sigma_2}(bx^\nu) \, J_{\sigma_1}(ax^\nu) = \frac{1}{\Gamma(1+\sigma_1)}$$

$$\times \left(\frac{a}{2}\right)^{\sigma_1} \left(\frac{b}{2}\right)^{-\frac{\gamma+1+\nu\sigma_1}{\nu}} \frac{1}{2\nu} \frac{\Gamma\left(\dfrac{1+\gamma+\nu(\sigma_1+\sigma_2)}{2\nu}\right)}{\Gamma\left(1+\sigma_2 - \dfrac{\gamma+1+\nu(\sigma_1+\sigma_2)}{2\nu}\right)} \tag{6.11}$$

$$\times F\left(\frac{1+\gamma+\nu(\sigma_1+\sigma_2)}{2\nu}, -\sigma_2 + \frac{1+\gamma+\nu(\sigma_1+\sigma_2)}{2\nu}; 1+\sigma_1, \frac{a^2}{b^2}\right),$$

$a/b \leq 1$. The function $F(\alpha, \beta; \gamma; z)$ entering into the formulas (6.10) and (6.11) is called the generalized hypergeometric function. Sometimes it is denoted by

$$F(\alpha, \beta; \gamma; z) =_2 F_1(\alpha, \beta; \gamma; z).$$

6.1.11 63rd *General Formula*

$$N_{63} = \int\limits_0^\infty dx x^\gamma \, J_{\sigma_1}(ax^{\nu_1}) \, J_{\sigma_2}(bx^{\nu_2})$$

$$= \left(\frac{a}{2}\right)^{\sigma_1} \left(\frac{b}{2}\right)^{-\frac{\gamma+1+\nu_1\sigma_1}{\nu_2}} \frac{1}{2\nu_2} \frac{1}{2i} \int\limits_{-\beta+i\infty}^{-\beta-i\infty} d\xi \frac{\left[\frac{a}{2}\left(\frac{2}{b}\right)^{\nu_1/\nu_2}\right]^{2\xi}}{\sin \pi\xi \, \Gamma(1+\xi)}$$

$$\times \frac{\Gamma\left(\dfrac{\gamma+1+\nu_1\sigma_1+\nu_2\sigma_2+2\nu_1\xi}{2\nu_2}\right)}{\Gamma\left(1+\sigma_2 - \dfrac{\gamma+1+\nu_1\sigma_1+\nu_2\sigma_2+2\nu_1\xi}{2\nu_2}\right)} \frac{1}{\Gamma(1+\sigma_1+\xi)} \tag{6.12}$$

or

$$N_{63} = \left(\frac{b}{2}\right)^{\sigma_2} \left(\frac{a}{2}\right)^{-\frac{\gamma+1+\nu_2\sigma_2}{\nu_1}} \frac{1}{2\nu_1} \frac{1}{2i} \int\limits_{-\beta+i\infty}^{-\beta-i\infty} d\xi \frac{\left[\frac{b}{2}\left(\frac{2}{a}\right)^{\nu_2/\nu_1}\right]^{2\xi}}{\sin \pi\xi \, \Gamma(1+\xi)}$$

$$\times \frac{\Gamma\left(\dfrac{\gamma+1+\nu_1\sigma_1+\nu_2\sigma_2+2\nu_2\xi}{2\nu_1}\right)}{\Gamma\left(1+\sigma_1 - \dfrac{\gamma+1+\nu_1\sigma_1+\nu_2\sigma_2+2\nu_2\xi}{2\nu_1}\right)} \frac{1}{\Gamma(1+\sigma_2+\xi)}. \tag{6.13}$$

6.2 Calculation of Concrete Integrals

1. The formula (6.2) with $\gamma = 0$, $\nu = 1$, where $\sigma > -1$ is an arbitrary number, gives

$$S_1 = \int\limits_0^\infty dx \, J_\sigma(bx) = \frac{1}{2}\left(\frac{b}{2}\right)^{-1} \frac{\Gamma\left(\frac{1+\sigma}{2}\right)}{\Gamma\left(1+\sigma - \frac{1+\sigma}{2}\right)} = \frac{1}{b}. \tag{6.14}$$

2. From the formula (6.2), it follows immediately

$$S_2 = \int\limits_0^\infty dx \, J_0(x) = \int\limits_0^\infty dx \frac{J_1(x)}{x} = 1. \tag{6.15}$$

3. Assuming $\gamma = 0$, $\nu = 1$ in (6.10), one gets

$$S_3 = \int_0^\infty dx \, J_{\sigma_1}(ax) \, J_{\sigma_2}(bx) = b^{\sigma_2} \, a^{-1-\sigma_2} \frac{\Gamma\left(\frac{1+\sigma_1+\sigma_2}{2}\right)}{\Gamma(1+\sigma_2) \, \Gamma\left(\frac{\sigma_1-\sigma_2+1}{2}\right)}$$

$$\times F\left(\frac{1+\sigma_1+\sigma_2}{2}, \frac{\sigma_2-\sigma_1+1}{2}; 1+\sigma_2; \frac{b^2}{a^2}\right), \tag{6.16}$$

where $a, b > 0$, $\mathrm{Re}(\sigma_1 + \sigma_2) > -1$, $b < a$. If $a < b$, then it should take interchanging constants $a \rightleftarrows b$.

4. From the formula (6.10), it follows immediately

$$S_4 = \int_0^\infty dx \, J_{\sigma_1}(ax) \, J_{\sigma_1-1}(bx) = \begin{cases} \dfrac{b^{\sigma_1-1}}{a^{\sigma_1}} & \text{for } b < a \\[2mm] \dfrac{1}{2b} & \text{for } b = a \\[2mm] 0 & \text{for } b > a. \end{cases} \tag{6.17}$$

5. The formula (6.2) with $\gamma = -1$, $\sigma = \nu = 2$ gives

$$S_5 = \int_0^\infty dx \, x^{-1} \, J_2(bx^2) = \frac{1}{4}. \tag{6.18}$$

6. The previous case with $\gamma = -2$, $\sigma = \nu = 2$ reads

$$S_6 = \int_0^\infty dx \, \frac{J_2(bx^2)}{x^2} = \frac{1}{4}\left(\frac{b}{2}\right)^{1/2} \frac{\Gamma\left(\frac{3}{4}\right)}{\Gamma\left(\frac{9}{4}\right)} = \frac{4\pi}{5}\sqrt{b}\,\frac{1}{\Gamma^2\left(\frac{1}{4}\right)}, \quad b > 0. \tag{6.19}$$

7. If we put $\gamma = -1$, $\sigma = 2$, $\nu = 1$ in (6.2), then

$$S_7 = \int_0^\infty dx \, x^{-1} \, J_2(bx) = \frac{1}{2}\frac{\Gamma(1)}{\Gamma(3-1)} = \frac{1}{2}. \tag{6.20}$$

8. The main formula (6.2) with $\gamma = -\frac{1}{4}$, $\sigma = 2$, $\nu = \frac{1}{4}$ gives

$$S_8 = \int_0^\infty dx \, x^{-1/4} \, J_2(bx^{1/4}) = 2\left(\frac{b}{2}\right)^{-3} \frac{\Gamma\left(\frac{5}{2}\right)}{\Gamma\left(\frac{1}{2}\right)} = 12\,\frac{1}{b^3}. \tag{6.21}$$

9. The formula (6.12) with

$$\gamma = 2, \; \sigma_1 = 2\rho, \; \sigma_2 = \rho + \frac{1}{2}, \; a \to 2a, \; b = 1, \; \nu_1 = 1, \; \nu_2 = 2$$

reads

$$S_9 = \int\limits_0^\infty dx x^2 \, J_{2\rho}(2ax) \, J_{\rho+\frac{1}{2}}(x^2)$$

$$= a^{2\rho} \left(\frac{1}{2}\right)^{-\frac{3+2\rho}{2}} \frac{1}{4} \frac{1}{2\pi i} \int\limits_{-\beta+i\infty}^{-\beta-i\infty} d\xi \frac{(\sqrt{2}a)^{2\xi}}{\sin \pi\xi \, \Gamma(1+\xi)}$$

$$\times \frac{\Gamma\left(\frac{3+2\rho+2\left(\rho+\frac{1}{2}\right)+2\xi}{4}\right)}{\Gamma\left(1+\rho+\frac{1}{2}-1-\rho-\frac{\xi}{2}\right)} \frac{1}{\Gamma(1+2\rho+\xi)}, \qquad (6.22)$$

where

$$\frac{\Gamma\left(1+\rho+\frac{\xi}{2}\right)}{\Gamma(1+2\rho+\xi)} = \frac{\left(\rho+\frac{\xi}{2}\right)\Gamma\left(\rho+\frac{\xi}{2}\right)}{(2\rho+\xi)\,\Gamma(2\rho+\xi)},$$

and

$$\Gamma\left(2\left(\rho+\frac{\xi}{2}\right)\right) = \frac{2^{2\left(\rho+\frac{\xi}{2}\right)-1}}{\sqrt{\pi}} \, \Gamma\left(\rho+\frac{\xi}{2}\right) \, \Gamma\left(\frac{1}{2}+\rho+\frac{\xi}{2}\right).$$

So that

$$\frac{\Gamma\left(1+\rho+\frac{\xi}{2}\right)}{\Gamma(1+2\rho+\xi)} \frac{1}{\Gamma\left(\frac{1}{2}(1-\xi)\right)} = \frac{\sqrt{\pi}}{2^{2\left(\rho+\frac{\xi}{2}\right)}} \frac{1}{\Gamma\left(\frac{1}{2}+\rho+\frac{\xi}{2}\right)} \frac{1}{\Gamma\left(\frac{1}{2}(1-\xi)\right)}.$$

Next we carry out the following transformation:

$$\Gamma\left(\frac{1}{2}-\frac{\xi}{2}\right) = \frac{\pi}{\cos\frac{\pi\xi}{2} \, \Gamma\left(\frac{1}{2}+\frac{\xi}{2}\right)}$$

and

$$\frac{\cos\frac{\pi\xi}{2}}{\sin\pi\xi} \frac{\Gamma\left(\frac{1}{2}+\frac{\xi}{2}\right)}{\Gamma(1+\xi)} = \frac{\Gamma\left(\frac{1}{2}+\frac{\xi}{2}\right)}{2\sin\frac{\pi\xi}{2}\,\xi} \frac{\sqrt{\pi}}{2^{\xi-1}\,\Gamma\left(\frac{\xi}{2}\right)\,\Gamma\left(\frac{1}{2}+\frac{\xi}{2}\right)}$$

$$= \frac{\sqrt{\pi}}{2} \frac{1}{2^\xi \, \sin\frac{\pi\xi}{2}\,\Gamma\left(1+\frac{\xi}{2}\right)}.$$

Now we change the integration variable $\frac{\xi}{2} \to x$ in (6.22) and obtain

$$S_9 = \frac{a}{2}\left(\frac{a^2}{2}\right)^{\rho-\frac{1}{2}} \sum_{n=0}^\infty \frac{(-1)^n \left(\frac{a^2}{2}\right)^{2n}}{n! \, \Gamma\left(1+\rho-\frac{1}{2}+n\right)} = \frac{a}{2} J_{\rho-\frac{1}{2}}(a^2). \qquad (6.23)$$

10. The integral (6.12) with $\gamma = 0$, $\nu_1 = -1$, $\sigma_1 = \sigma_2 = \lambda$, $\nu_2 = 1$ leads to the result

$$S_{10} = \int_0^\infty dx\, J_\lambda\left(\frac{a}{x}\right) J_\lambda(bx)$$

$$= (ab)^\lambda\, b^{-1}\, \frac{1}{2\pi i} \int_{-\beta+i\infty}^{-\beta-i\infty} d\xi\, \frac{\left(\frac{ab}{4}\right)^{2\xi}}{\sin \pi\xi\, \Gamma(1+\xi)}$$

$$\times \frac{\Gamma\left(\frac{1}{2}-\xi\right)}{\Gamma\left(\frac{1}{2}+\lambda+\xi\right)\Gamma(1+\lambda+\xi)}, \qquad (6.24)$$

where

- $\Gamma\left(\dfrac{1}{2}-\xi\right) = \dfrac{\pi}{\cos \pi\xi\, \Gamma\left(\frac{1}{2}+\xi\right)}$,

- $\sin \pi\xi\, \cos \pi\xi = \dfrac{1}{2}\sin 2\pi\xi$,

- $\Gamma(\xi)\, \Gamma\left(\dfrac{1}{2}+\xi\right) \dfrac{2^{2\xi-1}}{\sqrt{\pi}} = \Gamma(2\xi)$,

- $\Gamma\left(\dfrac{1}{2}+\lambda+\xi\right) \Gamma(\lambda+\xi) \dfrac{2^{2(\lambda+\xi)-1}}{\sqrt{\pi}} = \Gamma(2(\lambda+\xi))$.

After using these transformations and changing integration variable $2\xi = x$, one gets

$$S_{10} = b^{-1}(ab)^\lambda \sum_{n=0}^\infty \frac{(-1)^n\, (ab)^n}{n!\, \Gamma(1+2\lambda+n)} = b^{-1}\, J_{2\lambda}(2\sqrt{ab}), \qquad (6.25)$$

where $a, b > 0$, Re $\lambda > -1/2$.

11. Similar calculation of the integral (6.12) with

$$\gamma = 0, \quad \nu_1 = \frac{1}{2}, \quad \nu_2 = 1, \quad \sigma_1 = 2\lambda, \quad \sigma_2 = \lambda$$

gives

$$S_{11} = \int_0^\infty dx\, J_{2\lambda}(a\sqrt{x})\, J_\lambda(bx) = b^{-1}\, J_\lambda\left(\frac{a^2}{4b}\right), \qquad (6.26)$$

where $a > 0$, $b > 0$, Re $\lambda > -\frac{1}{2}$.

12. Also the main formula (6.12) with

$$\gamma = 1, \quad \sigma_1 = \frac{1}{2}\lambda, \quad \nu_1 = 2, \quad \sigma_2 = \lambda, \quad \nu_2 = 1$$

results

$$S_{12} = \int_0^\infty dx\, x\, J_{\lambda/2}(ax^2)\, J_\lambda(bx) = (2a)^{-1}\, J_{\lambda/2}\left(\frac{b^2}{4a}\right), \qquad (6.27)$$

where $a > 0$, $b > 0$, Re $\lambda > -1$. Notice that all these calculated integrals are presented in the textbook by Gradshteyn and Ryzhik, 1980 and coinciding these two results mean that our general method is valid for cylindrical functions.

13. If we put

$$a = b = 1, \ \nu = 1, \ \gamma = -\sigma_1 - \sigma_2$$

in the formula (6.10), then we have

$$S_{13} = \int_0^\infty dx \frac{J_{\sigma_1}(x) \, J_{\sigma_2}(x)}{x^{\sigma_1 + \sigma_2}} = \frac{1}{\Gamma(1 + \sigma_2)} \left(\frac{1}{2}\right)^{\sigma_1 + \sigma_2}$$

$$\times \frac{\Gamma\left(\frac{1}{2}\right)}{\Gamma\left(\frac{1}{2} + \sigma_1\right)} F\left(\frac{1}{2}, -\sigma_1 + \frac{1}{2}; \ 1 + \sigma_2; \ 1\right), \tag{6.28}$$

where

$$F(\alpha, \beta; \ \gamma, 1) = \frac{\Gamma(\gamma) \, \Gamma(\gamma - \alpha - \beta)}{\Gamma(\gamma - \alpha) \, \Gamma(\gamma - \beta)}.$$

So that

$$S_{13} = \sqrt{\pi} \left(\frac{1}{2}\right)^{\sigma_1 + \sigma_2} \frac{\Gamma(\sigma_1 + \sigma_2)}{\Gamma\left(\frac{1}{2} + \sigma_1\right) \Gamma\left(\frac{1}{2} + \sigma_2\right) \Gamma\left(\frac{1}{2} + \sigma_1 + \sigma_2\right)}. \tag{6.29}$$

This result coincides with the one mentioned by (11.308) in Chapter 3 of the book by A. D. Wheelon and J. T. Robacker, 1954.

14. Assuming

$$\gamma = -\frac{1}{2}, \ \sigma_1 = 0, \ a = 1, \ \nu = 1, \ \sigma_2 = \frac{1}{2}$$

in (6.10), one gets

$$S_{14} = \int_0^\infty \frac{dx}{\sqrt{x}} \, J_0(x) \, J_{1/2}(bx) = \frac{1}{\Gamma\left(\frac{3}{2}\right)} \frac{1}{2} \sqrt{\frac{a}{2}} \frac{\Gamma\left(\frac{1}{2}\right)}{\Gamma\left(\frac{1}{2}\right)}$$

$$\times F\left(\frac{1}{2}, \frac{1}{2}; \ \frac{3}{2}; \ b^2\right), \tag{6.30}$$

where

$$F\left(\frac{1}{2}, \frac{1}{2}; \ \frac{3}{2}; \ \sin^2 z\right) = \frac{z}{\sin z},$$

$$\sin^2 z = b^2, \ z = \arcsin b = \sin^{-1} b.$$

So that

$$S_{14} = \frac{1}{\sqrt{2\pi b}} \arcsin b, \ \text{where} \ 0 < b < 1. \tag{6.31}$$

15. Assuming

$$\delta = 1, \ æ = 2, \ \mu = -1, \ \sigma = 1, \ \nu = 1, \ \lambda = -\frac{1}{2}$$

in the main formula (6.3), one gets

$$S_{15} = \int_0^1 dx\, x(1-x^2)^{-1}\, J_1\left[bx(1-x^2)^{-1/2}\right]$$

$$= \frac{1}{2}\frac{1}{2i}\int_{-\beta+i\infty}^{-\beta-i\infty} d\xi \frac{(b/2)^{2\xi+1}}{\sin \pi\xi\, \Gamma(1+\xi)\, \Gamma(2+\xi)}$$

$$\times\; \frac{\Gamma\left(\frac{3}{2}+\xi\right)\Gamma\left(1-\frac{3}{2}-\xi\right)}{\Gamma(1)}, \tag{6.32}$$

where

$$\Gamma\left(\frac{3}{2}+\xi\right)\Gamma\left(1-\frac{3}{2}-\xi\right) = \frac{\pi}{\sin \pi\left(\frac{3}{2}+\xi\right)} = -\frac{\pi}{\cos \pi\xi}.$$

Thus

$$S_{15} = -\frac{\pi}{2}\left[\sum_{n=0}^{\infty}\frac{(b/2)^{2n+1}}{\Gamma(1+n)\,\Gamma(2+n)} - \sum_{n=0}^{\infty}\frac{(b/2)^{2n+2}}{\Gamma\left(\frac{3}{2}+n\right)\Gamma\left(\frac{3}{2}+n+1\right)}\right]$$

$$= -\frac{\pi}{2}\left[I_1(b) - \mathbf{L}_1(b)\right],\quad b > 0. \tag{6.33}$$

Here $I_1(b)$ and $\mathbf{L}_1(b)$ are called the modified Bessel function of the first kind and the Struve function, respectively.

16. We put

$$\delta = \frac{1}{2},\ \mu = -\frac{3}{2},\ \sigma = 2,\ \nu = -\frac{1}{2},\ \lambda = \frac{1}{2}$$

in (6.4) and then derive

$$S_{16} = \int_a^c dx\, (x-a)^{1/2}\,(c-x)^{-3/2}\, J_2\left[b(x-a)^{-1/2}(c-x)^{1/2}\right]$$

$$= \frac{1}{2i}\int_{\beta+i\infty}^{-\beta-i\infty} d\xi \frac{(b/2)^{2+2\xi}\,\Gamma\left(\frac{1}{2}-\xi\right)\Gamma\left(\frac{1}{2}+\xi\right)}{\sin \pi\xi\, \Gamma(1+\xi)\, \Gamma(3+\xi)}$$

$$= I_2(b) - \mathbf{L}_2(b). \tag{6.34}$$

17. Let $\mu = 1,\ \sigma = 1,\ \nu = -\frac{1}{2}$ be in (6.5), then we have

$$S_{17} = \int_0^{\infty} dx \frac{\sqrt{x}}{(x+a)(x+c)}\, J_1\left(b\sqrt{\frac{(x+a)(x+c)}{x}}\right)$$

$$= \frac{\sqrt{\pi}}{2i}\int_{-\beta+i\infty}^{-\beta-i\infty} d\xi \frac{(b/2)^{1+2\xi}\,A^{2\xi}}{\sin \pi\xi\, \Gamma(1+\xi)\, \Gamma(2+\xi)}\frac{\Gamma(-\xi)}{\Gamma\left(\frac{1}{2}-\xi\right)}, \tag{6.35}$$

where

$$\bullet \ \Gamma(-\xi) = -\frac{\pi}{\sin \pi \xi \ \Gamma(1 + \xi)}, \quad A = \sqrt{a} + \sqrt{c},$$

$$\bullet \ \Gamma\left(\frac{1}{2} - \xi\right) = \frac{\pi}{\cos \pi \xi \ \Gamma\left(\frac{1}{2} + \xi\right)}.$$

Thus

$$S_{17} = -\frac{\sqrt{\pi}}{2} b \frac{1}{2i} \int\limits_{-\beta+i\infty}^{-\beta-i\infty} d\xi \frac{\left(\frac{bA}{2}\right)^{2\xi} \cos \pi \xi \ \Gamma\left(\frac{1}{2} + \xi\right)}{\sin^2 \pi \xi \ \Gamma^3(1 + \xi)(1 + \xi)}$$

$$= -\frac{b}{2\sqrt{\pi}} \sum_{n=0}^{\infty} \frac{(-1)^n \left(\frac{bA}{2}\right)^{2n} \Gamma\left(\frac{1}{2} + n\right)}{\Gamma^3(1 + n)(1 + n)}$$

$$\times \left[2\ln\left(\frac{bA}{2}\right) - \frac{1}{1+n} - 3\Psi(1 + n) + \Psi\left(\frac{1}{2} + n\right)\right]. \tag{6.36}$$

18. From the formula (6.6) with

$$\ae = \frac{1}{2}, \ \mu = \frac{3}{2}, \ \lambda = \frac{1}{2}, \ \sigma = 3,$$

one gets

$$S_{18} = \int\limits_0^1 dx(1 - \sqrt{x})^{3/2} J_3\left[b(1 - \sqrt{x})^{1/2}\right] \tag{6.37}$$

$$= 2 \frac{1}{2i} \int\limits_{-\beta+i\infty}^{-\beta-i\infty} d\xi \frac{(b/2)^{3+2\xi} \ \Gamma(4 + \xi)}{\sin \pi \xi \ \Gamma(1 + \xi) \ \Gamma(4 + \xi) \ \Gamma(6 + \xi)}$$

$$= \frac{8}{b^2}\left[\left(\frac{b}{2}\right)^5 \sum_{n=0}^{\infty} \frac{(-1)^n \ (b/2)^{2n}}{\Gamma(1 + n) \ \Gamma(1 + 5 + n)}\right] = \frac{8}{b^2} J_5(b).$$

19. If we again consider the integral (6.6) with

$$\ae = \frac{1}{2}, \ \mu = 2, \ \sigma = 4, \ \lambda = \frac{1}{2},$$

then we obtain the same equivalent integral as in the previous case:

$$S_{19} = \int\limits_0^1 dx(1 - \sqrt{x})^2 J_4\left[b(1 - \sqrt{x})^{1/2}\right]$$

$$= 2 \frac{1}{2i} \int\limits_{-\beta+i\infty}^{-\beta-i\infty} d\xi \frac{(b/2)^{4+2\xi} \ \Gamma(2) \ \Gamma(5 + \xi)}{\sin \pi \xi \ \Gamma(1 + \xi) \ \Gamma(5 + \xi) \ \Gamma(7 + \xi)}$$

$$= \frac{8}{b^2} \sum_{n=0}^{\infty} \frac{(-1)^n \ (b/2)^{6+2\xi}}{\Gamma(1 + n) \ \Gamma(1 + 6 + \xi)} = \frac{8}{b^2} J_6(b). \tag{6.38}$$

Here, as before in Chapter 5, we have obtained the family of equivalent integrals.

20. The formula (6.9) with

$$\gamma = 0, \; p \to \Omega^2, \; t = 1, \; \text{æ} = 2, \; \lambda = 1, \; \sigma = 0, \; \nu = 1$$

reads

$$S_{20} = \int_0^\infty dx \, \frac{J_0(bx)}{\Omega^2 + x^2}$$

$$= \frac{1}{2\Omega} \frac{1}{2i} \int_{-\beta+i\infty}^{-\beta-i\infty} d\xi \, \frac{\left(\frac{\Omega b}{2}\right)^{2\xi} \Gamma\left(\frac{1}{2} + \xi\right) \Gamma\left(\frac{1}{2} - \xi\right)}{\sin \pi\xi \, \Gamma^2(1 + \xi) \Gamma(1)}. \tag{6.39}$$

Using the relation

$$\Gamma\left(\frac{1}{2} + \xi\right) \Gamma\left(\frac{1}{2} - \xi\right) = \frac{\pi}{\cos \pi\xi}$$

and taking the residues at the points $\xi = n$, $\xi = n + \frac{1}{2}$, one gets

$$S_{20} = \frac{\pi}{2\Omega} \left[\sum_{n=0}^\infty \frac{\left(\frac{b\Omega}{2}\right)^{2n}}{\Gamma^2(1 + n)} - \sum_{n=0}^\infty \frac{\left(\frac{b\Omega}{2}\right)^{2n+1}}{\Gamma^2\left(\frac{3}{2} + n\right)} \right]$$

$$= \frac{\pi}{2\Omega} \left[I_0(b\Omega) - \mathbf{L}_0(b\Omega) \right]. \tag{6.40}$$

21. Assuming

$$\gamma = 1, \; p \to \Omega^2, \; t = 1, \; \text{æ} = 2, \; \sigma = 0, \; \lambda = \frac{1}{2}, \; \nu = 1$$

in (6.9), one obtains

$$S_{21} = \int_0^\infty dx \, \frac{x}{(\Omega^2 + x^2)^{1/2}} \, J_0(bx)$$

$$= \frac{1}{2} \frac{\Omega}{\Gamma(1/2)} \frac{1}{2i} \int_{-\beta+i\infty}^{-\beta-i\infty} d\xi \, \frac{\left(\frac{b\Omega}{2}\right)^{2\xi} \Gamma\left(-\frac{1}{2} - \xi\right)}{\sin \pi\xi \, \Gamma(1 + \xi)}, \tag{6.41}$$

where

$$\frac{\Gamma\left(-\frac{1}{2} - \xi\right)}{\Gamma(1 + \xi)} = \frac{\Gamma\left(\frac{1}{2} - \xi\right)}{\left(-\frac{1}{2} - \xi\right) \Gamma(1 + \xi)} = \frac{-\pi}{\Gamma(1 + \xi) \Gamma\left(\frac{3}{2} + \xi\right)}$$

and

$$\Gamma(2 + 2\xi) = \frac{2^{2(1+\xi)-1}}{\sqrt{\pi}} \, \Gamma(1 + \xi) \, \Gamma\left(\frac{3}{2} + \xi\right).$$

Taking into account these relations and going to the integration variable $x = 1 + 2\xi$, we have

$$S_{21} = \frac{1}{b} \frac{1}{2i} \int_{-\beta+i\infty}^{-\beta-i\infty} dx \, \frac{(b\Omega)^x}{\sin \pi\xi \, \Gamma(1 + x)} = \frac{1}{b} e^{-b\Omega}. \tag{6.42}$$

22. We put

$$\gamma = 1 + \delta, \ \sigma = \delta, \ \lambda = 1 + \mu, \ p \to \omega^2, \ t = 1, \ \nu = 1, \ æ = 2$$

in (6.9) and obtain

$$S_{22} = \int\limits_0^\infty dx \frac{x^{1+\delta} \ J_\delta(bx)}{(\omega^2 + x^2)^{1+\mu}}$$

$$= \frac{1}{2} \frac{1}{\Gamma(1+\mu)} \frac{(b/2)^\delta}{(\omega^2)^{1+\mu}} (\omega^2)^{\frac{2+\delta+\delta}{2}} \frac{1}{2i} \int\limits_{-\beta+i\infty}^{-\beta-i\infty} d\xi \frac{\left(\frac{b\omega}{2}\right)^{2\xi}}{\sin \pi\xi \ \Gamma(1+\xi)}$$

$$\times \frac{\Gamma(1+\delta+\xi)}{\Gamma(1+\delta+\xi)} \Gamma(1+\mu-1-\delta-\xi)$$

$$= \frac{\omega^{\delta-\mu} \ b^\mu}{2^\mu \ \Gamma(1+\mu)} \ K_{\delta-\mu}(\omega b), \tag{6.43}$$

where we have denoted

$$K_{\delta-\mu}(\omega b) = \left(\frac{b\omega}{2}\right)^{\delta-\mu} \left(-\frac{\pi}{2}\right) \frac{1}{2i} \int\limits_{-\beta+i\infty}^{-\beta-i\infty} d\xi \frac{\left(\frac{b\omega}{2}\right)^{2\xi}}{\sin \pi\xi \ \Gamma(1+\xi)}$$

$$\times \frac{1}{\sin \pi(\delta-\mu+\xi) \ \Gamma(1+\delta-\mu+\xi)} \tag{6.44}$$

as the modified Bessel function of the second kind

$$K_\mu(z) = \frac{\pi}{2\sin(\mu\pi)} \left[I_{-\mu}(z) - I_\mu(z)\right]. \tag{6.45}$$

Here the case $\delta = \mu = 0$ gives exactly $K_0(z)$,

$$K_0(z) = -\ln\frac{z}{2} \ I_0(z) + \sum_{n=0}^\infty \frac{(z/2)^{2n}}{(n!)^2} \ \Psi(1+n)$$

as it should be. In the definition (6.44) we take into account a restriction of parameters

$$-1 < \text{Re } \delta < \text{Re}\left(2\mu + \frac{3}{2}\right), \ a > 0, \ b > 0.$$

23. From the integral (6.43) and the definition (6.44) it follows that

$$S_{23} = \int\limits_0^\infty dx \frac{x^{1+\delta}}{[\omega^2 + x^2]} \ J_\delta(bx) = S_{22}(\omega b)\Big|_{\mu=0} = \omega^\delta \ K_\delta(\omega b). \tag{6.46}$$

24. Let

$$\gamma = 1, \ p \to \omega^4, \ t = 1, \ æ = 4, \ \lambda = 1, \ \sigma = 0, \ \nu = 2$$

be in (6.9), then we have

$$S_{24} = \int\limits_0^\infty dx \frac{x}{\omega^4 + x^4} \, J_0(bx^2)$$

$$= \frac{1}{4} \frac{1}{\omega^2} \frac{1}{2i} \int\limits_{-\beta+i\infty}^{-\beta-i\infty} d\xi \frac{(b/2)^{2\xi} \, (\omega^2)^{2\xi}}{\sin \pi\xi \, \Gamma^2(1+\xi)} \, \Gamma\left(\frac{1}{2}+\xi\right) \Gamma\left(\frac{1}{2}-\xi\right)$$

$$= \frac{1}{2} \frac{\pi}{\omega^2} \frac{1}{2i} \int\limits_{-\beta+i\infty}^{-\beta-i\infty} d\xi \frac{\left(\frac{b\omega^2}{2}\right)^{2\xi}}{\sin 2\pi\xi \, \Gamma^2(1+\xi)}$$

$$= \frac{1}{4} \frac{\pi}{\omega^2} \left[I_0(b\omega^2) - \mathbf{L}_0(b\omega^2)\right]. \tag{6.47}$$

6.3   Integrals Containing $J_\mu(x)$ and Logarithmic Functions

6.3.1 64th *General Formula*

$$N_{64} = \int\limits_0^1 dx x^\gamma \, (1-x^{\ae})^\mu \, J_\sigma\left[bx^\nu(1-x^{\ae})^\delta\right] \ln x$$

$$= \frac{1}{\ae^2} \frac{1}{2i} \int\limits_{-\beta+i\infty}^{-\beta-i\infty} d\xi \frac{(b/2)^{\sigma+2\xi} \, B\left(\frac{1+\gamma+\nu(\sigma+2\xi)}{\ae}, \ \mu+1+\delta(\sigma+2\xi)\right)}{\sin \pi\xi \, \Gamma(1+\xi) \, \Gamma(1+\sigma+\xi)}$$

$$\times \left[\Psi\left(\frac{1+\gamma+\nu(\sigma+2\xi)}{\ae}\right) \right.$$

$$\left. -\Psi\left(\frac{1+\gamma+\nu(\sigma+2\xi)}{\ae} + \mu+1+\delta(\sigma+2\xi)\right)\right]. \tag{6.48}$$

6.3.2 65th *General Formula*

$$N_{65} = \int\limits_0^\infty dx\, x^\gamma\, \ln(1 + ax^{\ae})\, J_\sigma(bx^\nu)$$

$$= \pi\, \frac{1}{2i} \int\limits_{-\beta+i\infty}^{-\beta-i\infty} d\xi\, \frac{(b/2)^{\sigma+2\xi}\, a^{-1-\gamma-\nu(\sigma+2\xi)}}{\sin \pi\xi\, \Gamma(1+\xi)\, \Gamma(1+\sigma+\xi)} \qquad (6.49)$$

$$\times \frac{1}{\left[1 + \gamma + \nu(\sigma + 2\xi)\right]\, \sin\frac{\pi}{\ae}\left(1 + \gamma + \nu(\sigma + 2\xi)\right)},$$

where $\ae = 1,\ 2$.

6.3.3 *Examples of Concrete Integrals*

25. The formula (6.49) with

$$\gamma = 0,\ \ae = 1,\ \sigma = 0,\ \nu = 1$$

leads to an integral

$$S_{25} = \int\limits_0^\infty dx\, \ln(1 + ax)\, J_0(bx) \qquad (6.50)$$

$$= \frac{\pi}{a}\, \frac{1}{2i} \int\limits_{-\beta+i\infty}^{-\beta-i\infty} d\xi\, \frac{\left(\frac{b}{2a}\right)^{2\xi}}{\sin \pi\xi\, \Gamma^2(1+\xi)}\, \frac{1}{1+2\xi}\, \frac{1}{\sin \pi(1+2\xi)},$$

where

$$\sin \pi(1 + 2\xi) = -\sin 2\pi\xi.$$

Thus,

$$S_{25} = -\frac{\pi}{a}\, \frac{1}{2i} \int\limits_{-\beta+i\infty}^{-\beta-i\infty} d\xi\, \frac{\left(\frac{b}{2a}\right)^{2\xi}}{2\sin^2 \pi\xi\, \cos \pi\xi\, \Gamma^2(1+\xi)(1+2\xi)}$$

$$= -\frac{\pi}{2a}\left\{\frac{2}{\pi^2} \sum_{n=0}^\infty \frac{(-1)^n\, (b/2a)^{2n}}{(n!)^2(1+2n)}\left[\ln\frac{b}{2a} - \Psi(1+n) - \frac{1}{1+2n}\right]\right.$$

$$\left. - \frac{1}{2\pi} \sum_{n=0}^\infty (-1)^n\, \frac{(b/2a)^{2n}}{\Gamma^2\left(\frac{3}{2}+n\right)(2+n)}\right\}. \qquad (6.51)$$

26. Let

$$\gamma = 1, \ \text{æ} = 2, \ \mu = 1, \ \sigma = 0, \ \nu = 1, \ \sigma = \frac{1}{2}$$

be in (6.48), then we have

$$S_{26} = \int_0^1 dx\, x(1 - x^2)\, J_0\!\left[bx(1 - x^2)^{1/2}\right] \ln x$$

$$= \frac{1}{4}\,\frac{1}{2i} \int_{-\beta+i\infty}^{-\beta-i\infty} d\xi \frac{(b/2)^{2\xi}}{\sin \pi\xi\, \Gamma^2(1+\xi)}\, \frac{\Gamma(1+\xi)\, \Gamma(2+\xi)}{\Gamma(3+2\xi)}$$

$$\times \left[\Psi(1+\xi) - \Psi(3+2\xi)\right], \tag{6.52}$$

where

$$\bullet\ \Psi(1+\xi) = \frac{1}{\xi} + \Psi(\xi),$$

$$\bullet\ \Psi(2\xi) = \frac{1}{2} \sum_{k=0}^{1} \Psi\!\left(\xi + \frac{k}{2}\right) + \ln 2$$

$$= \frac{1}{2}\, \Psi(\xi) + \frac{1}{2}\, \Psi\!\left(\xi + \frac{1}{2}\right) + \ln 2$$

and

$$\Psi(3+2\xi) = \frac{1}{2}\, \Psi(1+\xi) + \frac{3+4\xi}{2(1+\xi)(1+2\xi)}$$

$$+ \frac{1}{2}\, \Psi(\xi + \frac{1}{2}) + \ln 2.$$

Finally, we have

$$S_{26} = \frac{1}{4b}\, \ln 2\, \sin\!\left(\frac{b}{2}\right) + \frac{1}{8b} \sum_{n=0}^{\infty} \frac{(-1)^n\, (b/2)^{2n+1}}{(1+2n)!}$$

$$\times \left[\Psi(1+n) - \frac{3+4n}{(1+n)(1+2n)} - \Psi\!\left(n + \frac{1}{2}\right)\right]. \tag{6.53}$$

6.4   Integrals Containing $J_\sigma(x)$ and Exponential Functions

6.4.1 66th *General Formula*

$$N_{66} = \int_0^\infty dx\, x^\gamma\, e^{-ax^\mu}\, J_\sigma(bx^\nu)$$

$$= \frac{1}{\mu}\, a^{-\frac{1+\gamma}{\mu}} \left(\frac{b}{2a^{\nu/\mu}}\right)^\sigma \frac{1}{2i} \int_{-\beta+i\infty}^{-\beta-i\infty} d\xi \frac{\left(\frac{b}{2a^{\nu/\mu}}\right)^{2\xi} \Gamma\!\left(\frac{1+\gamma+\nu(\sigma+2\xi)}{\mu}\right)}{\sin \pi\xi\, \Gamma(1+\xi)\, \Gamma(1+\sigma+\xi)} \tag{6.54}$$

or

$$N_{66} = \frac{1}{2\nu} \left(\frac{b}{2}\right)^{-\frac{\gamma+1}{\nu}} \frac{1}{2i} \int\limits_{-\beta'+i\infty}^{-\beta'-i\infty} d\eta \frac{\left[\dfrac{a}{(b/2)^{\mu/\nu}}\right]^{\eta}}{\sin \pi\eta \; \Gamma(1+\eta)}$$

$$\times \frac{\Gamma\left(\dfrac{1+\gamma+\mu\eta+\nu\sigma}{2\nu}\right)}{\Gamma\left(1+\sigma-\dfrac{1+\gamma+\mu\eta+\nu\sigma}{2\nu}\right)}.$$

(6.55)

6.4.2   67th *General Formula*

$$N_{67} = \int\limits_0^{\infty} dx\, x^{\gamma}\, e^{-ax^{\mu}}\, J_{\sigma_1}(b_1 x^{\nu})\, J_{\sigma_2}(b_2 x^{\nu})$$

$$= \frac{1}{\mu}\, a^{-\frac{1+\gamma}{\mu}} \left(\frac{b_1}{2a^{\nu/\mu}}\right)^{\sigma_1} \left(\frac{b_2}{2a^{\nu/\mu}}\right)^{\sigma_2} \times \frac{1}{\Gamma(1+\sigma_2)}$$

$$\times \frac{1}{2i} \int\limits_{-\beta+i\infty}^{-\beta-i\infty} d\xi \frac{\left(\dfrac{b_1}{2a^{\nu/\mu}}\right)^{2\xi}}{\sin \pi\xi \; \Gamma(1+\xi)\, \Gamma(1+\sigma_1+\xi)}$$

$$\times F\left(-\xi, -\sigma_1-\xi;\, 1+\sigma_2;\, \frac{b_2^2}{b_1^2}\right) \Gamma\left(\frac{1+\gamma+\nu\sigma_1+\nu\sigma_2+2\nu\xi}{\mu}\right)$$

(6.56)

or

$$N_{67} = \frac{1}{\Gamma(1+\sigma_2)} \left(\frac{b_1}{2}\right)^{-\frac{\gamma+1+\nu\sigma_2}{\nu}} \frac{1}{2\nu} \left(\frac{b_2}{2}\right)^{\sigma_2}$$

$$\times \frac{1}{2i} \int\limits_{-\beta'+i\infty}^{-\beta'-i\infty} d\eta \frac{\left[\dfrac{a}{(b_1/2)^{\mu/\nu}}\right]^{\eta}}{\sin \pi\eta \; \Gamma(1+\eta)}$$

$$\times \frac{\Gamma\left(\dfrac{1+\gamma+\mu\eta+\nu(\sigma_1+\sigma_2)}{2\nu}\right)}{\Gamma\left(1+\sigma_1-\dfrac{1+\gamma+\mu\eta+\nu(\sigma_1+\sigma_2)}{2\nu}\right)}$$

(6.57)

$$\times F\left(\frac{1+\gamma+\mu\eta+\nu(\sigma_1+\sigma_2)}{2\nu},\, -\sigma_1+\frac{1+\gamma+\mu\eta+\nu(\sigma_1+\sigma_2)}{2\nu};\right.$$

$$\left. 1+\sigma_2;\, \frac{b_1^2}{b_2^2}\right),$$

where $b_1/b_2 \leq 1$.

If $b_2/b_1 \leq 1$ then we should interchange $\sigma_1 \rightleftarrows \sigma_2$, $b_2 \rightleftarrows b_1$. As usual parameter ν is equal to 1, 2 in (6.56) and (6.57).

6.4.3 68th General Formula

$$
N_{68} = \int\limits_0^\infty dx x^\gamma \, e^{-ax^\mu} \, J_{\sigma_1}(b_1 x^{\nu_1}) \, J_{\sigma_2}(b_2 x^{\nu_2})
$$

$$
= \frac{1}{\mu} \, a^{-\frac{1+\gamma}{\mu}} \left(\frac{b_1}{2a^{\nu_1/\mu}}\right)^{\sigma_1} \left(\frac{b_2}{2a^{\nu_2/\mu}}\right)^{\sigma_2} \left(\frac{1}{2i}\right)^2
$$

$$
\times \int\limits_{-\beta+i\infty}^{-\beta-i\infty}\!\!\! d\xi_1 d\xi_2 \frac{\left(\dfrac{b_1}{2a^{\nu_1/\mu}}\right)^{2\xi} \left(\dfrac{b_2}{2a^{\nu_2/\mu}}\right)^{2\xi}}{\sin \pi\xi_1 \, \sin \pi\xi_2}
$$

$$
\times \frac{\Gamma\left(\dfrac{1+\gamma+\nu_1(\sigma_1+2\xi_1)+\nu_2(\sigma_2+2\xi_2)}{\mu}\right)}{\Gamma(1+\xi_1)\,\Gamma(1+\xi_2)\,\Gamma(1+\xi_1+\sigma_1)\,\Gamma(1+\xi_2+\sigma_2)}
$$

(6.58)

or

$$
N_{68} = \frac{1}{2\nu_2} \left(\frac{b_1}{2}\right)^{\sigma_1} \left(\frac{b_2}{2}\right)^{-\frac{1+\gamma+\nu_1\sigma_1}{\nu_2}} \frac{1}{2i} \int\limits_{-\beta'+i\infty}^{-\beta'-i\infty} d\eta \frac{\left[\dfrac{a}{(b_2/2)^{\mu/\nu_2}}\right]^\eta}{\sin \pi\eta \, \Gamma(1+\eta)} \frac{1}{2i}
$$

$$
\times \int\limits_{-\beta+i\infty}^{-\beta-i\infty} d\xi \frac{\left[\dfrac{b_1}{2}\left(\dfrac{2}{b_2}\right)^{\nu_1/\nu_2}\right]^{2\xi}}{\sin \pi\xi \, \Gamma(1+\xi)} \Gamma\left(\frac{1+\gamma+\mu\eta+\nu_1\sigma_1+\nu_2\sigma_2+2\nu_1\xi}{2\nu_2}\right)
$$

$$
\times \frac{1}{\Gamma\left(1+\sigma_2-\dfrac{1+\gamma+\mu\eta+\nu_1\sigma_1+\nu_2\sigma_2+2\nu_1\xi}{2\nu_2}\right)} \frac{1}{\Gamma(1+\sigma_1+\xi)}.
$$

(6.59)

Interchanges $\nu_1 \longleftrightarrow \nu_2$, $b_1 \longleftrightarrow b_2$ and $\sigma_1 \longleftrightarrow \sigma_2$ in formulas (6.58) and (6.59) are valid.

6.4.4 Calculation of Concrete Integrals

27. We put

$$
\gamma = \sigma, \; \mu = 1, \; \nu = 1
$$

in (6.54) and get

$$S_{27} = \int_0^\infty dx x^\sigma e^{-ax} J_\sigma(bx) \tag{6.60}$$

$$= a^{-(1+\sigma)} \left(\frac{b}{2a}\right)^\sigma \frac{1}{2i} \int_{-\beta+i\infty}^{-\beta-i\infty} d\xi \frac{\left(\frac{b}{2a}\right)^{2\xi} \Gamma(1 + 2\sigma + 2\xi)}{\sin \pi\xi \, \Gamma(1 + \xi) \, \Gamma(1 + \sigma + \xi)},$$

where

$$\Gamma(2 + 2\sigma + 2\xi) = \frac{2^{2(1+\sigma+\xi)-1}}{\sqrt{\pi}} \Gamma(1 + \sigma + \xi) \Gamma\left(\frac{3}{2} + \sigma + \xi\right).$$

Thus,

$$S_{27} = \frac{1}{a} \left(\frac{2b}{a^2}\right)^\sigma \frac{1}{\sqrt{\pi}} \sum_{n=0}^\infty \frac{(-1)^n (b^2/a^2)^n}{n!} \Gamma\left(\frac{1}{2} + \sigma + n\right)$$

$$= \frac{(2b)^\sigma}{\sqrt{\pi}} \frac{\Gamma\left(\sigma + \frac{1}{2}\right)}{(a^2 + b^2)^{\sigma + \frac{1}{2}}}. \tag{6.61}$$

28. The formula (6.56) with

$$\gamma = \lambda - 1, \ \mu = 1, \ \nu = 1$$

gives immediately:

$$S_{28} = \int_0^\infty dx x^{\lambda-1} e^{-ax} J_{\sigma_1}(b_1 x) J_{\sigma_2}(b_2 x)$$

$$= \frac{b_1^{\sigma_1} b_2^{\sigma_2}}{\Gamma(1 + \sigma_2)} 2^{-\sigma_1-\sigma_2} a^{-\lambda-\sigma_1-\sigma_2} \sum_{n=0}^\infty \frac{\Gamma(\lambda + \sigma_1 + \sigma_2 + 2n)}{n! \, \Gamma(1 + \sigma_1 + n)}$$

$$\times F\left(-n, -\sigma_1 - n; \ 1 + \sigma_2; \ \frac{b_2^2}{b_1^2}\right) \left(-\frac{b_1^2}{4a^2}\right)^n. \tag{6.62}$$

29. The formula (6.54) with

$$\gamma = 1 + \sigma, \ \mu = 2, \ \nu = 1$$

gives

$$S_{29} = \int_0^\infty dx x^{\sigma+1} e^{-ax^2} J_\sigma(bx)$$

$$= \frac{1}{2a} \left(\frac{b}{2a}\right)^\sigma \frac{1}{2i} \int_{-\beta+i\infty}^{-\beta-i\infty} d\xi \frac{(b^2/4a)^\xi}{\sin \pi\xi \, \Gamma(1 + \xi)} \frac{\Gamma(1 + \sigma + \xi)}{\Gamma(1 + \sigma + \xi)}$$

$$= \frac{b^\sigma}{(2a)^{1+\sigma}} \exp\left[-\frac{b^2}{4a}\right]. \tag{6.63}$$

30. Assuming

$$\gamma = -\frac{1}{2}, \ \mu = 1, \ \sigma = 1, \ \nu = \frac{1}{2}$$

in (6.54), one gets

$$S_{30} = \int\limits_0^\infty \frac{dx}{\sqrt{x}} \, e^{-ax} \, J_1(b\sqrt{x})$$

$$= \frac{1}{\sqrt{a}} \left(\frac{b}{2\sqrt{a}}\right) \frac{1}{2i} \int\limits_{-\beta+i\infty}^{-\beta-i\infty} d\xi \frac{\left(\frac{b}{2\sqrt{a}}\right)^{2\xi} \Gamma(1+\xi)}{\sin \pi\xi \, \Gamma(1+\xi) \, \Gamma(2+\xi)}.$$

Here we change the integration variable $1 + \xi = x$ and obtain after taking residues

$$S_{30} = \frac{2}{b} \left[1 - \exp\left(-\frac{b^2}{4a}\right)\right]. \tag{6.64}$$

31. The formula (6.54) with

$$\gamma = \frac{1}{2}, \ \mu = 1, \ \sigma = 1, \ \nu = \frac{1}{2}$$

reads

$$S_{31} = \int\limits_0^\infty dx \sqrt{x} \, e^{-ax} \, J_1(b\sqrt{x})$$

$$= \frac{1}{a\sqrt{a}} \left(\frac{b}{2\sqrt{a}}\right) \frac{1}{2i} \int\limits_{-\beta+i\infty}^{-\beta-i\infty} d\xi \frac{\left(\frac{b}{2\sqrt{a}}\right)^{2\xi} \Gamma(2+\xi)}{\sin \pi\xi \, \Gamma(1+\xi) \, \Gamma(2+\xi)}$$

$$= \frac{b}{2a^2} \exp\left[-\frac{b^2}{4a}\right]. \tag{6.65}$$

32. If

$$\gamma = 0, \ \mu = 1, \ \sigma = 0, \ \nu = 1$$

in (6.54), then we have

$$S_{32} = \int\limits_0^\infty dx \, e^{-ax} \, J_0(bx)$$

$$= \frac{1}{a} \frac{1}{2i} \int\limits_{-\beta+i\infty}^{-\beta-i\infty} d\xi \frac{(b/2a)^{2\xi} \Gamma(1+2\xi)}{\sin \pi\xi \, \Gamma^2(1+\xi)}, \tag{6.66}$$

where

$$\Gamma(1+2\xi) = 2\xi \, \Gamma(2\xi) = 2\xi \, \frac{2^{2\xi-1}}{\sqrt{\pi}} \, \Gamma(\xi) \, \Gamma\left(\frac{1}{2}+\xi\right)$$

$$= 2^{2\xi} \, \frac{1}{\sqrt{\pi}} \, \Gamma(1+\xi) \, \Gamma\left(\frac{1}{2}+\xi\right).$$

So that

$$S_{32} = \frac{1}{a} \frac{1}{\sqrt{\pi}} \sum_{n=0}^{\infty} \frac{(-1)^n (b^2/a^2)^n}{n!} \Gamma\left(\frac{1}{2} + n\right).$$ (6.67)

Since

$$\Gamma\left(\frac{1}{2} + n\right) = \frac{\sqrt{\pi}}{2^n}(2n-1)!!$$

and therefore

$$S_{32} = \frac{1}{\sqrt{a^2 + b^2}}.$$ (6.68)

6.5 Integrals Involving $J_\sigma(x)$, x^γ and Trigonometric Functions

6.5.1 69th General Formula

$$N_{69} = \int_0^{\infty} dx x^\gamma \ J_\sigma(bx^\nu) \ \sin^q(ax^\mu)$$

$$= \frac{1}{2\nu} \left(\frac{b}{2}\right)^{-\frac{\gamma+1}{\nu}} \frac{1}{2^{q-1}} \frac{1}{2i} \int_{\alpha+i\infty}^{\alpha-i\infty} d\xi \frac{\left[\frac{2a}{(b/2)^{\mu/\nu}}\right]^{2\xi}}{\sin \pi\xi \ \Gamma(1+2\xi)} \ I_q(\xi)$$

$$\times \frac{\Gamma\left(\frac{\gamma+1+\nu\sigma+2\mu\xi}{2\nu}\right)}{\Gamma\left(1+\sigma - \frac{\gamma+1+\nu\sigma+2\mu\xi}{2\nu}\right)}$$ (6.69)

or

$$N_{69} = \frac{\sqrt{\pi}}{2\mu} \frac{1}{2^{q-1}} a^{-\frac{\gamma+1}{\mu}} \left(\frac{b/2}{a^{\nu/\mu}}\right)^\sigma$$

$$\times \frac{1}{2i} \int_{-\beta+i\infty}^{-\beta-i\infty} d\eta \frac{\left(\frac{b/2}{a^{\nu/\mu}}\right)^{2\eta}}{\sin \pi\eta \ \Gamma(1+\eta) \ \Gamma(1+\sigma+\eta)}$$ (6.70)

$$\times I_q\left(\xi = -\frac{\gamma+1+\nu(\sigma+2\eta)}{2\mu}\right) \frac{\Gamma\left(\frac{1+\gamma+\nu(\sigma+2\eta)}{2\mu}\right)}{\Gamma\left(\frac{1}{2} - \frac{1+\gamma+\nu(\sigma+2\eta)}{2\mu}\right)},$$

where $I_q(\xi)$, $(q = 2, 4, 6, \ldots)$ is given by (1.32) in Chapter 1.

6.5.2   70th *General Formula*

$$N_{70} = \int\limits_0^\infty dx\, x^\gamma \; J_\sigma(bx^\nu) \; \sin^m(ax^\mu)$$

$$= \frac{1}{2\nu} \left(\frac{b}{2}\right)^{-\frac{1+\gamma+\mu}{\nu}} \frac{a}{2^{m-1}} \frac{1}{2i} \int\limits_{-\beta+i\infty}^{-\beta-i\infty} d\xi \frac{\left[\dfrac{a}{(b/2)^{\mu/\nu}}\right]^{2\xi}}{\sin \pi\xi \; \Gamma(2+2\xi)} \; N_m(\xi) \qquad (6.71)$$

$$\times \frac{\Gamma\left(\dfrac{\gamma+1+\mu+2\mu\xi+\nu\sigma}{2\nu}\right)}{\Gamma\left(1+\sigma - \dfrac{\gamma+1+\mu+\nu\sigma+2\mu\xi}{2\nu}\right)}$$

or

$$N_{70} = \frac{\sqrt{\pi}}{2\mu} \left(\frac{a}{2}\right)^{-\frac{1+\gamma}{\mu}} \left[\frac{b}{2} \Big/ \left(\frac{a}{2}\right)^{\nu/\mu}\right]^\sigma \frac{1}{2^{m-1}}$$

$$\times \frac{1}{2i} \int\limits_{-\beta+i\infty}^{-\beta-i\infty} d\eta \frac{\left[\dfrac{b}{2} \Big/ \left(\dfrac{a}{2}\right)^{\nu/\mu}\right]^{2\eta}}{\sin \pi\eta \; \Gamma(1+\eta) \; \Gamma(1+\sigma+\eta)}$$

$$\times N_m\left(\xi = -\frac{1}{2}\left[1 + \frac{\gamma+1+\nu\sigma+2\nu\eta}{\mu}\right]\right) \qquad (6.72)$$

$$\times \frac{\Gamma\left(\dfrac{1}{2}\left[1 + \dfrac{1+\gamma+\nu\sigma+2\nu\eta}{\mu}\right]\right)}{\Gamma\left(1 - \dfrac{1+\gamma+\nu\sigma+2\nu\eta}{2\mu}\right)},$$

$$m = 1,\, 3,\, 5,\, 7,\ldots.$$

6.5.3 71st *General Formula*

$$N_{71} = \int\limits_0^\infty dx x^\gamma \, J_\sigma(bx^\nu) \, \cos^m(ax^\mu)$$

$$= \frac{1}{2\nu} \left(\frac{b}{2}\right)^{-\frac{\gamma+1}{\nu}} \frac{1}{2^{m-1}} \frac{1}{2i} \int\limits_{-\beta+i\infty}^{-\beta-i\infty} d\xi \frac{\left[\dfrac{a}{(b/2)^{\mu/\nu}}\right]^{2\xi}}{\sin \pi\xi \, \Gamma(1+2\xi)} \, N_m'(\xi) \qquad (6.73)$$

$$\times \frac{\Gamma\left(\dfrac{1+\gamma+\nu\sigma+2\mu\xi}{2\nu}\right)}{\Gamma\left(1+\sigma-\dfrac{1+\gamma+\nu\sigma+2\mu\xi}{2\nu}\right)}$$

or

$$N_{71} = \frac{\sqrt{\pi}}{2\mu} \left(\frac{a}{2}\right)^{-\frac{1+\gamma}{\mu}} \left[\frac{b}{2}\Big/ \left(\frac{a}{2}\right)^{\nu/\mu}\right]^\sigma \frac{1}{2^{m-1}}$$

$$\times \frac{1}{2i} \int\limits_{-\beta+i\infty}^{-\beta-i\infty} d\eta \frac{\left[\dfrac{b}{2}\Big/ \left(\dfrac{a}{2}\right)^{\nu/\mu}\right]^{2\eta}}{\sin \pi\eta \, \Gamma(1+\eta) \, \Gamma(1+\sigma+\eta)}$$

$$\times N_m'\left(\xi = -\frac{\gamma+1+\nu\sigma+2\nu\eta}{2\mu}\right) \frac{\Gamma\left(\dfrac{1+\gamma+\nu\sigma+2\nu\eta}{2\mu}\right)}{\Gamma\left(\dfrac{1}{2}-\dfrac{1+\gamma+\nu\sigma+2\nu\eta}{2\mu}\right)},$$

$$m = 1, \, 3, \, 5, \, 7, \dots.$$

(6.74)

6.5.4 72nd *General Formula*

$$N_{72} = \int\limits_0^\infty dx x^\gamma \, J_\sigma(bx^\nu)\Big[\cos^q(ax^\mu) - 1\Big]$$

$$= \frac{1}{2\nu}\left(\frac{b}{2}\right)^{-\frac{1+\gamma}{\nu}} \frac{1}{2^{q-1}}\frac{1}{2i}\int\limits_{\alpha+i\infty}^{\alpha-i\infty} d\xi \frac{\left[\dfrac{2a}{(b/2)^{\mu/\nu}}\right]^{2\xi}}{\sin \pi\xi \, \Gamma(1+2\xi)} I'_q(\xi)$$

$$\times \frac{\Gamma\left(\dfrac{1+\gamma+\nu\sigma+2\mu\xi}{2\nu}\right)}{\Gamma\left(1+\sigma - \dfrac{\gamma+1+\nu\sigma+2\mu\xi}{2\nu}\right)}$$

(6.75)

or

$$N_{72} = \frac{\sqrt{\pi}}{2\mu}\frac{1}{2^{q-1}} a^{-\frac{1+\gamma}{\mu}}\left[\frac{b}{2}\Big/a^{\nu/\mu}\right]^\sigma$$

$$\times \frac{1}{2i}\int\limits_{-\beta+i\infty}^{-\beta-i\infty} d\eta \frac{\left[\dfrac{b}{2}\Big/a^{\nu/\mu}\right]^{2\eta}}{\sin \pi\eta \, \Gamma(1+\eta)\,\Gamma(1+\sigma+\eta)}$$

$$\times I'_q\left(\xi = -\frac{1+\gamma+\nu\sigma+2\nu\eta}{2\mu}\right)\frac{\Gamma\left(\dfrac{1+\gamma+\nu\sigma+2\nu\eta}{2\mu}\right)}{\Gamma\left(\dfrac{1}{2} - \dfrac{1+\gamma+\nu\sigma+2\nu\eta}{2\mu}\right)},$$

(6.76)

$$q = 2,\ 4,\ 6,\ldots.$$

In the formulas (6.71)-(6.76) quantities $N_m(\xi)$, $N'_m(\xi)$ and $I'_q(\xi)$ are defined by the expressions (1.34), (1.36) and (1.38), respectively.

6.6 Calculation of Particular Integrals

33. The formula (6.71) with

$$\gamma = 0,\ \sigma = 0,\ \nu = \mu = 1,\ m = 1$$

gives

$$S_{33}^a = \int\limits_0^\infty dx \; J_0(bx) \; \sin(ax) \tag{6.77}$$

$$= \frac{1}{2} \left(\frac{b}{2}\right)^{-2} \frac{1}{2i} \int\limits_{-\beta+i\infty}^{-\beta-i\infty} d\xi \frac{(2a/b)^{2\xi}}{\sin \pi\xi \; \Gamma(2+2\xi)} \frac{\Gamma(1+\xi)}{\Gamma(-\xi)}.$$

Taking into account relations

- $\Gamma(1+\xi) \; \Gamma(-\xi) = -\dfrac{\pi}{\sin \pi\xi}$,

- $\Gamma(2(1+\xi)) = \dfrac{2^{2\xi+1}}{\sqrt{\pi}} \; \Gamma(1+\xi) \; \Gamma\left(\dfrac{3}{2}+\xi\right)$,

one gets

$$S_{33}^a = -\frac{1}{4\sqrt{\pi}} \left(\frac{b}{2}\right)^{-2} \frac{1}{2i} \int\limits_{-\beta+i\infty}^{-\beta-i\infty} d\xi \frac{(a/b)^{2\xi}}{\Gamma\left(\frac{3}{2}+\xi\right)} \; \Gamma(1+\xi).$$

It turns out that this function has no poles in the right half plane and so it goes to zero, when $0 \le a \le b$.

The case $a > b$ is studied by means of the formula (6.72). The result reads

$$S_{33}^b = \int\limits_0^\infty dx \; J_0(bx) \; \sin(ax) \tag{6.78}$$

$$= \frac{\sqrt{\pi}}{2} \left(\frac{a}{2}\right)^{-1} \frac{1}{2i} \int\limits_{-\beta+i\infty}^{-\beta-i\infty} d\eta \frac{(b/a)^{2\eta}}{\sin \pi\eta \; \Gamma^2(1+\eta)} \frac{\Gamma(1+\eta)}{\Gamma\left(\frac{1}{2}-\eta\right)}.$$

Since

$$\Gamma\left(\frac{1}{2} - n\right) = (-1)^n \frac{2^n \sqrt{\pi}}{(2n-1)!!}$$

and therefore

$$S_{33}^b = \frac{1}{2} \left(\frac{a}{2}\right)^{-1} \sum_{n=0}^\infty \frac{(b/\sqrt{2}a)^{2n} \; (2n-1)!!}{n!} = \frac{1}{\sqrt{a^2-b^2}}, \tag{6.79}$$

where $0 < b < a$. Finally, we have

$$S_{33} = \int\limits_0^\infty dx \; J_0(bx) \; \sin(ax) = \begin{cases} 0 & \text{if} \quad 0 < a < b \\[2mm] \dfrac{1}{\sqrt{a^2-b^2}} & \text{if} \quad 0 < b < a. \end{cases}$$

34. We put

$$\gamma = 0, \; \sigma = 0, \; \nu = 1, \; m = 1, \; \mu = 1$$

in (6.73) and get

$$S_{34}^a = \int_0^\infty dx \; J_0(bx) \; \cos(ax) \tag{6.80}$$

$$= \frac{1}{2} \left(\frac{b}{2}\right)^{-1} \frac{1}{2i} \int_{-\beta+i\infty}^{-\beta-i\infty} d\xi \frac{\left(\frac{2a}{b}\right)^{2\xi}}{\sin \pi\xi \; \Gamma(1+2\xi)} \frac{\Gamma\left(\frac{1}{2}+\xi\right)}{\Gamma\left(\frac{1}{2}-\xi\right)},$$

where

$$\Gamma(1+2\xi) = 2\xi \, \frac{2^{2\xi-1}}{\sqrt{\pi}} \, \Gamma(\xi) \, \Gamma\left(\frac{1}{2}+\xi\right)$$

$$= \frac{2^{2\xi}}{\sqrt{\pi}} \, \Gamma(1+\xi) \, \Gamma\left(\frac{1}{2}+\xi\right).$$

Thus,

$$S_{34}^a = \frac{1}{2} \left(\frac{b}{2}\right)^{-1} \sum_{n=0}^\infty \frac{\left(\frac{a}{\sqrt{2b}}\right)^{2n} (2n-1)!!}{n!} = \frac{1}{\sqrt{b^2-a^2}}, \tag{6.81}$$

where $0 < a < b$. The case $0 < b < a$ is considered from (6.74). The result reads

$$S_{34}^b = \int_0^\infty dx \; J_0(bx) \; \cos(ax) \tag{6.82}$$

$$= \frac{\sqrt{\pi}}{2} \left(\frac{a}{2}\right)^{-1} \frac{1}{2i} \int_{-\beta+i\infty}^{-\beta-i\infty} d\eta \frac{(b/a)^{2\eta}}{\sin \pi\eta \; \Gamma^2(1+\eta)} \frac{\Gamma\left(\frac{1}{2}+\eta\right)}{\Gamma(-\eta)}.$$

It is obvious that this function has no poles in the right half plane and therefore displacement of the integration contour to the right gives zero. Collecting all results, we have

$$S_{34} = \int_0^\infty dx \; J_0(bx) \; \cos(ax) = \begin{cases} \dfrac{1}{\sqrt{b^2-a^2}} & \text{if} \quad 0 < a < b \\[2ex] \infty & \text{if} \quad a = b \\[2ex] 0 & \text{if} \quad 0 < b < a. \end{cases} \tag{6.83}$$

35. The formula (6.71) with

$$\gamma = 0, \; \sigma = 0, \; \nu = \frac{1}{2}, \; m = 1, \; \mu = 1$$

gives

$$S_{35} = \int_0^\infty dx \, J_0(b\sqrt{x}) \, \sin(ax) \tag{6.84}$$

$$= \left(\frac{b}{2}\right)^{-4} a \frac{1}{2i} \int_{-\beta+i\infty}^{-\beta-i\infty} d\xi \frac{\left(a\sqrt{\frac{2}{b}}\right)^{2\xi}}{\sin \pi\xi \, \Gamma(2+2\xi)} \frac{\Gamma(2+2\xi)}{\Gamma(-1-2\xi)}.$$

It is obvious that this variant representation for the given integral gives zero result. Therefore we would like to consider a second variant Mellin representation (6.72) for this integral:

$$S_{35} = \frac{\sqrt{\pi}}{2} \left(\frac{a}{2}\right)^{-1} \frac{1}{2i} \int_{-\beta+i\infty}^{-\beta-i\infty} d\eta \frac{\left(\frac{b}{2}\sqrt{\frac{2}{a}}\right)^{2\eta}}{\sin \pi\eta \, \Gamma^2(1+\eta)} \frac{\Gamma\left(1+\frac{1}{2}\eta\right)}{\Gamma\left(\frac{1}{2}-\frac{1}{2}\eta\right)},$$

where

$$\frac{\Gamma\left(1+\frac{1}{2}\eta\right)}{\Gamma(1+\eta) \, \Gamma\left(\frac{1}{2}-\frac{1}{2}\eta\right)} = \frac{\cos\frac{\pi}{2}\eta}{\sqrt{\pi} \, 2^\eta}.$$

So that by changing integration variable $\frac{\eta}{2} = x$, one gets

$$S_{35} = \frac{1}{a} \frac{1}{2i} \int_{-\beta+i\infty}^{-\beta-i\infty} dx \frac{\left(\frac{b}{2\sqrt{a}}\right)^{4x}}{\sin \pi x \, \Gamma(1+2x)} = \frac{1}{a} \cos\left(\frac{b^2}{4a}\right). \tag{6.85}$$

36. Similarly if we put

$$\gamma = 0, \; \sigma = 0, \; \nu = \frac{1}{2}, \; m = 1, \; \mu = 1$$

in (6.74), we have

$$S_{36} = \int_0^\infty dx \, J_0(b\sqrt{x}) \, \cos(ax)$$

$$= \frac{\sqrt{\pi}}{2} \left(\frac{a}{2}\right)^{-1} \frac{1}{2i} \int_{-\beta+i\infty}^{-\beta-i\infty} d\eta \frac{\left[\frac{b}{2}\sqrt{\frac{2}{a}}\right]^{2\eta}}{\sin \pi\eta \, \Gamma^2(1+\eta)} \frac{\Gamma\left(\frac{1}{2}+\frac{1}{2}\eta\right)}{\Gamma\left(-\frac{1}{2}\eta\right)},$$

where

$$\frac{\Gamma\left(\frac{1}{2}+\frac{1}{2}\eta\right)}{\Gamma(1+\eta) \, \Gamma\left(-\frac{1}{2}\eta\right)} = -\frac{\sin\frac{\pi\eta}{2}}{\sqrt{\pi} \, 2^\eta}.$$

After changing the integration variable $\eta \to 1 + 2x$ and taking into account the identity

$$\cos\left(\frac{\pi}{2} + \pi x\right) = -\sin \pi x,$$

one gets

$$S_{36} = \frac{1}{a} \frac{1}{2i} \int_{-\beta+i\infty}^{-\beta-i\infty} dx \frac{\left(\frac{b^2}{4a}\right)^{2x+1}}{\sin \pi x \, \Gamma(2+2x)} = \frac{1}{a} \sin\left(\frac{b^2}{4a}\right). \tag{6.86}$$

37. Let

$$\gamma = 0, \ \sigma = 1, \ \nu = 1, \ \mu = 2, \ m = 1$$

be in (6.72). Then

$$S_{37} = \int_0^\infty dx \ J_1(bx) \ \sin(ax^2)$$

$$= \frac{\sqrt{\pi}}{4} \left(\frac{a}{2}\right)^{-\frac{1}{2}} \left(\frac{b}{2}\sqrt{\frac{2}{a}}\right) \frac{1}{2i} \int_{-\beta+i\infty}^{-\beta-i\infty} d\eta \frac{\left[\frac{b}{2}\sqrt{\frac{2}{a}}\right]^{2\eta}}{\sin \pi\eta \ \Gamma(1+\eta) \ \Gamma(2+\eta)}$$

$$\times \frac{\Gamma\left(1+\frac{\eta}{2}\right)}{\Gamma\left(\frac{1}{2}-\frac{\eta}{2}\right)}.$$

After some elementary calculations, we have

$$S_{37} = \frac{1}{b} \frac{1}{2i} \int_{-\beta+i\infty}^{-\beta-i\infty} dx \frac{\left(\frac{b^2}{4a}\right)^{2x+1}}{\sin \pi x \ \Gamma(2+2x)} = \frac{1}{b} \sin\left(\frac{b^2}{4a}\right). \tag{6.87}$$

38. The formula (6.74) with

$$\gamma = 0, \ \sigma = 1, \ \mu = 2, \ m = 1, \ \nu = 1$$

reads

$$S_{38} = \int_0^\infty dx \ J_1(bx) \ \cos(ax^2) = \frac{\sqrt{\pi}}{4} \left(\frac{a}{2}\right)^{-1/2} \left(\frac{b}{2}\sqrt{\frac{2}{a}}\right)$$

$$\times \frac{1}{2i} \int_{-\beta+i\infty}^{-\beta-i\infty} d\eta \frac{\left[\frac{b}{2}\sqrt{\frac{2}{a}}\right]^{2\eta}}{\sin \pi\eta \ \Gamma(1+\eta) \ \Gamma(2+\eta)} \frac{\Gamma\left(\frac{1}{2}+\frac{\eta}{2}\right)}{\Gamma\left(-\frac{\eta}{2}\right)}.$$

As before, here

$$\frac{\Gamma\left(\frac{1}{2}+\frac{1}{2}\eta\right)}{\Gamma(1+\eta) \ \Gamma\left(-\frac{1}{2}\eta\right)} = -\frac{\sin \frac{\pi\eta}{2}}{\sqrt{\pi} \ 2^\eta}$$

and changing the integration variable $\eta \to 2x - 1$, one gets

$$S_{38} = \frac{1}{b}(-1) \frac{1}{2i} \int_{-\beta+i\infty}^{-\beta-i\infty} dx \frac{\left(2 \frac{b^2}{8a}\right)^{2x}}{\sin \pi x \ \Gamma(1+2x)} = \frac{2}{b} \sin^2\left(\frac{b^2}{8a}\right). \tag{6.88}$$

39. If we put

$$\gamma = 0, \ \sigma = 1, \ \nu = 1, \ q = 2, \ \mu = 2$$

in (6.70), we obtain

$$S_{39} = \int\limits_0^\infty dx \, J_1(bx) \, \sin^2(ax^2) = -\frac{\sqrt{\pi}}{4} \frac{1}{2} a^{-\frac{1}{2}} \left(\frac{b}{2} \frac{1}{\sqrt{a}} \right)$$

$$\times \frac{1}{2i} \int\limits_{-\beta+i\infty}^{-\beta-i\infty} d\eta \frac{\left[\frac{b}{2} \frac{1}{\sqrt{a}} \right]^{2\eta}}{\sin \pi\eta \, \Gamma(1+\eta) \, \Gamma(2+\eta)} \frac{\Gamma\left(\frac{2+2\eta}{4} \right)}{\Gamma\left(-\frac{\eta}{2} \right)}, \tag{6.89}$$

where

$$\frac{\Gamma\left(\frac{1}{2} + \frac{1}{2}\eta \right)}{\Gamma(1+\eta) \, \Gamma\left(-\frac{1}{2}\eta \right)} = -\frac{\sin\left(\frac{\pi\eta}{2} \right)}{\sqrt{\pi} \, 2^\eta}.$$

After changing the integration variable $\eta \to 2x - 1$, where

$$\cos \frac{\pi}{2}(2x - 1) = \sin \pi x,$$

one gets

$$S_{39} = \frac{1}{2b} \frac{1}{2i} \int\limits_{-\beta+i\infty}^{-\beta-i\infty} d\eta \frac{\left[\frac{b^2}{8a^2} \right]^{2x}}{\sin \pi x \, \Gamma(1+2x)} = \frac{1}{2b} \cos\left(\frac{b^2}{8a} \right). \tag{6.90}$$

40. The formula (6.72) with

$$\gamma = 0, \ \sigma = 0, \ \nu = 1, \ m = 1, \ \mu = 2$$

leads to the following integral

$$S_{40} = \int\limits_0^\infty dx \, J_0(bx) \, \sin(ax^2)$$

$$= \frac{\sqrt{\pi}}{4} \left(\frac{a}{2} \right)^{-1/2} \frac{1}{2i} \int\limits_{-\beta+i\infty}^{-\beta-i\infty} d\eta \frac{\left(\frac{b}{2} \sqrt{\frac{2}{a}} \right)^{2\eta}}{\sin \pi\eta \, \Gamma(1+\eta) \, \Gamma(1+\eta)} \frac{\Gamma\left(\frac{3}{4} + \frac{\eta}{2} \right)}{\Gamma\left(\frac{3}{4} - \frac{\eta}{2} \right)}.$$

After some similar transformations as above, we have

$$S_{40} = \frac{1}{2a} \cos \frac{b^2}{4a}. \tag{6.91}$$

41. Let

$$\gamma = 0, \ \sigma = 0, \ \nu = 1, \ m = 1, \ \mu = 2$$

be in the formula (6.74), then we have

$$S_{41} = \int\limits_0^\infty dx \, J_0(bx) \, \cos(ax^2) \tag{6.92}$$

$$= \frac{\sqrt{\pi}}{4} \left(\frac{a}{2} \right)^{-1/2} \frac{1}{2i} \int\limits_{-\beta+i\infty}^{-\beta-i\infty} d\eta \frac{\left[\frac{b}{2} \sqrt{\frac{2}{a}} \right]^{2\eta}}{\sin \pi\eta \, \Gamma^2(1+\eta)} \frac{\Gamma\left(\frac{1}{4} + \frac{\eta}{2} \right)}{\Gamma\left(\frac{1}{4} - \frac{\eta}{2} \right)}.$$

After some elementary transformations with connected gamma-functions, one gets

$$S_{41} = \frac{1}{2a} \sin \frac{b^2}{4a}. \tag{6.93}$$

6.7   Integrals Containing Two $J_\sigma(x)$, x^γ and Trigonometric Functions

6.7.1 73rd *General Formula*

$$N_{73} = \int\limits_0^\infty dx\, x^\gamma \, J_{\sigma_1}(b_1 x^\nu)\, J_{\sigma_2}(b_2 x^\nu)\, \sin^q(ax^\mu)$$

$$= \frac{1}{\Gamma(1+\sigma_2)} \left(\frac{b_1}{2}\right)^{-\frac{\gamma+1+\nu\sigma_2}{\nu}} \left(\frac{b_2}{2}\right)^{\sigma_2} \frac{1}{2\nu}\frac{1}{2^{q-1}}$$

$$\times \frac{1}{2i} \int\limits_{\alpha+i\infty}^{\alpha-i\infty} d\xi \, \frac{\left[\dfrac{2a}{(b_1/2)^{\mu/\nu}}\right]^{2\xi}}{\sin \pi\xi\, \Gamma(1+2\xi)} \, I_q(\xi)$$

$$\times \frac{\Gamma\left(\dfrac{1+\gamma+\nu(\sigma_1+\sigma_2)+2\mu\xi}{2\nu}\right)}{\Gamma\left(1+\sigma_1 - \dfrac{1+\gamma+\nu(\sigma_1+\sigma_2)+2\mu\xi}{2\nu}\right)}$$

$$\times F\left(\frac{1+\gamma+\nu(\sigma_1+\sigma_2)+2\mu\xi}{2\nu}, -\sigma_1 + \frac{\gamma+1+\nu(\sigma_1+\sigma_2)+2\mu\xi}{2\nu}; \right.$$

$$\left. ;1+\sigma_2;\ \frac{b_2^2}{b_1^2}\right)$$

$$(6.94)$$

or

$$N_{73} = \frac{\sqrt\pi}{2\mu}\frac{1}{2^{q-1}}\, a^{-\frac{1+\gamma}{\mu}} \left[\frac{b_1}{2}\Big/ a^{\nu/\mu}\right]^{\sigma_1} \left[\frac{b_2}{2}\Big/ a^{\nu/\mu}\right]^{\sigma_2} \frac{1}{\Gamma(1+\sigma_2)}$$

$$\times \frac{1}{2i} \int\limits_{-\beta+i\infty}^{-\beta-i\infty} d\eta\, \frac{\left[\dfrac{b_1}{2}\Big/ a^{\nu/\mu}\right]^{2\eta}}{\sin \pi\eta\, \Gamma(1+\eta)\, \Gamma(1+\sigma_1+\eta)}$$

$$\times I_q\left(\xi = -\frac{1+\gamma+\nu(\sigma_1+\sigma_2)+2\nu\eta}{2\mu}\right)$$

$$\times \frac{\Gamma\left(\dfrac{1+\gamma+\nu(\sigma_1+\sigma_2)+2\nu\eta}{2\mu}\right)}{\Gamma\left(\dfrac{1}{2} - \dfrac{1+\gamma+\nu(\sigma_1+\sigma_2)+2\nu\eta}{2\mu}\right)}\, F\left(-\eta, -\sigma_1-\eta;\ 1+\sigma_2;\ \frac{b_2^2}{b_1^2}\right),$$

$$(6.95)$$

where $q = 2,\ 4,\ 6,\ldots$ and $I_q(\xi)$ is given by (1.32) in Chapter 1.

6.7.2   74th *General Formula*

$$N_{74} = \int\limits_0^\infty dx\, x^\gamma\, J_{\sigma_1}(b_1 x^\nu)\, J_{\sigma_2}(b_2 x^\nu)\, \sin^m(ax^\mu)$$

$$= \frac{1}{\Gamma(1+\sigma_2)}\, \left(\frac{b_1}{2}\right)^{-\frac{1+\gamma+\mu+\nu\sigma_2}{\nu}}\, \frac{a}{2\nu}\, \left(\frac{b_2}{2}\right)^{\sigma_2}\, \frac{1}{2^{m-1}}$$

$$\times \frac{1}{2i} \int\limits_{-\beta+i\infty}^{-\beta-i\infty} d\xi\, \frac{\left[\frac{a}{(b_1/2)^{\mu/\nu}}\right]^{2\xi}}{\sin \pi\xi\, \Gamma(2+2\xi)}\, N_m(\xi)$$

$$\times \frac{\Gamma\left(\dfrac{1+\gamma+\mu+\nu(\sigma_1+\sigma_2)+2\mu\xi}{2\nu}\right)}{\Gamma\left(1+\sigma_1-\dfrac{1+\gamma+\mu+\nu(\sigma_1+\sigma_2)+2\mu\xi}{2\nu}\right)}$$

$$\times F\left(\frac{1+\gamma+\mu+\nu(\sigma_1+\sigma_2)+2\mu\xi}{2\nu}, \right.$$

$$\left. -\sigma_1 + \frac{\gamma+1+\mu+\nu(\sigma_1+\sigma_2)+2\mu\xi}{2\nu};\ 1+\sigma_2;\ \frac{b_2^2}{b_1^2}\right) \tag{6.96}$$

or

$$N_{74} = \frac{\sqrt{\pi}}{2\mu}\, \frac{1}{2^{m-1}}\, \left(\frac{a}{2}\right)^{-\frac{1+\gamma}{\mu}}$$

$$\times \left[\frac{b_1}{2}\Big/(a/2)^{\nu/\mu}\right]^{\sigma_1}\, \left[\frac{b_2}{2}\Big/(a/2)^{\nu/\mu}\right]^{\sigma_2}$$

$$\times \frac{1}{\Gamma(1+\sigma_2)}\, \frac{1}{2i} \int\limits_{-\beta+i\infty}^{-\beta-i\infty} d\eta\, \frac{\left[\frac{b_1}{2}\Big/(a/2)^{\nu/\mu}\right]^{2\eta}}{\sin \pi\eta\, \Gamma(1+\eta)\, \Gamma(1+\sigma_1+\eta)}$$

$$\times N_m\left(\xi = -\frac{1}{2}\left[1 + \frac{1+\gamma+\nu(\sigma_1+\sigma_2)+2\nu\eta}{\mu}\right]\right)$$

$$\times \frac{\Gamma\left(\dfrac{1}{2}\left[1+\dfrac{1+\gamma+\nu(\sigma_1+\sigma_2)+2\nu\eta}{\mu}\right]\right)}{\Gamma\left(1-\dfrac{1+\gamma+\nu(\sigma_1+\sigma_2)+2\nu\eta}{2\mu}\right)}$$

$$\times F\left(-\eta, -\sigma_1-\eta;\ 1+\sigma_2;\ \frac{b_2^2}{b_1^2}\right), \tag{6.97}$$

where $m = 1,\ 3,\ 5,\ 7,\ldots$ and $N_m(\xi)$ is given by (1.34) in Chapter 1.

6.7.3　75th *General Formula*

$$
N_{75} = \int\limits_{0}^{\infty} dx\, x^{\gamma}\, J_{\sigma_1}(b_1 x^{\nu})\, J_{\sigma_2}(b_2 x^{\nu})\, \cos^m(a x^{\mu})
$$

$$
= \frac{1}{\Gamma(1+\sigma_2)} \left(\frac{b_1}{2}\right)^{-\frac{1+\gamma+\nu\sigma_2}{\nu}} \frac{1}{2\nu} \left(\frac{b_2}{2}\right)^{\sigma_2} \frac{1}{2^{m-1}}
$$

$$
\times \frac{1}{2i} \int\limits_{-\beta+i\infty}^{-\beta-i\infty} d\xi\, \frac{\left[\dfrac{a}{(b_1/2)^{\mu/\nu}}\right]^{2\xi}}{\sin \pi\xi\ \Gamma(1+2\xi)}\ N'_m(\xi)
$$

$$
\times \frac{\Gamma\left(\dfrac{1+\gamma+\nu(\sigma_1+\sigma_2)+2\nu\xi}{2\nu}\right)}{\Gamma\left(1+\sigma_1 - \dfrac{1+\gamma+\nu(\sigma_1+\sigma_2)+2\mu\xi}{2\nu}\right)}
$$

$$
\times F\left(\frac{1+\gamma+\nu(\sigma_1+\sigma_2)+2\mu\xi}{2\nu},\right.
$$

$$
\left. -\sigma_1 + \frac{\gamma+1+\nu(\sigma_1+\sigma_2)+2\mu\xi}{2\nu};\ 1+\sigma_2;\ \frac{b_2^2}{b_1^2}\right)
$$

(6.98)

or

$$
N_{75} = \frac{\sqrt{\pi}}{2\mu}\, \frac{1}{2^{m-1}} \left(\frac{a}{2}\right)^{-\frac{1+\gamma}{\mu}} \left[\frac{b_1}{2}\Big/(a/2)^{\nu/\mu}\right]^{\sigma_1} \left[\frac{b_2}{2}\Big/(a/2)^{\nu/\mu}\right]^{\sigma_2}
$$

$$
\times \frac{1}{\Gamma(1+\sigma_2)}\, \frac{1}{2i} \int\limits_{-\beta+i\infty}^{-\beta-i\infty} d\eta\, \frac{\left[\dfrac{b_1}{2}\Big/(a/2)^{\nu/\mu}\right]^{2\eta}}{\sin \pi\eta\ \Gamma(1+\eta)\ \Gamma(1+\sigma_1+\eta)}
$$

$$
\times N'_m\left(\xi = -\frac{1+\gamma+\nu(\sigma_1+\sigma_2)+2\nu\eta}{2\mu}\right)
$$

$$
\times \frac{\Gamma\left(\dfrac{1+\gamma+\nu(\sigma_1+\sigma_2)+2\nu\eta}{2\mu}\right)}{\Gamma\left(\dfrac{1}{2} - \dfrac{1+\gamma+\nu(\sigma_1+\sigma_2)+2\nu\eta}{2\mu}\right)}
$$

$$
\times F\left(-\eta, -\sigma_1-\eta;\ 1+\sigma_2;\ \frac{b_2^2}{b_1^2}\right),
$$

(6.99)

where $m = 1,\ 3,\ 5,\ 7,\dots$ and $N'_m(\xi)$ is given by (1.36) in Chapter 1.

6.7.4 76th *General Formula*

$$N_{76} = \int\limits_0^\infty dx\, x^\gamma\, J_{\sigma_1}(b_1 x^\nu)\, J_{\sigma_2}(b_2 x^\nu)\, \left[\cos^q(a x^\mu) - 1\right]$$

$$= \frac{1}{\Gamma(1+\sigma_2)} \left(\frac{b_1}{2}\right)^{-\frac{\gamma+1+\nu\sigma_2}{\nu}} \left(\frac{b_2}{2}\right)^{\sigma_2} \frac{1}{2\nu}\, \frac{1}{2^{q-1}}$$

$$\times \frac{1}{2i} \int\limits_{\alpha+i\infty}^{\alpha-i\infty} d\xi\, \frac{\left[\dfrac{2a}{(b_1/2)^{\mu/\nu}}\right]^{2\xi}}{\sin\pi\xi\,\Gamma(1+2\xi)}\, I_q'(\xi)$$

$$\times \frac{\Gamma\left(\dfrac{1+\gamma+\nu(\sigma_1+\sigma_2)+2\mu\xi}{2\nu}\right)}{\Gamma\left(1+\sigma_1 - \dfrac{1+\gamma+\nu(\sigma_1+\sigma_2)+2\mu\xi}{2\nu}\right)}$$

$$\times F\left(\frac{1+\gamma+\nu(\sigma_1+\sigma_2)+2\mu\xi}{2\nu}, -\sigma_1 + \frac{\gamma+1+\nu(\sigma_1+\sigma_2)+2\mu\xi}{2\nu}\right.$$

$$\left.; 1+\sigma_2;\ \frac{b_2^2}{b_1^2}\right)$$

$$(6.100)$$

or

$$N_{76} = \frac{\sqrt{\pi}}{2\mu}\, \frac{1}{2^{q-1}}\, a^{-\frac{1+\gamma}{\mu}} \left[\frac{b_1}{2}\Big/ a^{\nu/\mu}\right]^{\sigma_1} \left[\frac{b_2}{2}\Big/ a^{\nu/\mu}\right]^{\sigma_2}$$

$$\times \frac{1}{\Gamma(1+\sigma_2)}\, \frac{1}{2i} \int\limits_{-\beta+i\infty}^{-\beta-i\infty} d\eta\, \frac{\left[\dfrac{b_1}{2}\Big/ a^{\nu/\mu}\right]^{2\eta}}{\sin\pi\eta\,\Gamma(1+\eta)\,\Gamma(1+\sigma_1+\eta)}$$

$$\times I_q'\left(\xi = -\frac{1+\gamma+\nu(\sigma_1+\sigma_2)+2\nu\eta}{2\mu}\right)$$

$$\times \frac{\Gamma\left(\dfrac{1+\gamma+\nu(\sigma_1+\sigma_2)+2\nu\eta}{2\mu}\right)}{\Gamma\left(\dfrac{1}{2} - \dfrac{1+\gamma+\nu(\sigma_1+\sigma_2)+2\nu\eta}{2\mu}\right)}$$

$$\times F\left(-\eta, -\sigma_1 - \eta;\ 1+\sigma_2;\ \frac{b_2^2}{b_1^2}\right),$$

$$(6.101)$$

where $q = 2,\ 4,\ 6,\ldots$ and $I_q'(\xi)$ is given by (1.38) in Chapter 1.

6.7.5   77th General Formula

$$N_{77} = \int_0^\infty dx\, x^\gamma\, J_{\sigma_1}(b_1 x^{\nu_1})\, J_{\sigma_2}(b_2 x^{\nu_2})\, \sin^q(ax^\mu)$$

$$= \left(\frac{b_1}{2}\right)^{\sigma_1} \left(\frac{b_2}{2}\right)^{-\frac{\gamma+1+\nu_1\sigma_1}{\nu_2}} \frac{1}{2\nu_2}\, \frac{1}{2^{q-1}}$$

$$\times \frac{1}{2i} \int_{\alpha+i\infty}^{\alpha-i\infty} d\xi \frac{\left[2a\Big/(b_2/2)^{\nu_1/\nu_2}\right]^{2\xi}}{\sin \pi\xi\, \Gamma(1+2\xi)}\, I_q(\xi)$$

$$\times \frac{1}{2i} \int_{-\beta+i\infty}^{-\beta-i\infty} d\eta \frac{\left[\frac{b_1}{2}\Big/(b_2/2)^{\nu_1/\nu_2}\right]^{2\eta}}{\sin \pi\eta\, \Gamma(1+\eta)\, \Gamma(1+\sigma_1+\eta)}$$

$$\times \frac{\Gamma\left(\dfrac{1+\gamma+2\mu\xi+\nu_1\sigma_1+\nu_2\sigma_2+2\nu_1\eta}{2\nu_2}\right)}{\Gamma\left(1+\sigma_2-\dfrac{1+\gamma+2\mu\xi+\nu_1\sigma_1+\nu_2\sigma_2+2\nu_1\eta}{2\nu_2}\right)}$$

(6.102)

or

$$N_{77} = \frac{\sqrt{\pi}}{2\mu}\, \frac{1}{2^{q-1}}\, a^{-\frac{1+\gamma}{\mu}} \left(\frac{b_1}{2}\Big/a^{\nu_1/\mu}\right)^{\sigma_1} \left(\frac{b_2}{2}\Big/a^{\nu_2/\mu}\right)^{\sigma_2}$$

$$\times \frac{1}{(2i)^2} \iint_{-\beta+i\infty}^{-\beta-i\infty} d\eta_1 d\eta_2 \frac{\left(\frac{b_1}{2}\Big/a^{\nu_1/\mu}\right)^{2\eta_1} \left(\frac{b_2}{2}\Big/a^{\nu_2/\mu}\right)^{2\eta_2}}{\sin \pi\eta_1\, \sin \pi\eta_2\, \Gamma(1+\eta_1)\, \Gamma(1+\eta_2)}$$

$$\times \frac{I_q\left(\xi=-\dfrac{1+\gamma+\nu_1\sigma_1+\nu_2\sigma_2+2\nu_1\eta_1+2\nu_2\eta_2}{2\mu}\right)}{\Gamma(1+\sigma_1+\eta_1)\, \Gamma(1+\sigma_2+\eta_2)}$$

(6.103)

$$\times \frac{\Gamma\left(\dfrac{1+\gamma+\nu_1\sigma_1+\nu_2\sigma_2+2\nu_1\eta_1+2\nu_2\eta_2}{2\mu}\right)}{\Gamma\left(\dfrac{1}{2}-\dfrac{1+\gamma+\nu_1\sigma_1+\nu_2\sigma_2+2\nu_1\eta_1+2\nu_2\eta_2}{2\mu}\right)},$$

where $q = 2,\, 4,\, 6,\ldots$ and $I_q(\xi)$ is defined by the formula (1.32) in Chapter 1.

6.7.6 78th *General Formula*

$$N_{78} = \int_0^\infty dx\, x^\gamma\, J_{\sigma_1}(b_1 x^{\nu_1})\, J_{\sigma_2}(b_2 x^{\nu_2})\, \sin^m(a x^\mu)$$

$$= \left(\frac{b_1}{2}\right)^{\sigma_1} \left(\frac{b_2}{2}\right)^{-\frac{\gamma+1+\mu+\nu_1\sigma_1}{\nu_2}} \frac{1}{2\nu_2}\, \frac{1}{2^{m-1}}$$

$$\times \frac{1}{2i} \int_{-\beta+i\infty}^{-\beta-i\infty} d\xi \frac{\left[\dfrac{a}{(b_2/2)^{\nu_1/\nu_2}}\right]^{2\xi}}{\sin\pi\xi\,\Gamma(2+2\xi)}\, N_m(\xi)$$

$$\times \frac{1}{2i} \int_{-\beta'+i\infty}^{-\beta'-i\infty} d\eta \frac{\left[\dfrac{b_1}{2}\Big/(b_2/2)^{\nu_1/\nu_2}\right]^{2\eta}}{\sin\pi\eta\,\Gamma(1+\eta)\,\Gamma(1+\sigma_1+\eta)}$$

$$\times \frac{\Gamma\left(\dfrac{1+\gamma+\mu+2\mu\xi+\nu_1\sigma_1+\nu_2\sigma_2+2\nu_1\eta}{2\nu_2}\right)}{\Gamma\left(1+\sigma_2 - \dfrac{1+\gamma+\mu+2\mu\xi+\nu_1\sigma_1+\nu_2\sigma_2+2\nu_1\eta}{2\nu_2}\right)}$$

(6.104)

or

$$N_{78} = \frac{\sqrt{\pi}}{2\mu}\, \frac{1}{2^{m-1}} \left(\frac{a}{2}\right)^{-\frac{1+\gamma}{\mu}} \left(\frac{b_1/2}{(a/2)^{\nu_1/\mu}}\right)^{\sigma_1} \left(\frac{b_2/2}{(a/2)^{\nu_2/\mu}}\right)^{\sigma_2}$$

$$\times \frac{1}{(2i)^2} \iint_{-\beta+i\infty}^{-\beta-i\infty} d\eta_1 d\eta_2 \frac{\left(\dfrac{b_1}{2}\Big/(a/2)^{\nu_1/\mu}\right)^{2\eta_1} \left(\dfrac{b_2}{2}\Big/(a/2)^{\nu_2/\mu}\right)^{2\eta_2}}{\sin\pi\eta_1\,\sin\pi\eta_2\,\Gamma(1+\eta_1)\,\Gamma(1+\eta_2)}$$

$$\times \frac{N_m\left(\xi = -\dfrac{1}{2}\left[1+\dfrac{1+\gamma+\nu_1\sigma_1+\nu_2\sigma_2+2\nu_1\eta_1+2\nu_2\eta_2}{\mu}\right]\right)}{\Gamma(1+\sigma_1+\eta_1)\,\Gamma(1+\sigma_2+\eta_2)}$$

$$\times \frac{\Gamma\left(\dfrac{1}{2}\left[1+\dfrac{1+\gamma+\nu_1\sigma_1+\nu_2\sigma_2+2\nu_1\eta_1+2\nu_2\eta_2}{\mu}\right]\right)}{\Gamma\left(1-\dfrac{1+\gamma+\nu_1\sigma_1+\nu_2\sigma_2+2\nu_1\eta_1+2\nu_2\eta_2}{2\mu}\right)},$$

(6.105)

where $m = 1, 3, 5, 7, \ldots$ and $N_m(\xi)$ is given by (1.34) in Chapter 1.

6.7.7 79th General Formula

$$N_{79} = \int\limits_0^\infty dx x^\gamma \; J_{\sigma_1}(b_1 x^{\nu_1}) \; J_{\sigma_2}(b_2 x^{\nu_2}) \; \cos^m(ax^\mu)$$

$$= \left(\frac{b_1}{2}\right)^{\sigma_1} \left(\frac{b_2}{2}\right)^{-\frac{\gamma+1+\nu_1\sigma_1}{\nu_2}} \frac{1}{2\nu_2} \frac{1}{2^{m-1}}$$

$$\times \frac{1}{2i} \int\limits_{-\beta+i\infty}^{-\beta-i\infty} d\xi \frac{\left[\dfrac{a}{(b_2/2)^{\nu_1/\nu_2}}\right]^{2\xi}}{\sin\pi\xi \; \Gamma(1+2\xi)} \; N_m'(\xi)$$

$$\times \frac{1}{2i} \int\limits_{-\beta'+i\infty}^{-\beta'-i\infty} d\eta \frac{\left[\dfrac{b_1}{2}\Big/(b_2/2)^{\nu_1/\nu_2}\right]^{2\eta}}{\sin\pi\eta \; \Gamma(1+\eta) \; \Gamma(1+\sigma_1+\eta)}$$

$$\times \frac{\Gamma\left(\dfrac{1+\gamma+2\mu\xi+\nu_1\sigma_1+\nu_2\sigma_2+2\nu_1\eta}{2\nu_2}\right)}{\Gamma\left(1+\sigma_2-\dfrac{1+\gamma+2\mu\xi+\nu_1\sigma_1+\nu_2\sigma_2+2\nu_1\eta}{2\nu_2}\right)}$$

(6.106)

or

$$N_{79} = \frac{\sqrt{\pi}}{2\mu} \frac{1}{2^{m-1}} \left(\frac{a}{2}\right)^{-\frac{1+\gamma}{\mu}} \left(\frac{b_1/2}{(a/2)^{\nu_1/\mu}}\right)^{\sigma_1} \left(\frac{b_2/2}{(a/2)^{\nu_2/\mu}}\right)^{\sigma_2}$$

$$\times \frac{1}{(2i)^2} \iint\limits_{-\beta+i\infty}^{-\beta-i\infty} d\eta_1 d\eta_2 \frac{\left(\dfrac{b_1}{2}\Big/(a/2)^{\nu_1/\mu}\right)^{2\eta_1} \left(\dfrac{b_2}{2}\Big/(a/2)^{\nu_2/\mu}\right)^{2\eta_2}}{\sin\pi\eta_1 \; \sin\pi\eta_2 \; \Gamma(1+\eta_1) \; \Gamma(1+\eta_2)}$$

$$\times \frac{N_m'\left(\xi=-\dfrac{1+\gamma+\nu_1\sigma_1+\nu_2\sigma_2+2\nu_1\eta_1+2\nu_2\eta_2}{2\mu}\right)}{\Gamma(1+\sigma_1+\eta_1) \; \Gamma(1+\sigma_2+\eta_2)}$$

$$\times \frac{\Gamma\left(\dfrac{1+\gamma+\nu_1\sigma_1+\nu_2\sigma_2+2\nu_1\eta_1+2\nu_2\eta_2}{2\mu}\right)}{\Gamma\left(\dfrac{1}{2}-\dfrac{1+\gamma+\nu_1\sigma_1+\nu_2\sigma_2+2\nu_1\eta_1+2\nu_2\eta_2}{2\mu}\right)},$$

(6.107)

where $m = 1, 3, 5, 7, \ldots$ and $N_m'(\xi)$ is determined by the formula (1.36) in Chapter 1.

6.7.8 80th *General Formula*

$$N_{80} = \int\limits_0^\infty dx x^\gamma \, J_{\sigma_1}(b_1 x^{\nu_1}) \, J_{\sigma_2}(b_2 x^{\nu_2}) \left[\cos^q(ax^\mu) - 1 \right]$$

$$= \left(\frac{b_1}{2}\right)^{\sigma_1} \left(\frac{b_2}{2}\right)^{-\frac{\gamma+1+\nu_1\sigma_1}{\nu_2}} \frac{1}{2\nu_2} \frac{1}{2^{q-1}}$$

$$\times \frac{1}{2i} \int\limits_{\alpha+i\infty}^{\alpha-i\infty} d\xi \frac{\left[\frac{2a}{(b_2/2)^{\nu_1/\nu_2}}\right]^{2\xi}}{\sin \pi\xi \, \Gamma(1+2\xi)} \, I_q'(\xi)$$

$$\times \frac{1}{2i} \int\limits_{-\beta+i\infty}^{-\beta-i\infty} d\eta \frac{\left[\frac{b_1}{2} \Big/ (b_2/2)^{\nu_1/\nu_2}\right]^{2\eta}}{\sin \pi\eta \, \Gamma(1+\eta) \, \Gamma(1+\sigma_1+\eta)}$$

$$\times \frac{\Gamma\left(\dfrac{1+\gamma+2\mu\xi+\nu_1\sigma_1+\nu_2\sigma_2+2\nu_1\eta}{2\nu_2}\right)}{\Gamma\left(1+\sigma_2 - \dfrac{1+\gamma+2\mu\xi+\nu_1\sigma_1+\nu_2\sigma_2+2\nu_1\eta}{2\nu_2}\right)}$$

(6.108)

or

$$N_{80} = \frac{\sqrt{\pi}}{2\mu} \frac{1}{2^{q-1}} \left(\frac{a}{2}\right)^{-\frac{1+\gamma}{\mu}} \left(\frac{b_1/2}{a^{\nu_1/\mu}}\right)^{\sigma_1} \left(\frac{b_2/2}{a^{\nu_2/\mu}}\right)^{\sigma_2}$$

$$\times \frac{1}{(2i)^2} \iint\limits_{-\beta+i\infty}^{-\beta-i\infty} d\eta_1 d\eta_2 \frac{\left(\frac{b_1}{2}\Big/a^{\nu_1/\mu}\right)^{2\eta_1} \left(\frac{b_2}{2}\Big/a^{\nu_2/\mu}\right)^{2\eta_2}}{\sin \pi\eta_1 \, \sin \pi\eta_2 \, \Gamma(1+\eta_1) \, \Gamma(1+\eta_2)}$$

$$\times \frac{I_q'\left(\xi = -\dfrac{1+\gamma+\nu_1\sigma_1+\nu_2\sigma_2+2\nu_1\eta_1+2\nu_2\eta_2}{2\mu}\right)}{\Gamma(1+\sigma_1+\eta_1) \, \Gamma(1+\sigma_2+\eta_2)}$$

$$\times \frac{\Gamma\left(\dfrac{1+\gamma+\nu_1\sigma_1+\nu_2\sigma_2+2\nu_1\eta_1+2\nu_2\eta_2}{2\mu}\right)}{\Gamma\left(\dfrac{1}{2} - \dfrac{1+\gamma+\nu_1\sigma_1+\nu_2\sigma_2+2\nu_1\eta_1+2\nu_2\eta_2}{2\mu}\right)},$$

(6.109)

where $q = 2, \, 4, \, 6, \ldots$ and $I_q'(\xi)$ is derived from the formula (1.38) in Chapter 1.

6.8   Exercises 1

By using formulas (6.94)–(6.109), calculate the following integrals

$$S_{42} = \int_0^\infty dx \; J_\sigma(a\sqrt{x}) \, J_\sigma(b\sqrt{x}) \, \sin(cx), \tag{6.110}$$

$$S_{43} = \int_0^\infty dx \; J_\sigma(a\sqrt{x}) \, J_\sigma(b\sqrt{x}) \, \cos(cx), \tag{6.111}$$

$$S_{44} = \int_0^\infty dx\, x^{\sigma_2-\sigma_1-2} \, J_{\sigma_1}(ax) \, J_{\sigma_2}(bx) \, \sin(cx), \tag{6.112}$$

$$S_{45} = \int_0^\infty dx\, x^{\sigma_2-\sigma_1-1} \, J_{\sigma_1}(ax) \, J_{\sigma_2}(bx) \, \cos(cx), \tag{6.113}$$

$$S_{46} = \int_0^\infty dx\, x \, J_\sigma(bx) \, J_\sigma(cx) \, \sin(ax^2), \tag{6.114}$$

$$S_{47} = \int_0^\infty dx\, x \, J_\sigma(bx) \, J_\sigma(cx) \, \cos(ax^2). \tag{6.115}$$

Answers

$$S_{42} = \frac{1}{c} \, J_\sigma\left(\frac{ab}{2c}\right) \, \cos\left(\frac{a^2+b^2}{4c} - \frac{\sigma\pi}{2}\right),$$
where a, b, $c > 0$, $\operatorname{Re}\sigma > -2$.

$$S_{43} = \frac{1}{c} \, J_\sigma\left(\frac{ab}{2c}\right) \, \sin\left(\frac{a^2+b^2}{4c} - \frac{\sigma\pi}{2}\right),$$
where a, b, $c > 0$, $\operatorname{Re}\sigma > -1$.

$$S_{44} = 2^{\sigma_2-\sigma_1-1} \, a^{\sigma_1} \, b^{-\sigma_2} \, c \, \frac{\Gamma(\sigma_1)}{\Gamma(1+\sigma_2)},$$
where $0 < a$, b, $0 < c < b - a$, $0 < \operatorname{Re}\sigma_2 < \operatorname{Re}\sigma_1 + 3$.

$$S_{45} = 2^{\sigma_2-\sigma_1-1} \, a^{\sigma_1} \, b^{-\sigma_2} \, c \, \frac{\Gamma(\sigma_2)}{\Gamma(1+\sigma_1)},$$
where $0 < a$, b, $0 < c < b - a$, $0 < \operatorname{Re}\sigma_2 < \operatorname{Re}\sigma_1 + 2$.

$$S_{46} = \frac{1}{2a} \, \cos\left(\frac{b^2 + c^2}{4a} - \frac{\sigma\pi}{2}\right) J_\sigma\left(\frac{bc}{2a}\right),$$

where a, b, $c > 0$, Re $\sigma > -2$.

$$S_{47} = \frac{1}{2a} \, \sin\left(\frac{b^2 + c^2}{4a} - \frac{\sigma\pi}{2}\right) J_\sigma\left(\frac{bc}{2a}\right),$$

where a, b, $c > 0$, Re $\sigma > -1$.

6.9 Integrals Containing $J_\sigma(x)$, x^γ, Trigonometric and Exponential Functions

Here we use formulas like (5.41), (5.40), (6.54), (6.55), (6.69) and (6.70) and obtain universal formulas for the following integrals.

6.9.1 81st *General Formula*

$$N_{81} = \int_0^\infty dx \, x^\gamma \, J_\sigma(bx^\nu) \, \sin^q(ax^\nu) \, e^{-cx^\delta}$$

$$= \left(\frac{b}{2} \Big/ c^{\nu/\delta}\right)^\sigma \frac{1}{2i} \int_{-\beta'+i\infty}^{-\beta'-i\infty} d\eta \frac{\left(\dfrac{b/2}{c^{\nu/\delta}}\right)^{2\eta}}{\sin\pi\eta \, \Gamma(1+\eta) \, \Gamma(1+\sigma+\eta)} \tag{6.116}$$

$$\times \frac{1}{\delta} \, c^{-\frac{1+\gamma}{\delta}} \frac{1}{2^{q-1}} \frac{1}{2i} \int_{\alpha+i\infty}^{\alpha-i\infty} d\xi \frac{\left(\dfrac{2a}{c^{\nu/\delta}}\right)^{2\xi}}{\sin\pi\xi \, \Gamma(1+2\xi)} I_q(\xi)$$

$$\times \Gamma\left(\frac{1+\gamma+\nu\sigma+2\nu\eta+2\mu\xi}{\delta}\right)$$

or

$$N_{81} = \left(\frac{b/2}{a^{\nu/\mu}}\right)^{\sigma} \frac{\sqrt{\pi}}{2\mu} \frac{1}{2^{q-1}}$$

$$\times \frac{1}{2i} \int\limits_{-\beta'+i\infty}^{-\beta'-i\infty} d\eta \frac{\left(\dfrac{b/2}{a^{\nu/\mu}}\right)^{2\eta}}{sin\pi\eta\ \Gamma(1+\eta)\ \Gamma(1+\sigma+\eta)}\ a^{-\frac{1+\gamma}{\mu}}$$

$$\times \frac{1}{2i} \int\limits_{-\beta+i\infty}^{-\beta-i\infty} d\xi \frac{\left(c/a^{\delta/\mu}\right)^{\xi}}{\sin\pi\xi\ \Gamma(1+\xi)}\ I_q\left(\xi = -\frac{1+\gamma+\nu\sigma+2\nu\eta+\delta\xi}{2\mu}\right)$$

$$\times \frac{\Gamma\left(\dfrac{1+\gamma+\nu\sigma+2\nu\eta+\delta\xi}{2\mu}\right)}{\Gamma\left(\dfrac{1}{2} - \dfrac{1+\gamma+\nu\sigma+2\nu\eta+\delta\xi}{2\mu}\right)}$$

(6.117)

or

$$N_{81}$$

$$= \frac{1}{2^{q-1}} \frac{1}{2i} \int\limits_{\alpha+i\infty}^{\alpha-i\infty} d\xi \frac{\left[2a/c^{\mu/\delta}\right]^{2\xi}}{\sin\pi\xi\ \Gamma(1+2\xi)}\ \frac{I_q(\xi)\ c^{-\frac{1+\gamma}{\delta}}\ (b/2c^{\nu/\delta})^{\sigma}}{\delta}$$

$$\times \frac{1}{2i} \int\limits_{-\beta+i\infty}^{-\beta-i\infty} d\eta \frac{(b/2c^{\nu/\delta})^{2\eta}}{\sin\pi\eta\ \Gamma(1+\eta)\ \Gamma(1+\sigma+\eta)}$$

$$\times \Gamma\left(\frac{1+\gamma+2\mu\xi+\nu(\sigma+2\eta)}{\delta}\right)$$

(6.118)

or

$$
N_{81} = \frac{1}{2^{q-1}} \frac{1}{2i} \int\limits_{\alpha+i\infty}^{\alpha-i\infty} d\xi \frac{\left[\dfrac{2a}{(b/2)^{\mu/\nu}}\right]^{2\xi}}{\sin \pi\xi \; \Gamma(1+2\xi)} \frac{I_q(\xi)}{2\nu} \left(\frac{b}{2}\right)^{-\frac{1+\gamma}{\nu}}
$$

$$
\times \frac{1}{2i} \int\limits_{-\beta+i\infty}^{-\beta-i\infty} d\eta \frac{\left(\dfrac{c}{(b/2)^{\delta/\nu}}\right)^{2\eta}}{\sin \pi\eta \; \Gamma(1+\eta)}
$$

$$
\times \frac{\Gamma\left(\dfrac{1+\gamma+2\mu\xi+\delta\eta+\nu\sigma}{2\nu}\right)}{\Gamma\left(1+\sigma - \dfrac{1+\gamma+2\mu\xi+\delta\eta+\nu\sigma}{2\nu}\right)}
$$

(6.119)

or

$$
N_{81} = \frac{1}{2i} \int\limits_{-\beta+i\infty}^{-\beta-i\infty} d\eta \frac{\left(\dfrac{c}{(b/2)^{\delta/\nu}}\right)^{\eta}}{\sin \pi\eta \; \Gamma(1+\eta)} \frac{1}{2\nu} \left(\frac{b}{2}\right)^{-\frac{1+\gamma}{\nu}} \frac{1}{2^{q-1}}
$$

$$
\times \frac{1}{2i} \int\limits_{\alpha+i\infty}^{\alpha-i\infty} d\xi \frac{\left[\dfrac{2a}{(b/2)^{\mu/\nu}}\right]^{2\xi}}{\sin \pi\xi \; \Gamma(1+2\xi)} I_q(\xi)
$$

$$
\times \frac{\Gamma\left(\dfrac{1+\gamma+\nu\sigma+\delta\eta+2\mu\xi}{2\nu}\right)}{\Gamma\left(1+\sigma - \dfrac{1+\gamma+\nu\sigma+\delta\eta+2\mu\xi}{2\nu}\right)}
$$

(6.120)

or

$$
N_{81} = \frac{1}{2i} \int_{-\beta+i\infty}^{-\beta-i\infty} d\xi \frac{\left[c/a^{\delta/\mu} \right]^{\xi}}{\sin \pi \xi \ \Gamma(1+\xi)} \frac{\sqrt{\pi}}{2\mu} \frac{1}{2^{q-1}} a^{-\frac{1+\gamma}{\mu}} \left(\frac{b/2}{a^{\nu/\mu}} \right)^{\sigma}
$$

$$
\times \frac{1}{2i} \int_{-\beta'+i\infty}^{-\beta'-i\infty} d\eta \frac{\left[\frac{b}{2} / a^{\nu/\mu} \right]^{2\eta}}{\sin \pi \eta \ \Gamma(1+\eta) \ \Gamma(1+\sigma+\eta)}
$$

$$
\times I_q \left(\xi = -\frac{1+\gamma+\delta\xi+\nu(\sigma+2\eta)}{2\mu} \right)
$$

$$
\times \frac{\Gamma\left(\dfrac{1+\gamma+\delta\xi+\nu(\sigma+2\eta)}{2\mu} \right)}{\Gamma\left(\dfrac{1}{2} - \dfrac{1+\gamma+\delta\xi+\nu(\sigma+2\eta)}{2\mu} \right)}.
$$

(6.121)

From these formulas we see that

$$\text{the formula} \quad (6.116) = \text{the formula} \quad (6.118),$$

$$\text{the formula} \quad (6.117) = \text{the formula} \quad (6.121)$$

and

$$\text{the formula} \quad (6.119) = \text{the formula} \quad (6.120)$$

as expected. In these formulas, $q = 2, 4, 6, \ldots$ and I_q is given by expression (1.32) in Chapter 1.

6.9.2 82nd *General Formula*

$$
N_{82} = \int_0^\infty dx \, x^\gamma \, J_\sigma(bx^\nu) \, \sin^m(ax^\mu) \, e^{-cx^\delta}
$$

$$
= \left(\frac{b/2}{c^{\nu/\delta}} \right)^{\sigma} \frac{1}{2i} \int_{-\beta'+i\infty}^{-\beta'-i\infty} d\eta \frac{\left[\frac{b}{2} / c^{\nu/\delta} \right]^{2\eta}}{\sin \pi \eta \ \Gamma(1+\eta) \ \Gamma(1+\sigma+\eta)} \frac{1}{\delta}
$$

(6.122)

$$
\times c^{-\frac{1+\gamma+\mu}{\delta}} \frac{a}{2^{m-1}} \frac{1}{2i} \int_{-\beta+i\infty}^{-\beta-i\infty} d\xi \frac{(a/c^{\nu/\delta})^{2\xi}}{\sin \pi \xi \ \Gamma(2+2\xi)} N_m(\xi)
$$

$$
\times \Gamma\left(\frac{1+\gamma+\mu+\nu\delta+2\nu\eta+2\mu\xi}{\delta} \right)
$$

or

$$N_{82} = \left(\frac{b/2}{(a/2)^{\nu/\mu}} \right)^{\sigma} \frac{\sqrt{\pi}}{2\mu} \frac{1}{2^{m-1}}$$

$$\times \frac{1}{2i} \int\limits_{-\beta'+i\infty}^{-\beta'-i\infty} d\eta \frac{\left[\frac{b/2}{(a/2)^{\nu/\mu}} \right]^{2\eta}}{\sin \pi\eta \; \Gamma(1+\eta) \; \Gamma(1+\sigma+\eta)}$$

$$\times \left(\frac{a}{2} \right)^{-\frac{1+\gamma}{\mu}} \frac{1}{2i} \int\limits_{-\beta+i\infty}^{-\beta-i\infty} d\xi \frac{\left[\frac{c}{(a/2)^{\delta/\mu}} \right]^{\xi}}{\sin \pi\xi \; \Gamma(1+\xi)} \qquad (6.123)$$

$$\times N_m \left(\xi = -\frac{1}{2} \left[1 + \frac{1+\gamma+\nu\sigma+2\nu\eta+\delta\xi}{\mu} \right] \right)$$

$$\times \frac{\Gamma \left(\frac{1}{2} \left[1 + \frac{1+\gamma+\nu\sigma+2\nu\eta+\delta\xi}{\mu} \right] \right)}{\Gamma \left(1 - \frac{1+\gamma+\nu\sigma+2\nu\eta+\delta\xi}{2\mu} \right)}$$

or

$$N_{82} = \frac{a}{2^{m-1}} \frac{1}{2i} \int\limits_{-\beta+i\infty}^{-\beta-i\infty} d\xi \frac{\left[\frac{a}{(b/2)^{\mu/\nu}} \right]^{2\xi}}{\sin \pi\xi \; \Gamma(2+2\xi)} \frac{N_m(\xi)}{2\nu} \left(\frac{b}{2} \right)^{-\frac{1+\gamma+\mu}{\nu}}$$

$$\times \frac{1}{2i} \int\limits_{-\beta'+i\infty}^{-\beta'-i\infty} d\eta \frac{\left[\frac{c}{(b/2)^{\delta/\nu}} \right]^{\eta}}{\sin \pi\eta \; \Gamma(1+\eta)} \qquad (6.124)$$

$$\times \frac{\Gamma \left(\frac{1+\gamma+\mu+2\mu\xi+\delta\eta+\nu\sigma}{2\nu} \right)}{\Gamma \left(1+\sigma - \frac{1+\gamma+\mu+2\mu\xi+\delta\eta+\nu\sigma}{2\nu} \right)},$$

where $m = 1, 3, 5, 7, \ldots$ and $N_m(\xi)$ is defined by the expressions (1.34) in Chapter 1.

6.9.3 83rd General Formula

$$N_{83} = \int\limits_0^\infty dx\, x^\gamma\, J_\sigma(bx^\nu) \cos^m(ax^\mu)\, e^{-cx^\delta}$$

$$= \left(\frac{b/2}{c^{\nu/\delta}}\right)^\sigma \frac{1}{2i} \int\limits_{-\beta'+i\infty}^{-\beta'-i\infty} d\eta\, \frac{\left[\frac{b}{2}\Big/c^{\nu/\delta}\right]^{2\eta}}{\sin\pi\eta\ \Gamma(1+\eta)\ \Gamma(1+\sigma+\eta)} \frac{1}{\delta}$$

$$\times c^{-\frac{1+\gamma}{\delta}} \frac{1}{2^{m-1}} \frac{1}{2i} \int\limits_{-\beta+i\infty}^{-\beta-i\infty} d\xi\, \frac{\left(a/c^{\nu/\delta}\right)^{2\xi}}{\sin\pi\xi\ \Gamma(1+2\xi)}\, N_q'(\xi)$$

$$\times \Gamma\left(\frac{1+\gamma+\nu\delta+2\nu\eta+2\mu\xi}{\delta}\right)$$

(6.125)

or

$$N_{83} = \left(\frac{b/2}{(a/2)^{\nu/\mu}}\right)^\sigma \frac{\sqrt{\pi}}{2\mu} \frac{1}{2^{m-1}}$$

$$\times \frac{1}{2i} \int\limits_{-\beta'+i\infty}^{-\beta'-i\infty} d\eta\, \frac{\left[\frac{b/2}{(a/2)^{\nu/\mu}}\right]^{2\eta}}{\sin\pi\eta\ \Gamma(1+\eta)\ \Gamma(1+\sigma+\eta)}$$

$$\times \left(\frac{a}{2}\right)^{-\frac{1+\gamma}{\mu}} \frac{1}{2i} \int\limits_{-\beta+i\infty}^{-\beta-i\infty} d\xi\, \frac{\left[\frac{c}{(a/2)^{\delta/\mu}}\right]^\xi}{\sin\pi\xi\ \Gamma(1+\xi)}$$

$$\times N_q'\left(\xi = -\frac{1+\gamma+\nu\sigma+2\nu\eta+\delta\xi}{2\mu}\right)$$

$$\times \frac{\Gamma\left(\dfrac{1+\gamma+\nu\sigma+2\nu\eta+\delta\xi}{2\mu}\right)}{\Gamma\left(\dfrac{1}{2}-\dfrac{1+\gamma+\nu\sigma+2\nu\eta+\delta\xi}{2\mu}\right)}$$

(6.126)

or

$$N_{83}$$

$$= \frac{1}{2^{m-1}} \frac{1}{2i} \int\limits_{-\beta+i\infty}^{-\beta-i\infty} d\xi \frac{\left[\dfrac{a}{(b/2)^{\mu/\nu}}\right]^{2\xi}}{\sin \pi\xi \; \Gamma(1+2\xi)} \frac{N'_m(\xi)}{2\nu} \left(\frac{b}{2}\right)^{-\frac{1+\gamma}{\nu}}$$

$$\times \frac{1}{2i} \int\limits_{-\beta+i\infty}^{-\beta-i\infty} d\eta \frac{\left[\dfrac{c}{(b/2)^{\delta/\nu}}\right]^{\eta}}{\sin \pi\eta \; \Gamma(1+\eta)} \tag{6.127}$$

$$\times \frac{\Gamma\left(\dfrac{1+\gamma+2\mu\xi+\delta\eta+\nu\sigma}{2\nu}\right)}{\Gamma\left(1+\sigma-\dfrac{1+\gamma+2\mu\xi+\delta\eta+\nu\sigma}{2\nu}\right)}.$$

6.9.4 84th *General Formula*

$$N_{84}$$

$$= \int\limits_{0}^{\infty} dx\, x^{\gamma}\; J_{\sigma}(bx^{\nu}) \left[\cos^{q}(ax^{\mu}) - 1\right] e^{-cx^{\delta}}$$

$$= \left(\frac{b/2}{c^{\nu/\delta}}\right)^{\sigma} \frac{1}{2i} \int\limits_{-\beta'+i\infty}^{-\beta'-i\infty} d\eta \frac{\left[\dfrac{b}{2}\Big/c^{\nu/\delta}\right]^{2\eta}}{\sin \pi\eta \; \Gamma(1+\eta)\;\Gamma(1+\sigma+\eta)} \frac{1}{\delta} \tag{6.128}$$

$$\times c^{-\frac{1+\gamma}{\delta}} \frac{1}{2^{q-1}} \frac{1}{2i} \int\limits_{\alpha+i\infty}^{\alpha-i\infty} d\xi \frac{\left(2a/c^{\nu/\delta}\right)^{2\xi}}{\sin \pi\xi \; \Gamma(1+2\xi)} I'_q(\xi)$$

$$\times \Gamma\left(\frac{1+\gamma+\nu\delta+2\nu\eta+2\mu\xi}{\delta}\right)$$

or

$$N_{84} = \left(\frac{b/2}{a^{\nu/\mu}}\right)^{\sigma} \frac{\sqrt{\pi}}{2\mu} \frac{1}{2^{q-1}}$$

$$\times \frac{1}{2i} \int_{-\beta'+i\infty}^{-\beta'-i\infty} d\eta \frac{\left[\frac{b/2}{a^{\nu/\mu}}\right]^{2\eta}}{\sin \pi\eta \; \Gamma(1+\eta) \; \Gamma(1+\sigma+\eta)}$$

$$\times a^{-\frac{1+\gamma}{\mu}} \frac{1}{2i} \int_{-\beta+i\infty}^{-\beta-i\infty} d\xi \frac{\left[\frac{c}{a^{\delta/\mu}}\right]^{\xi}}{\sin \pi\xi \; \Gamma(1+\xi)}$$

$$\times I_q'\left(\xi = -\frac{1+\gamma+\nu\sigma+2\nu\eta+\delta\xi}{2\mu}\right)$$

$$\times \frac{\Gamma\left(\dfrac{1+\gamma+\nu\sigma+2\nu\eta+\delta\xi}{2\mu}\right)}{\Gamma\left(\dfrac{1}{2} - \dfrac{1+\gamma+\nu\sigma+2\nu\eta+\delta\xi}{2\mu}\right)}$$

$$(6.129)$$

or

$$N_{84} = \frac{1}{2^{q-1}} \frac{1}{2i} \int_{\alpha+i\infty}^{\alpha-i\infty} d\xi \frac{\left[\frac{2a}{(b/2)^{\mu/\nu}}\right]^{2\xi}}{\sin \pi\xi \; \Gamma(1+2\xi)} \frac{I_q'(\xi)}{2\nu} \left(\frac{b}{2}\right)^{-\frac{1+\gamma}{\nu}}$$

$$\times \frac{1}{2i} \int_{-\beta+i\infty}^{-\beta-i\infty} d\eta \frac{\left[\frac{c}{(b/2)^{\delta/\nu}}\right]^{\eta}}{\sin \pi\eta \; \Gamma(1+\eta)} \frac{\Gamma\left(\dfrac{1+\gamma+2\mu\xi+\delta\eta+\nu\sigma}{2\nu}\right)}{\Gamma\left(1+\sigma - \dfrac{1+\gamma+2\mu\xi+\delta\eta+\nu\sigma}{2\nu}\right)},$$

$$(6.130)$$

where $m = 1, 3, 5, 7, \ldots$, $q = 2, 4, 6, \ldots$ and $N_m'(\xi)$ and $I_q'(\xi)$ are determined by the relations (1.36) and (1.38), respectively.

6.10   Exercises 2

Taking into account the formulas (6.116)-(6.130), calculate the following integrals

$$S_{48} = \int\limits_0^\infty \frac{dx}{x} \, e^{-cx} \, J_1(bx) \, \sin(ax), \qquad (6.131)$$

$$S_{49} = \int\limits_0^\infty \frac{dx}{x} \, e^{-cx} \, J_1(bx) \, \cos(ax). \qquad (6.132)$$

Answers

$$S_{48} = \frac{a}{b}(1 - r),$$

$$\text{where } a^2 = \frac{b^2}{1 - r^2} - \frac{c^2}{r^2}, \ b > 0,$$

$$S_{49} = \arcsin\left(\frac{2a}{\sqrt{c^2 + (a + b)^2} + \sqrt{c^2 + (a - b)^2}}\right)$$

$$a > 0, \ c > b.$$

Chapter 7

Integrals Involving the Neumann Function $N_\sigma(x)$

7.1 Definition of the Neumann Function

The Bessel function of the second kind or the Neumann function, also denoted by $Y_\sigma(x)$, is defined as

$$N_\sigma(x) = \frac{J_\sigma(x)\,\cos(\sigma\pi) - J_{-\sigma}(x)}{\sin(\sigma\pi)}$$

$$= \frac{-2^{1+\sigma}}{\Gamma\left(\frac{1}{2}\right)\Gamma\left(-\sigma+\frac{1}{2}\right)x^\sigma} \int\limits_1^\infty dy\, \frac{\cos(xy)}{(y^2-1)^{\sigma+1/2}}, \qquad (7.1)$$

where σ is not equal to integers.

7.2 The Mellin Representation of $N_\sigma(x)$

$$N_\sigma(x) = \left[\sin(\sigma\pi)\right]^{-1} \left\{ \frac{1}{2i} \int\limits_{-\beta+i\infty}^{-\beta-i\infty} d\xi \frac{(x/2)^{2\xi}}{\sin\pi\xi\,\Gamma(1+\xi)} \right. \qquad (7.2)$$

$$\left. \times \left[\cos(\pi\sigma)\left(\frac{x}{2}\right)^\sigma \frac{1}{\Gamma(1+\xi+\sigma)} - \left(\frac{x}{2}\right)^{-\sigma} \frac{1}{\Gamma(1+\xi-\sigma)}\right]\right\}.$$

7.3 85th General Formula

$$N_{85} = \int\limits_0^\infty dx\, x^\gamma\, N_\sigma(bx^\nu)$$

$$= \frac{1}{\sin(\pi\sigma)} \left\{ \cos(\pi\sigma) \left[\frac{1}{2\nu} \left(\frac{b}{2}\right)^{-\frac{1+\gamma}{\nu}} \frac{\Gamma\left(\dfrac{1+\gamma+\nu\sigma}{2\nu}\right)}{\Gamma\left(1+\sigma - \dfrac{1+\gamma+\nu\sigma}{2\nu}\right)} \right] \right.$$

$$\left. - \left[\frac{1}{2\nu} \left(\frac{b}{2}\right)^{-\frac{1+\gamma}{\nu}} \frac{\Gamma\left(\dfrac{1+\gamma-\nu\sigma}{2\nu}\right)}{\Gamma\left(1-\sigma - \dfrac{1+\gamma-\nu\sigma}{2\nu}\right)} \right] \right\}. \tag{7.3}$$

7.4 86th General Formula

$$N_{86} = \int\limits_0^\infty dx \frac{x^\gamma}{\left[p + tx^{\text{æ}}\right]^\lambda}\, N_\sigma(bx^\nu)$$

$$= \frac{1}{\text{æ}} \frac{1}{\Gamma(\lambda)} \frac{1}{p^\lambda} \left(\frac{p}{t}\right)^{\frac{1+\gamma}{\text{æ}}} \frac{1}{\sin(\pi\sigma)} \frac{1}{2i} \int\limits_{-\beta+i\infty}^{-\beta-i\infty} d\xi \frac{\left(\dfrac{b}{2}\right)^{2\xi} \left(\dfrac{p}{t}\right)^{\frac{2\nu\xi}{\text{æ}}}}{\sin \pi\xi\, \Gamma(1+\xi)}$$

$$\times \left\{ \cos(\pi\sigma) \left(\frac{p}{t}\right)^{\frac{\nu\sigma}{\text{æ}}} \frac{(b/2)^\sigma}{\Gamma(1+\sigma+\xi)} \Gamma\left(\frac{1+\gamma+\nu(\sigma+2\xi)}{\text{æ}}\right) \right.$$

$$\times \Gamma\left(\lambda - \frac{1+\gamma+\nu(\sigma+2\xi)}{\text{æ}}\right) - \left(\frac{p}{t}\right)^{-\frac{\nu\sigma}{\text{æ}}} \frac{(b/2)^{-\sigma}}{\Gamma(1-\sigma+\xi)}$$

$$\left. \times \Gamma\left(\frac{1+\gamma-\nu\sigma+2\nu\xi}{\text{æ}}\right) \Gamma\left(\lambda - \frac{1+\gamma-\nu\sigma+2\nu\xi}{\text{æ}}\right) \right\}. \tag{7.4}$$

7.5 87th General Formula

$$N_{87} = \int_0^\infty dx\, x^\gamma\, e^{-ax^\mu}\, N_\sigma(bx^\nu)$$

$$= \frac{1}{\mu}\, a^{-\frac{1+\gamma}{\mu}}\, \frac{1}{\sin(\pi\sigma)}\, \frac{1}{2i} \int_{-\beta+i\infty}^{-\beta-i\infty} d\xi \frac{\left(\dfrac{b}{2a^{\nu/\mu}}\right)^{2\xi}}{\sin \pi\xi\, \Gamma(1+\xi)}$$

$$\times \left\{ \frac{\left(\dfrac{b}{2a^{\nu/\mu}}\right)^\sigma}{\Gamma(1+\sigma+\xi)}\, \Gamma\left(\frac{1+\gamma+\nu(\sigma+2\xi)}{\mu}\right)\, \cos(\pi\sigma) \right.$$

$$\left. - \frac{\left(\dfrac{b}{2a^{\nu/\mu}}\right)^{-\sigma}}{\Gamma(1-\sigma+\xi)}\, \Gamma\left(\frac{1+\gamma-\nu\sigma+2\nu\xi}{\mu}\right) \right\}$$

$$\tag{7.5}$$

or

$$N_{87} = \frac{1}{2\nu}\, \left(\frac{b}{2}\right)^{-\frac{1+\gamma}{\nu}}\, \frac{1}{\sin(\pi\sigma)}\, \frac{1}{2i} \int_{-\beta+i\infty}^{-\beta-i\infty} d\eta \frac{\left[\dfrac{a}{(b/2)^{\mu/\nu}}\right]^\eta}{\sin \pi\eta\, \Gamma(1+\eta)}$$

$$\times \left\{ \frac{\cos(\pi\sigma)\, \Gamma\left(\dfrac{1+\gamma+\mu\eta+\nu\sigma}{2\nu}\right)}{\Gamma\left(1+\sigma-\dfrac{1+\gamma+\mu\eta+\nu\sigma}{2\nu}\right)} - \frac{\Gamma\left(\dfrac{1+\gamma+\mu\eta-\nu\sigma}{2\nu}\right)}{\Gamma\left(1-\sigma-\dfrac{1+\gamma+\mu\eta-\nu\sigma}{2\nu}\right)} \right\}.$$

$$\tag{7.6}$$

7.6 88th General Formula

$$N_{88} = \int\limits_0^\infty dx x^\gamma \, N_\sigma(bx^\nu) \, \sin^q(ax^\mu)$$

$$= \frac{1}{2\nu} \left(\frac{b}{2}\right)^{-\frac{1+\gamma}{\nu}} \frac{1}{2^{q-1}} \frac{1}{\sin(\pi\sigma)} \frac{1}{2i} \int\limits_{\alpha+i\infty}^{\alpha-i\infty} d\xi \frac{\left[\dfrac{2a}{(b/2)^{\mu/\nu}}\right]^{2\xi}}{\sin\pi\xi \, \Gamma(1+2\xi)} \, I_q(\xi)$$

$$\times \left\{ \frac{\cos(\pi\sigma) \, \Gamma\left(\dfrac{1+\gamma+\nu\sigma+2\mu\xi}{2\nu}\right)}{\Gamma\left(1+\sigma - \dfrac{1+\gamma+\nu\sigma+2\mu\xi}{2\nu}\right)} - \frac{\Gamma\left(\dfrac{1+\gamma-\nu\sigma+2\mu\xi}{2\nu}\right)}{\Gamma\left(1-\sigma - \dfrac{1+\gamma-\nu\sigma+2\mu\xi}{2\nu}\right)} \right\}$$

(7.7)

or

$$N_{88} = \frac{\sqrt{\pi}}{2\mu} \frac{1}{2^{q-1}} a^{-\frac{1+\gamma}{\mu}} \frac{1}{\sin(\pi\sigma)} \frac{1}{2i} \int\limits_{-\beta+i\infty}^{-\beta-i\infty} d\eta \frac{\left[\dfrac{b/2}{a^{\nu/\mu}}\right]^{2\eta}}{\sin\pi\eta \, \Gamma(1+\eta)}$$

$$\times \left\{ \frac{\cos(\pi\sigma) \left(\dfrac{b/2}{a^{\nu/\mu}}\right)^\sigma}{\Gamma\left(\dfrac{1}{2} - \dfrac{1+\gamma+\nu\sigma+2\nu\eta}{2\mu}\right)} \frac{I_q\left(\xi = -\dfrac{\gamma+1+\nu(\sigma+2\eta)}{2\mu}\right)}{\Gamma(1+\sigma+\eta)} \right.$$

$$\times \Gamma\left(\dfrac{1+\gamma+\nu(\sigma+2\eta)}{2\mu}\right) - \frac{\left(\dfrac{b/2}{a^{\nu/\mu}}\right)^{-\sigma}}{\Gamma\left(\dfrac{1}{2} - \dfrac{1+\gamma-\nu\sigma+2\nu\eta}{2\mu}\right)}$$

$$\left. \times \frac{I_q\left(\xi = -\dfrac{1+\gamma-\nu\sigma+2\nu\eta}{2\mu}\right)}{\Gamma(1-\sigma+\eta)} \Gamma\left(\dfrac{1+\gamma-\nu\sigma+2\nu\eta}{2\mu}\right) \right\},$$

(7.8)

where $I_q(\xi)$ ($q = 2, 4, 6, \ldots$) is given by (1.32) in Chapter 1.

7.7 89th General Formula

$$N_{89} = \int_0^\infty dx\, x^\gamma \, N_\sigma(bx^\nu) \, \sin^m(ax^\mu) = \frac{1}{2\nu} \left(\frac{b}{2}\right)^{-\frac{1+\gamma+\mu}{\nu}}$$

$$\times \frac{a}{2^{m-1}} \frac{1}{\sin(\pi\sigma)} \frac{1}{2i} \int_{-\beta+i\infty}^{-\beta-i\infty} d\xi \frac{\left[\dfrac{a}{(b/2)^{\mu/\nu}}\right]^{2\xi}}{\sin \pi\xi \, \Gamma(2+2\xi)} N_m(\xi)$$

$$\times \left\{ \frac{\cos(\pi\sigma)\, \Gamma\left(\dfrac{1+\gamma+\mu+\nu\sigma+2\mu\xi}{2\nu}\right)}{\Gamma\left(1+\sigma - \dfrac{1+\gamma+\mu+\nu\sigma+2\mu\xi}{2\nu}\right)} \right.$$

$$\left. - \frac{\Gamma\left(\dfrac{1+\gamma+\mu-\nu\sigma+2\mu\xi}{2\nu}\right)}{\Gamma\left(1-\sigma - \dfrac{1+\gamma+\mu-\nu\sigma+2\mu\xi}{2\nu}\right)} \right\}$$

(7.9)

or

$$N_{89} = \frac{\sqrt{\pi}}{2\mu} \left(\frac{a}{2}\right)^{-\frac{1+\gamma}{\mu}} \frac{1}{2^{m-1}} \frac{1}{\sin(\pi\sigma)} \frac{1}{2i} \int_{-\beta+i\infty}^{-\beta-i\infty} d\eta \frac{\left[\dfrac{b/2}{(a/2)^{\nu/\mu}}\right]^{2\eta}}{\sin \pi\eta \, \Gamma(1+\eta)}$$

$$\times \left\{ \frac{\cos(\pi\sigma) \left(\dfrac{b/2}{(a/2)^{\nu/\mu}}\right)^\sigma N_m\left(\xi = -\dfrac{1}{2}\left[1 + \dfrac{\gamma+1+\nu(\sigma+2\eta)}{\mu}\right]\right)}{\Gamma\left(1 - \dfrac{1+\gamma+\nu\sigma+2\nu\eta}{2\mu}\right) \qquad \Gamma(1+\sigma+\eta)} \right.$$

$$\times \Gamma\left(\frac{1}{2}\left[1 + \frac{1+\gamma+\nu(\sigma+2\eta)}{\mu}\right]\right) - \frac{\left(\dfrac{b/2}{(a/2)^{\nu/\mu}}\right)^{-\sigma}}{\Gamma\left(1 - \dfrac{1+\gamma-\nu\sigma+2\nu\eta}{2\mu}\right)}$$

$$\times \frac{N_m\left(\xi = -\dfrac{1}{2}\left[1 + \dfrac{1+\gamma-\nu\sigma+2\nu\eta}{\mu}\right]\right)}{\Gamma(1-\sigma+\eta)}$$

$$\left. \times \Gamma\left(\frac{1}{2}\left[1 + \frac{1+\gamma-\nu\sigma+2\nu\eta}{\mu}\right]\right) \right\},$$

(7.10)

where $N_m(\xi)$ ($m = 1, 3, 5, 7, \ldots$) is determined by (1.34) in Chapter 1.

7.8   90th General Formula

$$N_{90} = \int\limits_{0}^{\infty} dx\, x^{\gamma}\, N_{\sigma}(bx^{\nu})\, \cos^{m}(ax^{\mu})$$

$$= \frac{1}{2\nu}\left(\frac{b}{2}\right)^{-\frac{1+\gamma}{\nu}} \frac{1}{2^{m-1}}\frac{1}{2i} \int\limits_{-\beta+i\infty}^{-\beta-i\infty} d\xi\, \frac{\left[\dfrac{a}{(b/2)^{\mu/\nu}}\right]^{2\xi}}{\sin\pi\xi\, \Gamma(1+2\xi)}\, N_{m}'(\xi)$$

$$\times \frac{1}{\sin(\pi\sigma)}\left\{ \frac{\cos(\pi\sigma)\, \Gamma\left(\dfrac{1+\gamma+\nu\sigma+2\mu\xi}{2\nu}\right)}{\Gamma\left(1+\sigma-\dfrac{1+\gamma+\nu\sigma+2\mu\xi}{2\nu}\right)} \right.$$

$$\left. - \frac{\Gamma\left(\dfrac{1+\gamma-\nu\sigma+2\mu\xi}{2\nu}\right)}{\Gamma\left(1-\sigma-\dfrac{1+\gamma-\nu\sigma+2\mu\xi}{2\nu}\right)} \right\} \tag{7.11}$$

or

$$N_{90} = \frac{\sqrt{\pi}}{2\mu}\left(\frac{a}{2}\right)^{-\frac{1+\gamma}{\mu}} \frac{1}{2^{m-1}}\frac{1}{\sin(\pi\sigma)}\frac{1}{2i} \int\limits_{-\beta+i\infty}^{-\beta-i\infty} d\eta\, \frac{\left[\dfrac{b/2}{(a/2)^{\nu/\mu}}\right]^{2\eta}}{\sin\pi\eta\, \Gamma(1+\eta)}$$

$$\times \left\{ \frac{\cos(\pi\sigma)\left(\dfrac{b/2}{(a/2)^{\nu/\mu}}\right)^{\sigma}}{\Gamma\left(\dfrac{1}{2}-\dfrac{1+\gamma+\nu\sigma+2\nu\eta}{2\mu}\right)} \frac{N_{m}'\left(\xi=-\dfrac{\gamma+1+\nu(\sigma+2\eta)}{2\mu}\right)}{\Gamma(1+\sigma+\eta)} \right.$$

$$\times \Gamma\left(\dfrac{1+\gamma+\nu(\sigma+2\eta)}{2\mu}\right) - \frac{\left(\dfrac{b/2}{(a/2)^{\nu/\mu}}\right)^{-\sigma}}{\Gamma\left(\dfrac{1}{2}-\dfrac{1+\gamma-\nu\sigma+2\nu\eta}{2\mu}\right)} \tag{7.12}$$

$$\left. \times \frac{N_{m}'\left(\xi=-\dfrac{\gamma+1+\nu(-\sigma+2\eta)}{2\mu}\right)}{\Gamma(1-\sigma+\eta)} \Gamma\left(\dfrac{1+\gamma-\nu\sigma+2\nu\eta}{2\mu}\right) \right\},$$

where $N_{m}'(\xi)$ $(m = 1,\, 3,\, 5,\, 7,\ldots)$ is defined by (1.36) in Chapter 1.

7.9 91st General Formula

$$N_{91} = \int\limits_0^\infty dx\, x^\gamma\, N_\sigma(bx^\nu) \left[\cos^q(ax^\mu) - 1 \right]$$

$$= \frac{1}{2\nu} \left(\frac{b}{2} \right)^{-\frac{1+\gamma}{\nu}} \frac{1}{2^{q-1}} \frac{1}{2i} \int\limits_{\alpha+i\infty}^{\alpha-i\infty} d\xi \frac{\left[\dfrac{2a}{(b/2)^{\mu/\nu}} \right]^{2\xi}}{\sin \pi\xi\, \Gamma(1+2\xi)} I_q'(\xi)$$

$$\times \frac{1}{\sin(\pi\sigma)} \left\{ \frac{\cos(\pi\sigma)\, \Gamma\left(\dfrac{1+\gamma+\nu\sigma+2\mu\xi}{2\nu} \right)}{\Gamma\left(1+\sigma - \dfrac{1+\gamma+\nu\sigma+2\mu\xi}{2\nu} \right)} \right.$$

$$\left. - \frac{\Gamma\left(\dfrac{1+\gamma-\nu\sigma+2\mu\xi}{2\nu} \right)}{\Gamma\left(1-\sigma - \dfrac{1+\gamma-\nu\sigma+2\mu\xi}{2\nu} \right)} \right\} \tag{7.13}$$

or

$$N_{91} = \frac{\sqrt{\pi}}{2\mu}\, a^{-\frac{1+\gamma}{\mu}} \frac{1}{2^{q-1}} \frac{1}{\sin(\pi\sigma)} \frac{1}{2i} \int\limits_{-\beta+i\infty}^{-\beta-i\infty} d\eta \frac{\left[\dfrac{b/2}{a^{\nu/\mu}} \right]^{2\eta}}{\sin \pi\eta\, \Gamma(1+\eta)}$$

$$\times \left\{ \frac{\cos(\pi\sigma) \left(\dfrac{b/2}{a^{\nu/\mu}} \right)^\sigma}{\Gamma\left(\dfrac{1}{2} - \dfrac{1+\gamma+\nu\sigma+2\nu\eta}{2\mu} \right)} \frac{I_q'\left(\xi = -\dfrac{\gamma+1+\nu(\sigma+2\eta)}{2\mu} \right)}{\Gamma(1+\sigma+\eta)} \right.$$

$$\times \Gamma\left(\frac{1+\gamma+\nu(\sigma+2\eta)}{2\mu} \right) - \frac{\left(\dfrac{b/2}{a^{\nu/\mu}} \right)^{-\sigma}}{\Gamma\left(\dfrac{1}{2} - \dfrac{1+\gamma-\nu\sigma+2\nu\eta}{2\mu} \right)}$$

$$\left. \times \frac{I_m'\left(\xi = -\dfrac{\gamma+1+\nu(-\sigma+2\eta)}{2\mu} \right)}{\Gamma(1-\sigma+\eta)} \Gamma\left(\frac{1+\gamma-\nu\sigma+2\nu\eta}{2\mu} \right) \right\}, \tag{7.14}$$

where $I_q'(\xi)$ ($q = 2, 4, 6, \ldots$) is given by (1.38) in Chapter 1.

7.10   Calculation of Concrete Integrals

1. The formula (7.3) with $\gamma = 0$, $\nu = 1$ gives

$$S_{50} = \int_0^\infty dx \, N_\sigma(bx) = -\frac{1}{b} \frac{[1 - \cos(\pi\sigma)]}{\sin \pi\sigma} = -\frac{1}{b} \tan\left(\frac{\pi\sigma}{2}\right). \tag{7.15}$$

2. Assuming $\sigma = \frac{1}{4}$, $\gamma = -\frac{1}{2}$, $\nu = \frac{1}{4}$ in (7.3), one gets

$$S_{51} = \int_0^\infty dx \frac{1}{\sqrt{x}} \, N_{\frac{1}{4}}(bx^{1/4}) = \frac{\sqrt{2}}{b^2} \left(\frac{1}{\sqrt{2}} + 1\right). \tag{7.16}$$

3. We put $\gamma = 0$, $p \to a^2$, $t = 1$, $æ = 2$, $\lambda = 1$, $\sigma = 0$, $\nu = 1$ in (7.4) and obtain

$$S_{52} = \int_0^\infty dx \frac{N_0(bx)}{x^2 + a^2} = -\frac{1}{a} K_0(ab), \tag{7.17}$$

   where $a > 0$, $b > 0$.

4. Let $\gamma = \sigma$, $\nu = 1$, $\lambda = 1$, $t = -1$, $p \to a^2$ be in (7.4) then we have

$$S_{53} = \int_0^\infty dx x^\sigma \, N_\sigma(bx) \frac{1}{a^2 - x^2} = -\frac{\pi}{2} a^{\sigma-1} \, J_\sigma(ab), \tag{7.18}$$

   where $a, b > 0$, $-\frac{1}{2} < \text{Re } \sigma < \frac{5}{2}$.

5. Assuming $\gamma = 0$, $\mu = 1$, $\sigma = 0$, $\nu = 1$ in (7.5), one gets

$$S_{54} = \int_0^\infty dx \, e^{-ax} \, N_0(bx) = \frac{-2}{\pi\sqrt{a^2 + b^2}} \, \ln \frac{a + \sqrt{a^2 + b^2}}{b}. \tag{7.19}$$

6. The formulas (7.11) and (7.12) with $\gamma = 0$, $\sigma = 0$, $\nu = 1$, $m = 1$, $\mu = 1$ give

$$S_{55} = \int_0^\infty dx \, N_0(bx) \, \cos(ax) = \begin{cases} 0 & \text{if} \quad 0 < a < b \\[2mm] -\dfrac{1}{\sqrt{a^2 - b^2}} & \text{if} \quad 0 < b < a. \end{cases} \tag{7.20}$$

7. We put $\gamma = 0$, $\sigma = 0$, $\nu = 1$, $m = 1$, $\mu = 1$ in formulas (7.9) and (7.10) and obtain

$$S_{56} = \int_0^\infty dx \, N_0(bx) \, \sin(ax)$$

$$= \begin{cases} \dfrac{2 \arcsin\left(\frac{a}{b}\right)}{\pi\sqrt{b^2 - a^2}} & \text{if} \quad 0 < a < b \\[4mm] \dfrac{2}{\pi\sqrt{a^2 - b^2}} \, \ln\left[\dfrac{a}{b} - \sqrt{\dfrac{a^2}{b^2} - 1}\right] & \text{if} \quad 0 < b < a. \end{cases} \tag{7.21}$$

8. The formulas (7.9) and (7.10) with $\gamma = \lambda$, $\sigma = \lambda - 1$, $\nu = 1$, $m = 1$, $\mu = 1$ read

$$S_{57} = \int_0^\infty dx\, x^\lambda\, N_{\lambda-1}(bx)\, \sin(ax)$$

$$= \begin{cases} 0 & \text{if } 0 < a < b, \ |\mathrm{Re}\, \lambda| < \dfrac{1}{2} \\[2mm] \dfrac{2^\lambda \sqrt{\pi}\, b^{\lambda-1}\, a}{\Gamma\left(\frac{1}{2} - \lambda\right)}(a^2 - b^2)^{-\lambda-\frac{1}{2}} & \text{if } 0 < b < a, \ |\mathrm{Re}\, \lambda| < \dfrac{1}{2}. \end{cases} \tag{7.22}$$

9. While the formulas (7.11) and (7.12) with $\gamma = \sigma$, $\nu = 1$, $\mu = 1$, $m = 1$ lead

$$S_{58} = \int_0^\infty dx\, x^\sigma\, N_\sigma(bx)\, \cos(ax)$$

$$= \begin{cases} 0 & \text{if } 0 < a < b, \ |\mathrm{Re}\, \sigma| < \dfrac{1}{2} \\[2mm] \dfrac{-2^\sigma \sqrt{\pi}\, b^\nu}{\Gamma\left(\frac{1}{2} - \sigma\right)}(a^2 - b^2)^{-\sigma-\frac{1}{2}} & \text{if } 0 < b < a, \ |\mathrm{Re}\, \sigma| < \dfrac{1}{2}. \end{cases} \tag{7.23}$$

10. Assuming $\gamma = \dfrac{1}{2}$, $\sigma = \dfrac{1}{4}$, $\nu = 2$, $m = 1$, $\mu = 1$, $b \to b^2$ in (7.9), one gets

$$S_{59} = \int_0^\infty dx\, \sqrt{x}\, N_{\frac{1}{4}}(b^2 x^2)\, \sin(ax)$$

$$= -2^{-\frac{3}{2}} \sqrt{\pi a}\, b^{-2}\, \mathbf{H}_{\frac{1}{4}}\left(\frac{a^2}{4b^2}\right), \tag{7.24}$$

where

$$\mathbf{H}_\nu(x) = \sum_{m=0}^\infty (-1)^m \frac{\left(\frac{x}{2}\right)^{2m+\nu+1}}{\Gamma\left(m + \frac{3}{2}\right)\Gamma\left(\nu + m + \frac{3}{2}\right)}$$

is called the Struve function.

11. Similarly the formula (7.11) with $\gamma = \dfrac{1}{2}$, $\sigma = -\dfrac{1}{4}$, $b \to b^2$, $\nu = 2$, $m = 1$, $\mu = 1$ gives

$$S_{60} = \int_0^\infty dx\, \sqrt{x}\, N_{-\frac{1}{4}}(b^2 x^2)\, \cos(ax)$$

$$= -2^{-\frac{3}{2}} \sqrt{\pi a}\, b^{-2}\, \mathbf{H}_{-\frac{1}{4}}\left(\frac{a^2}{4b^2}\right). \tag{7.25}$$

Chapter 8

Integrals Containing Other Cylindrical and Special Functions

8.1 Integrals Involving Modified Bessel Function of the Second Kind

$$K_\mu(x) = \frac{\pi}{2\sin(\mu\pi)} \left[I_{-\mu}(x) - I_\mu(x) \right], \tag{8.1}$$

where

$$I_\lambda(x) = \frac{x^\lambda}{\Gamma\left(\frac{1}{2}\right) \Gamma\left(\lambda + \frac{1}{2}\right) 2^\lambda} \int_0^\pi dy \ \cosh[x \ \cos y] \ \sin^{2\lambda} y. \tag{8.2}$$

Sometimes $K_\delta(x)$ is called the Macdonald function.

8.1.1 *Mellin Representations of $K_\delta(x)$ and $I_\lambda(x)$*

$$K_\delta(x) = -\frac{\pi}{2} \left(\frac{x}{2}\right)^\delta \frac{1}{2i} \int_{-\beta+i\infty}^{-\beta-i\infty} d\xi \frac{\left(\frac{x}{2}\right)^{2\xi}}{\sin \pi\xi \ \Gamma(1+\xi)}$$

$$\times \frac{1}{\sin \pi(\delta+\xi) \ \Gamma(1+\delta+\xi)} \tag{8.3}$$

and

$$I_\lambda(x) = \frac{1}{2i} \int_{-\beta+i\infty}^{-\beta-i\infty} d\xi \frac{(-1)^\xi \left(\frac{x}{2}\right)^{\lambda+2\xi}}{\sin \pi\xi \ \Gamma(\lambda+\xi+1)}, \tag{8.4}$$

where

$$\sin\left[(\delta+\xi)\pi\right] \Gamma(1+\delta+\xi) = -\frac{\pi}{\Gamma(-\delta-\xi)}.$$

8.1.2   92nd *General Formula*

$$N_{92} = \int\limits_0^\infty dx x^\gamma \ K_\sigma(bx^\nu) = -\frac{\pi}{2} \frac{1}{2\nu} \left(\frac{b}{2}\right)^{-\frac{1+\gamma}{\nu}}$$

$$\times \frac{\Gamma\left(\dfrac{1+\gamma+\nu\sigma}{2\nu}\right)}{\Gamma\left(1+\sigma - \dfrac{1+\gamma+\nu\sigma}{2\nu}\right)} \frac{1}{\sin \pi \left(\delta - \dfrac{1+\gamma+\nu\sigma}{2\nu}\right)}.$$

(8.5)

8.1.3   93rd *General Formula*

$$N_{93} = \int\limits_0^\infty dx x^\gamma \ J_{\sigma_1}(ax^{\nu_1}) \ K_{\sigma_2}(bx^{\nu_2})$$

$$= -\frac{\pi}{2} \left(\frac{a}{2}\right)^{\sigma_1} \left(\frac{b}{2}\right)^{-\frac{1+\gamma+\nu_1\sigma_1}{\nu_2}} \frac{1}{2\nu_2} \frac{1}{2i} \int\limits_{-\beta+i\infty}^{-\beta-i\infty} d\xi \frac{\left[\dfrac{a/2}{(b/2)^{\nu_1/\nu_2}}\right]^{2\xi}}{\sin \pi\xi \ \Gamma(1+\xi)}$$

$$\times \frac{\Gamma\left(\dfrac{1+\gamma+\nu_1\sigma_1+\nu_2\sigma_2+2\nu_1\xi}{2\nu_2}\right)}{\Gamma\left(1+\sigma_2 - \dfrac{1+\gamma+\nu_1\sigma_1+\nu_2\sigma_2+2\nu_1\xi}{2\nu_2}\right)}$$

$$\times \frac{1}{\sin \pi \left(\sigma_2 - \dfrac{1+\gamma+\nu_1\sigma_1+\nu_2\sigma_2+2\nu_1\xi}{2\nu_2}\right)} \frac{1}{\Gamma(1+\sigma_1+\xi)}$$

(8.6)

or

$$N_{93} = -\frac{\pi}{2} \left(\frac{b}{2}\right)^{\sigma_2} \left(\frac{a}{2}\right)^{-\frac{1+\gamma+\nu_2\sigma_2}{\nu_1}} \frac{1}{2\nu_1} \frac{1}{2i} \int\limits_{-\beta+i\infty}^{-\beta-i\infty} d\xi \frac{\left[\dfrac{b/2}{(a/2)^{\nu_2/\nu_1}}\right]^{2\xi}}{\sin \pi\xi \ \Gamma(1+\xi)}$$

$$\times \frac{\Gamma\left(\dfrac{1+\gamma+\nu_1\sigma_1+\nu_2\sigma_2+2\nu_2\xi}{2\nu_1}\right)}{\Gamma\left(1+\sigma_1 - \dfrac{1+\gamma+\nu_1\sigma_1+\nu_2\sigma_2+2\nu_2\xi}{2\nu_1}\right)}$$

$$\times \frac{1}{\sin \pi (\sigma_2+\xi)} \frac{1}{\Gamma(1+\sigma_2+\xi)}.$$

(8.7)

8.1.4   94th *General Formula*

$$N_{94} = \int\limits_0^\infty dx \frac{x^\gamma}{\left[p + tx^{\text{æ}}\right]^\lambda} \, K_\sigma(bx^\nu)$$

$$= -\frac{\pi}{2} \frac{1}{\text{æ}} \frac{1}{\Gamma(\lambda)} \frac{1}{p^\lambda} \left(\frac{p}{t}\right)^{\frac{\gamma+1+\nu\sigma}{\text{æ}}}$$

$$\times \frac{1}{2i} \int\limits_{-\beta+i\infty}^{-\beta-i\infty} d\xi \frac{(b/2)^{\sigma+2\xi}}{\sin \pi\xi \, \Gamma(1+\xi) \, \Gamma(1+\sigma+\xi)} \frac{(p/t)^{\frac{2\nu\xi}{\text{æ}}}}{\sin \pi(\sigma+\xi)}$$

$$\times \Gamma\left(\frac{1+\gamma+\nu(\sigma+2\xi)}{\text{æ}}\right) \Gamma\left(\lambda - \frac{1+\gamma+\nu(\sigma+2\xi)}{\text{æ}}\right).$$

(8.8)

8.1.5   95th *General Formula*

$$N_{95} = \int\limits_0^\infty dx x^\gamma \, K_{\sigma_1}(ax^{\nu_1}) \, K_{\sigma_2}(bx^{\nu_2})$$

$$= \frac{\pi^2}{4} \left(\frac{a}{2}\right)^{\sigma_1} \left(\frac{b}{2}\right)^{-\frac{1+\gamma+\nu_1\sigma_1}{\nu_2}} \frac{1}{2\nu_2} \frac{1}{2i} \int\limits_{-\beta+i\infty}^{-\beta-i\infty} d\xi \frac{\left[\frac{a/2}{(b/2)^{\nu_1/\nu_2}}\right]^{2\xi}}{\sin \pi\xi \, \Gamma(1+\xi)}$$

$$\times \frac{\Gamma\left(\dfrac{1+\gamma+\nu_1\sigma_1+\nu_2\sigma_2+2\nu_1\xi}{2\nu_2}\right)}{\Gamma\left(1+\sigma_2 - \dfrac{1+\gamma+\nu_1\sigma_1+\nu_2\sigma_2+2\nu_1\xi}{2\nu_2}\right)}$$

$$\times \frac{1}{\sin \pi\left(\sigma_2 - \dfrac{1+\gamma+\nu_1\sigma_1+\nu_2\sigma_2+2\nu_1\xi}{2\nu_2}\right)} \frac{1}{\Gamma(1+\sigma_1+\xi)} \frac{1}{\sin \pi(\sigma_1+\xi)}$$

(8.9)

or

$$N_{95} = \frac{\pi^2}{4} \left(\frac{b}{2}\right)^{\sigma_2} \left(\frac{a}{2}\right)^{-\frac{1+\gamma+\nu_2\sigma_2}{\nu_1}} \frac{1}{2\nu_1} \frac{1}{2i} \int\limits_{-\beta+i\infty}^{-\beta-i\infty} d\xi \frac{\left[\frac{b/2}{(a/2)^{\nu_2/\nu_1}}\right]^{2\xi}}{\sin \pi\xi \; \Gamma(1+\xi)}$$

$$\times \frac{\Gamma\left(\dfrac{1+\gamma+\nu_1\sigma_1+\nu_2\sigma_2+2\nu_2\xi}{2\nu_1}\right)}{\Gamma\left(1+\sigma_1 - \dfrac{1+\gamma+\nu_1\sigma_1+\nu_2\sigma_2+2\nu_2\xi}{2\nu_1}\right)} \; \frac{1}{\sin \pi\,(\sigma_2+\xi)} \qquad (8.10)$$

$$\times \frac{1}{\Gamma(1+\sigma_2+\xi)} \; \frac{1}{\sin \pi\left(\sigma_1 - \dfrac{1+\gamma+\nu_1\sigma_1+\nu_2\sigma_2+2\nu_2\xi}{2\nu_1}\right)}.$$

8.1.6 96th General Formula

$$N_{96} = \int\limits_0^\infty dx\, x^\gamma \; e^{-ax^\mu} \; K_\sigma(bx^\nu)$$

$$= -\frac{\pi}{2}\, a^{-\frac{1+\gamma}{\mu}} \left(\frac{b/2}{a^{\nu/\mu}}\right)^\sigma \frac{1}{2i} \int\limits_{-\beta+i\infty}^{-\beta-i\infty} d\xi \frac{\left[\dfrac{b/2}{a^{\nu/\mu}}\right]^{2\xi}}{\sin \pi\xi \; \Gamma(1+\xi)} \qquad (8.11)$$

$$\times \frac{\Gamma\left(\dfrac{1+\gamma+\nu(\sigma+2\xi)}{\mu}\right)}{\Gamma(1+\sigma+\xi)\; \sin \pi(\sigma+\xi)}$$

or

$$N_{96} = -\frac{\pi}{2} \left(\frac{b}{2}\right)^{-\frac{1+\gamma}{\nu}} \frac{1}{2i} \int\limits_{-\beta'+i\infty}^{-\beta'-i\infty} d\eta \frac{\left[\dfrac{a}{(b/2)^{\mu/\nu}}\right]^\eta}{\sin \pi\eta \; \Gamma(1+\eta)}$$

$$\times \frac{\Gamma\left(\dfrac{1+\gamma+\mu\eta+\nu\sigma}{2\nu}\right)}{\Gamma\left(1+\sigma - \dfrac{1+\gamma+\mu\eta+\nu\sigma}{2\nu}\right)} \; \frac{1}{\sin \pi\left(\sigma - \dfrac{1+\gamma+\mu\eta+\nu\sigma}{2\nu}\right)}. \qquad (8.12)$$

8.1.7   97th *General Formula*

$$N_{97} = \int\limits_0^\infty dx\, x^\gamma\, K_\sigma(bx^\nu)\, \sin^q(ax^\mu)$$

$$= -\frac{\pi}{2}\frac{1}{2\nu}\left(\frac{b}{2}\right)^{-\frac{1+\gamma}{\nu}}\frac{1}{2^{q-1}}\frac{1}{2i}\int\limits_{\alpha+i\infty}^{\alpha-i\infty} d\xi \frac{\left[\dfrac{2a}{(b/2)^{\mu/\nu}}\right]^{2\xi}}{\sin\pi\xi\,\Gamma(1+2\xi)}\, I_q(\xi)$$

$$\times \frac{\Gamma\left(\dfrac{1+\gamma+\nu\sigma+2\mu\xi}{2\nu}\right)}{\Gamma\left(1+\sigma-\dfrac{1+\gamma+\nu\sigma+2\mu\xi}{2\nu}\right)}\frac{1}{\sin\pi\left(\sigma-\dfrac{1+\gamma+\nu\sigma+2\mu\xi}{2\nu}\right)}$$

(8.13)

or

$$N_{97} = -\frac{\pi}{2}\frac{\sqrt{\pi}}{2\mu}\frac{1}{2^{q-1}}\, a^{-\frac{1+\gamma}{\mu}}\left(\frac{b/2}{a^{\nu/\mu}}\right)^\sigma$$

$$\times\frac{1}{2i}\int\limits_{-\beta+i\infty}^{-\beta-i\infty} d\eta \frac{\left[\dfrac{b/2}{a^{\nu/\mu}}\right]^{2\eta}}{\sin\pi\eta\,\Gamma(1+\eta)\,\Gamma(1+\sigma+\eta)}\frac{1}{\sin\pi(\sigma+\eta)}$$

$$\times I_q\left(\xi=-\frac{1+\gamma+\nu(\sigma+2\eta)}{2\mu}\right)\frac{\Gamma\left(\dfrac{1+\gamma+\nu(\sigma+2\eta)}{2\nu}\right)}{\Gamma\left(\dfrac{1}{2}-\dfrac{1+\gamma+\nu(\sigma+2\eta)}{2\mu}\right)},$$

(8.14)

where $I_q(\xi)$ $(q = 2,\, 4,\, 6, \ldots)$ is defined by (1.32) in Chapter 1.

8.1.8   98th *General Formula*

$$N_{98} = \int_0^\infty dx\, x^\gamma\, K_\sigma(bx^\nu)\, \sin^m(ax^\mu)$$

$$= -\frac{\pi}{2} \left(\frac{b}{2}\right)^{-\frac{1+\gamma+\mu}{\nu}} \frac{a}{2^{m-1}} \frac{1}{2i} \int_{-\beta+i\infty}^{-\beta-i\infty} d\xi \frac{\left[\dfrac{a}{(b/2)^{\mu/\nu}}\right]^{2\xi}}{\sin \pi\xi\, \Gamma(2+2\xi)}\, N_m(\xi)$$

$$\times \frac{\Gamma\left(\dfrac{1+\gamma+\mu+\nu\sigma+2\mu\xi}{2\nu}\right)}{\Gamma\left(1+\sigma-\dfrac{1+\gamma+\mu+\nu\sigma+2\mu\xi}{2\nu}\right)\sin \pi\left(\sigma-\dfrac{1+\gamma+\mu+\nu\sigma+2\mu\xi}{2\nu}\right)}$$

(8.15)

or

$$N_{98} = -\frac{\pi}{2} \frac{\sqrt{\pi}}{2\mu} \frac{1}{2^{m-1}} \left(\frac{a}{2}\right)^{-\frac{1+\gamma}{\mu}} \left(\frac{b/2}{(a/2)^{\nu/\mu}}\right)^\sigma$$

$$\times \frac{1}{2i} \int_{-\beta+i\infty}^{-\beta-i\infty} d\eta \frac{\left[\dfrac{b/2}{(a/2)^{\nu/\mu}}\right]^{2\eta}}{\sin \pi\eta\, \Gamma(1+\eta)\, \Gamma(1+\sigma+\eta)\, \sin \pi(\sigma+\eta)}$$

$$\times N_m\left(\xi = -\frac{1}{2}\left[1 + \frac{1+\gamma+\nu(\sigma+2\eta)}{\mu}\right]\right)$$

$$\times \frac{\Gamma\left(\dfrac{1}{2}\left[1+\dfrac{1+\gamma+\nu(\sigma+2\eta)}{\mu}\right]\right)}{\Gamma\left(1-\dfrac{1+\gamma+\nu(\sigma+2\eta)}{2\mu}\right)},$$

(8.16)

where $N_m(\xi)$ ($m = 1, 3, 5, \ldots$) is given by (1.34) in Chapter 1.

8.1.9 99th *General Formula*

$$N_{99} = \int\limits_0^\infty dx\, x^\gamma\, K_\sigma(bx^\nu)\, \cos^m(ax^\mu)$$

$$= -\frac{\pi}{2}\frac{1}{2\nu}\left(\frac{b}{2}\right)^{-\frac{1+\gamma}{\nu}}\frac{1}{2^{m-1}}\frac{1}{2i}\int\limits_{-\beta+i\infty}^{-\beta-i\infty} d\xi \frac{\left[\dfrac{a}{(b/2)^{\mu/\nu}}\right]^{2\xi}}{\sin\pi\xi\,\Gamma(1+2\xi)}\, N'_m(\xi)$$

$$\times \frac{\Gamma\left(\dfrac{1+\gamma+\nu\sigma+2\mu\xi}{2\nu}\right)}{\Gamma\left(1+\sigma-\dfrac{1+\gamma+\nu\sigma+2\mu\xi}{2\nu}\right)}\frac{1}{\sin\pi\left(\sigma-\dfrac{1+\gamma+\nu\sigma+2\mu\xi}{2\nu}\right)}$$

(8.17)

or

$$N_{99} = -\frac{\pi}{2}\frac{\sqrt{\pi}}{2\mu}\frac{1}{2^{m-1}}\left(\frac{a}{2}\right)^{-\frac{1+\gamma}{\mu}}\left(\frac{b/2}{(a/2)^{\nu/\mu}}\right)^{\sigma}$$

$$\times\frac{1}{2i}\int\limits_{-\beta+i\infty}^{-\beta-i\infty} d\eta \frac{\left[\dfrac{b/2}{(a/2)^{\nu/\mu}}\right]^{2\eta}}{\sin\pi\eta\,\Gamma(1+\eta)\,\Gamma(1+\sigma+\eta)}\frac{1}{\sin\pi(\sigma+\eta)}$$

$$\times N'_m\left(\xi=-\frac{1+\gamma+\nu(\sigma+2\eta)}{2\mu}\right)\frac{\Gamma\left(\dfrac{1+\gamma+\nu(\sigma+2\eta)}{2\nu}\right)}{\Gamma\left(\dfrac{1}{2}-\dfrac{1+\gamma+\nu(\sigma+2\eta)}{2\mu}\right)},$$

(8.18)

where $N'_m(\xi)$ ($m = 1, 3, 5, \dots$) is defined by (1.36) in Chapter 1.

8.1.10 100th *General Formula*

$$N_{100} = \int_0^\infty dx\, x^\gamma \; K_\sigma(bx^\nu) \left[\cos^q(ax^\mu) - 1 \right]$$

$$= -\frac{\pi}{2} \frac{1}{2\nu} \left(\frac{b}{2}\right)^{-\frac{1+\gamma}{\nu}} \frac{1}{2^{q-1}} \frac{1}{2i} \int_{\alpha+i\infty}^{\alpha-i\infty} d\xi \frac{\left[\dfrac{2a}{(b/2)^{\mu/\nu}}\right]^{2\xi}}{\sin \pi\xi \; \Gamma(1+2\xi)} \, I_q'(\xi)$$

$$\times \frac{\Gamma\left(\dfrac{1+\gamma+\nu\sigma+2\mu\xi}{2\nu}\right)}{\Gamma\left(1+\sigma-\dfrac{1+\gamma+\nu\sigma+2\mu\xi}{2\nu}\right)} \frac{1}{\sin \pi\left(\sigma - \dfrac{1+\gamma+\nu\sigma+2\mu\xi}{2\nu}\right)}$$

$$\tag{8.19}$$

or

$$N_{100} = -\frac{\pi}{2} \frac{\sqrt{\pi}}{2\mu} \frac{1}{2^{q-1}} \, a^{-\frac{1+\gamma}{\mu}} \left(\frac{b/2}{(a/2)^{\nu/\mu}}\right)^\sigma$$

$$\times \frac{1}{2i} \int_{-\beta+i\infty}^{-\beta-i\infty} d\eta \frac{\left[\dfrac{b/2}{a^{\nu/\mu}}\right]^{2\eta}}{\sin \pi\eta \; \Gamma(1+\eta) \; \Gamma(1+\sigma+\eta)} \frac{1}{\sin \pi(\sigma+\eta)}$$

$$\times I_q'\left(\xi = -\frac{1+\gamma+\nu(\sigma+2\eta)}{2\mu}\right) \frac{\Gamma\left(\dfrac{1+\gamma+\nu(\sigma+2\eta)}{2\mu}\right)}{\Gamma\left(\dfrac{1}{2} - \dfrac{1+\gamma+\nu(\sigma+2\eta)}{2\mu}\right)},$$

$$\tag{8.20}$$

where I_q' $(q = 2, 4, 6, \ldots)$ is given by (1.38) in Chapter 1.

8.1.11   *Calculation of Concrete Integrals*

(1) We put $\nu = 1$ in (8.5) and take into account

$$\Gamma\left(1+\sigma-\frac{1+\gamma+\nu\sigma}{2\nu}\right) \sin \pi\left(\sigma - \frac{1+\gamma+\nu\sigma}{2\nu}\right) = -\frac{\pi}{\Gamma\left(-\sigma+\frac{1+\gamma+\nu\sigma}{2\nu}\right)}$$

and get

$$S_{61} = \int\limits_0^\infty dx x^\gamma \, K_\sigma(bx)$$

$$= \frac{1}{4} \left(\frac{b}{2}\right)^{-1-\gamma} \Gamma\left(\frac{1+\gamma+\sigma}{2}\right) \Gamma\left(\frac{1+\gamma-\sigma}{2}\right)$$

$$= 2^{\gamma-1} b^{-\gamma-1} \Gamma\left(\frac{1+\gamma+\sigma}{2}\right) \Gamma\left(\frac{1+\gamma-\sigma}{2}\right), \qquad (8.21)$$

where $b > 0$, $\mathrm{Re}(\gamma + 1 \pm \sigma) > 0$.

(2) The formula (8.8) with

$$\gamma = \sigma, \ \nu = 1, \ p \to a^2, \ t = 1, \ \text{æ} = 2, \ \lambda = 1$$

gives

$$S_{62} = \int\limits_0^\infty dx x^\sigma \, K_\sigma(bx) \, \frac{1}{a^2 + x^2}$$

$$= -\frac{\pi}{4} \frac{1}{a^2} (a^2)^{\frac{\sigma+1+\sigma}{2}} \frac{1}{2i} \int\limits_{-\beta+i\infty}^{-\beta-i\infty} d\xi \frac{(b/2)^{\sigma+2\xi} (a^2)^\xi}{\sin \pi\xi \, \Gamma(1+\xi)} \frac{\Gamma(-\sigma - \xi)}{-\pi}$$

$$\times \Gamma\left(\frac{1+\sigma+\sigma+2\xi}{2}\right) \Gamma\left(1 - \frac{1+2\sigma+2\xi}{2}\right),$$

where we have used the identity

$$\Gamma(1 + x) \, \Gamma(-x) = -\frac{\pi}{\sin \pi x}.$$

After some elementary calculations, one gets

$$S_{62} = \frac{\pi^2 \, a^{\sigma-1}}{4 \cos \pi\sigma} \left[\mathbf{H}_{-\sigma}(ab) - N_{-\sigma}(ab)\right], \qquad (8.22)$$

where $a, b > 0$, $\mathrm{Re} \, \sigma > -\frac{1}{2}$.

(3) Similarly from (8.8), one gets

$$S_{63} = \int\limits_0^\infty dx x^{-\sigma} \, K_\sigma(bx) \, \frac{1}{a^2 + x^2}$$

$$= \frac{\pi^2}{4a^{\sigma+1} \, \cos \pi\sigma} \left[\mathbf{H}_\sigma(ab) - N_\sigma(ab)\right]. \qquad (8.23)$$

(4) If we put

$$\nu_1 = \nu_2 = 1, \ \sigma_1 = \sigma_2 = \sigma, \ \gamma = 1$$

in the formula (8.6) or (8.7) and obtain after some transformations:

$$S_{64} = \int\limits_0^\infty dx x \, J_\sigma(ax) \, K_\sigma(bx) = \frac{a^\sigma}{b^\sigma \, (a^2 + b^2)}, \qquad (8.24)$$

where $\mathrm{Re} \, a > 0$, $b > 0$, $\mathrm{Re} \, \sigma > -1$.

(5) The formula (8.6) with

$$\gamma = 1, \ \sigma_1 = \sigma, \ \nu_1 = 1, \ \nu_2 = 2, \ \sigma_2 = \frac{1}{2}\sigma$$

gives

$$S_{65} = \int\limits_0^\infty dx x \ J_\sigma(ax) \ K_{\frac{1}{2}\sigma}(bx^2)$$

$$= -\frac{\pi}{2} \left(\frac{a}{2}\right)^\sigma \left(\frac{b}{2}\right)^{-\frac{2+\sigma}{2}} \frac{1}{4} \frac{1}{2i} \int\limits_{-\beta+i\infty}^{-\beta-i\infty} d\xi \frac{\left[\frac{a/2}{(b/2)^{1/2}}\right]^{2\xi}}{\sin \pi\xi \ \Gamma(1+\xi)}$$

$$\times \frac{\Gamma\left(\frac{2+\sigma+\sigma+2\xi}{4}\right)}{\Gamma\left(1+\frac{1}{2}\sigma - \frac{1}{2} - \frac{\sigma}{2} - \frac{\xi}{2}\right)} \frac{1}{\sin \pi\left(\frac{1}{2} - \frac{1}{2} - \frac{\sigma}{2} - \frac{\xi}{2}\right)} \frac{1}{\Gamma(1+\sigma+\xi)}.$$

After some elementary calculations, one gets

$$S_{65} = \frac{\pi}{4b} \left[I_{\frac{1}{2}\sigma}\left(\frac{a^2}{4b}\right) - \mathbf{L}_{\frac{1}{2}\sigma}\left(\frac{a^2}{4b}\right) \right], \tag{8.25}$$

where Re $a > 0$, $b > 0$, Re $\sigma > -1$, and $\mathbf{L}_\sigma(x)$ is the Struve function

$$\mathbf{L}_\sigma(x) = \sum_{n=0}^\infty \frac{(x/2)^{2n+\sigma+1}}{\Gamma\left(n+\frac{3}{2}\right)\Gamma\left(\sigma+n+\frac{3}{2}\right)} \tag{8.26}$$

and $I_\sigma(x)$ is the modified Bessel function of the first kind (8.2).

(6) The formula (8.8) with

$$\gamma = 0, \ p \to a^2, \ t = 1, \ æ = 2, \ \lambda = \frac{1}{2}, \ \nu = 1$$

reads

$$S_{66} = \int\limits_0^\infty dx \frac{1}{(a^2 + x^2)^{1/2}} \ K_\sigma(bx) \tag{8.27}$$

$$= \frac{\pi^2}{8} \sec\left(\frac{1}{2}\sigma\pi\right) \left\{ \left[J_{\frac{1}{2}\sigma}\left(\frac{1}{2}ab\right) \right]^2 + \left[N_{\frac{1}{2}\sigma}\left(\frac{1}{2}ab\right) \right]^2 \right\},$$

where Re $a > 0$, $b > 0$, $|\text{Re } \sigma| < 1$.

(7) Let $\gamma = 0$, $\mu = 1$, $\sigma = 0$, $\nu = 1$ be in (8.11), then we have

$$S_{67} = \int\limits_0^\infty dx \ e^{-ax} \ K_0(bx) = -\frac{\pi}{2} a^{-1} \frac{1}{2i} \int\limits_{-\beta+i\infty}^{-\beta-i\infty} d\xi \frac{(b/2a)^{2\xi}}{\sin \pi\xi \ \Gamma(1+\xi)}$$

$$\times \frac{\Gamma(1+2\xi)}{-\pi} \Gamma(-\xi). \tag{8.28}$$

After some elementary calculations, one gets

$$S_{67} = \frac{\arccos\frac{a}{b}}{\sqrt{b^2 - a^2}}, \tag{8.29}$$

where $0 < a < b$, Re$(a + b) > 0$.

(8) The formula (8.11) with $\gamma = \frac{1}{2}$, $\sigma = \pm\frac{1}{2}$, $\mu = \nu = 1$ reads

$$S_{68} = \int_0^\infty dx \sqrt{x}\, e^{-ax}\, K_{\pm\frac{1}{2}}(bx) = \sqrt{\frac{\pi}{2b}}\, \frac{1}{a+b}. \tag{8.30}$$

(9) Let $\gamma = 0$, $\sigma = 0$, $\mu = \nu = 1$, $m = 1$ be in (8.15), then we have

$$S_{69} = \int_0^\infty dx\, K_0(bx)\, \sin(ax) = \frac{1}{\sqrt{a^2+b^2}}\, \ln\left[\frac{b}{a} + \sqrt{\frac{b^2}{a^2}+1}\right], \tag{8.31}$$

where $a > 0$, $b > 0$.

(10) If $\gamma = 0$, $\sigma = 0$, $\mu = \nu = 1$, $m = 1$ in (8.17), then we have

$$S_{70} = \int_0^\infty dx\, K_0(bx)\, \cos(ax) = \frac{\pi}{2\sqrt{a^2+b^2}}. \tag{8.32}$$

(11) Let us put $\gamma = 1$, $\sigma = 0$, $\nu = \mu = 1$, $m = 1$ in (8.15). Then the result reads

$$S_{71} = \int_0^\infty dx\, x\, K_0(bx)\, \sin(ax) = \frac{a\pi}{2}(a^2+b^2)^{-3/2}. \tag{8.33}$$

(12) After some tedious calculations from the formulas (8.15)-(8.18), one gets the following two formulas

$$S_{72} = \int_0^\infty dx\, x^\gamma\, K_\sigma(bx)\, \sin(ax)$$

$$= \frac{2^\gamma\, a\, \Gamma\left(\frac{2+\sigma+\gamma}{2}\right)\, \Gamma\left(\frac{2+\gamma-\sigma}{2}\right)}{b^{2+\gamma}}$$

$$\times F\left(\frac{2+\sigma+\gamma}{2}, \frac{2+\gamma-\sigma}{2}; \frac{3}{2}; -\frac{a^2}{b^2}\right), \tag{8.34}$$

where $\mathrm{Re}(-\gamma \pm \sigma) < 2$, $\mathrm{Re}\, b > 0$, $a > 0$ and

(13)

$$S_{73} = \int_0^\infty dx\, x^\gamma\, K_\sigma(bx)\, \cos(ax)$$

$$= 2^{\gamma-1}\, b^{-1-\gamma}\, \Gamma\left(\frac{\sigma+\gamma+1}{2}\right)\, \Gamma\left(\frac{1+\gamma-\sigma}{2}\right)$$

$$\times F\left(\frac{\gamma+\sigma+1}{2}, \frac{1+\gamma-\sigma}{2}; \frac{1}{2}; -\frac{a^2}{b^2}\right), \tag{8.35}$$

where $\mathrm{Re}(-\gamma \pm \sigma) < 1$, $\mathrm{Re}\, b > 0$, $a > 0$.

(14) Let $\gamma = 1 + \sigma$, $\nu = \mu = 1$, $m = 1$ be in (8.15), then one gets

$$S_{74} = \int\limits_0^\infty dx\, x^{1+\sigma}\, K_\sigma(bx)\, \sin(ax)$$

$$= \sqrt{\pi}\,(2b)^\sigma\, \Gamma\left(\frac{3}{2} + \sigma\right)\, a\, (a^2 + b^2)^{-\frac{1}{2} - \sigma}, \tag{8.36}$$

where $a, b > 0$, $\mathrm{Re}\,\sigma > -\dfrac{3}{2}$.

(15) Similarly, we assume $\gamma = \sigma$, $\nu = \mu = 1$, $m = 1$ in (8.17) and obtain

$$S_{75} = \int\limits_0^\infty dx\, x^\sigma\, K_\sigma(bx)\, \cos(ax)$$

$$= \frac{1}{2}\,\sqrt{\pi}\,(2b)^\sigma\, \Gamma\left(\sigma + \frac{1}{2}\right)(a^2 + b^2)^{-\sigma - \frac{1}{2}}, \tag{8.37}$$

where $a, b > 0$, $\mathrm{Re}\,\sigma > -\dfrac{1}{2}$.

(16) Assuming $\gamma = \dfrac{1}{2}$, $\sigma = \dfrac{1}{4}$, $b \to b^2$, $\nu = 2$, $\mu = 1$, $m = 1$ in (8.15), one gets

$$S_{76} = \int\limits_0^\infty dx\, \sqrt{x}\, K_{\frac{1}{4}}(b^2 x^2)\, \sin(ax)$$

$$= 2^{-\frac{5}{2}}\sqrt{\pi^2 a}\, b^{-2}\left[I_{-\frac{1}{4}}\left(\frac{a^2}{4b^2}\right) - \mathbf{L}_{\frac{1}{4}}\left(\frac{a^2}{4b^2}\right)\right], \tag{8.38}$$

where $a > 0$.

(17) The formula (8.17) with $\gamma = \dfrac{1}{2}$, $\sigma = -\dfrac{1}{4}$, $b \to b^2$, $\nu = 2$, $\mu = 1$, $m = 1$ leads to

$$S_{77} = \int\limits_0^\infty dx\, \sqrt{x}\, K_{-\frac{1}{4}}(b^2 x^2)\, \cos(ax)$$

$$= 2^{-\frac{5}{2}}\sqrt{\pi^2 a}\, b^{-2}\left[I_{-\frac{1}{4}}\left(\frac{a^2}{4b^2}\right) - \mathbf{L}_{-\frac{1}{4}}\left(\frac{a^2}{4b^2}\right)\right], \tag{8.39}$$

where $a > 0$.

8.1.12 101st *General Formula*

$$N_{101} = \int\limits_0^\infty dx\, x^\gamma\, K_\sigma(bx^\nu)\, \sin^q(ax^\mu)\, e^{-cx^\delta}$$

$$= \frac{1}{2}\left(\frac{b/2}{c^{\nu/\delta}}\right)^\sigma \frac{1}{2i}\int\limits_{-\beta'+i\infty}^{-\beta'-i\infty} d\eta\, \frac{\left[\frac{b/2}{c^{\nu/\delta}}\right]^{2\eta}\Gamma(-\sigma-\eta)}{\sin\pi\eta\,\Gamma(1+\eta)}\, c^{-\frac{1+\gamma}{\delta}}\frac{1}{\delta}$$

$$\times \frac{1}{2^{q-1}}\frac{1}{2i}\int\limits_{\alpha+i\infty}^{\alpha-i\infty} d\xi\, \frac{\left[\frac{2a}{c^{\nu/\delta}}\right]^{2\xi}}{\sin\pi\xi\,\Gamma(1+2\xi)}\, I_q(\xi)$$

$$\times \Gamma\left(\frac{1+\gamma+\nu\sigma+2\nu\eta+2\mu\xi}{\delta}\right)$$

(8.40)

or

$$N_{101} = \frac{1}{2}\left[\frac{b/2}{a^{\nu/\mu}}\right]^\sigma \frac{\sqrt{\pi}}{2\mu}\frac{1}{2^{q-1}}\frac{1}{2i}\int\limits_{-\beta'+i\infty}^{-\beta'-i\infty} d\eta\, \frac{\left[\frac{b/2}{a^{\nu/\mu}}\right]^{2\eta}\Gamma(-\sigma-\eta)}{\sin\pi\eta\,\Gamma(1+\eta)}$$

$$\times a^{-\frac{1+\gamma}{\mu}}\frac{1}{2i}\int\limits_{-\beta+i\infty}^{-\beta-i\infty} d\xi\, \frac{\left[\frac{c}{a^{\delta/\mu}}\right]^\xi}{\sin\pi\xi\,\Gamma(1+\xi)}$$

$$\times I_q\left(\xi=-\frac{1+\gamma+\nu\delta+2\nu\eta+\delta\xi}{2\mu}\right)\frac{\Gamma\left(\dfrac{1+\gamma+\nu\sigma+2\nu\eta+\delta\xi}{2\mu}\right)}{\Gamma\left(\dfrac{1}{2}-\dfrac{1+\gamma+\nu\sigma+2\nu\eta+\delta\xi}{2\mu}\right)}$$

(8.41)

or

$$
N_{101} = \frac{1}{2} \frac{1}{2^{q-1}} \frac{1}{2i} \int_{\alpha+i\infty}^{\alpha-i\infty} d\xi \frac{\left[\dfrac{2a}{(b/2)^{\mu/\nu}}\right]^{2\xi}}{\sin \pi\xi \; \Gamma(1+2\xi)} \frac{I_q(\xi)}{2\nu} \left(\frac{b}{2}\right)^{-\frac{1+\gamma}{\nu}}
$$

$$
\times \frac{1}{2i} \int_{-\beta+i\infty}^{-\beta-i\infty} d\eta \frac{\left[\dfrac{c}{(b/2)^{\delta/\nu}}\right]^{\eta}}{\sin \pi\eta \; \Gamma(1+\eta)} \; \Gamma\left(-\sigma + \frac{1+\gamma+2\mu\xi+\delta\eta+\nu\sigma}{2\nu}\right)
$$

$$
\times \Gamma\left(\frac{1+\gamma+2\mu\xi+\delta\eta+\nu\sigma}{2\nu}\right),
$$

(8.42)

where $I_q(\xi)$ $(q = 2, \, 4, \, 6, \dots)$ is derived from (1.32) in Chapter 1.

8.1.13 102nd *General Formula*

$$
N_{102} = \int_0^{\infty} dx x^{\gamma} \; K_{\sigma}(bx^{\nu}) \; \sin^m(ax^{\mu}) \; e^{-cx^{\delta}}
$$

$$
= \frac{1}{2} \left(\frac{b/2}{c^{\nu/\delta}}\right)^{\sigma} \frac{1}{2i} \int_{-\beta'+i\infty}^{-\beta'-i\infty} d\eta \frac{\left[\dfrac{b/2}{c^{\nu/\delta}}\right]^{2\eta} \Gamma(-\sigma-\eta)}{\sin \pi\eta \; \Gamma(1+\eta)} \frac{1}{\delta} c^{-\frac{1+\gamma+\mu}{\delta}}
$$

$$
\times \frac{a}{2^{m-1}} \frac{1}{2i} \int_{-\beta+i\infty}^{-\beta-i\infty} d\xi \frac{\left[\dfrac{a}{c^{\nu/\delta}}\right]^{2\xi}}{\sin \pi\xi \; \Gamma(2+2\xi)} N_m(\xi)
$$

$$
\times \Gamma\left(\frac{1+\gamma+\mu+\nu\sigma+2\nu\eta+2\mu\xi}{\delta}\right)
$$

(8.43)

or

$$N_{102} = \frac{1}{2} \left[\frac{b/2}{(a/2)^{\nu/\mu}} \right]^{\sigma} \frac{\sqrt{\pi}}{2\mu} \frac{1}{2^{m-1}}$$

$$\times \frac{1}{2i} \int\limits_{-\beta'+i\infty}^{-\beta'-i\infty} d\eta \frac{\left[\dfrac{b/2}{(a/2)^{\nu/\mu}} \right]^{2\eta} \Gamma(-\sigma-\eta)}{\sin \pi \eta \; \Gamma(1+\eta)}$$

$$\times \left(\frac{a}{2} \right)^{-\frac{1+\gamma}{\mu}} \frac{1}{2i} \int\limits_{-\beta+i\infty}^{-\beta-i\infty} d\xi \frac{\left[\dfrac{c}{(a/2)^{\delta/\mu}} \right]^{\xi}}{\sin \pi \xi \; \Gamma(1+\xi)}$$

$$\times N_m \left(\xi = -\frac{1}{2} \left[1 + \frac{1+\gamma+\nu\delta+2\nu\eta+\delta\xi}{\mu} \right] \right)$$

$$\times \frac{\Gamma\left[\dfrac{1}{2} \left(1 + \dfrac{1+\gamma+\nu\sigma+2\nu\eta+\delta\xi}{\mu} \right) \right]}{\Gamma\left(1 - \dfrac{1+\gamma+\nu\sigma+2\nu\eta+\delta\xi}{2\mu} \right)}$$

(8.44)

or

$$N_{102} = \frac{1}{2} \frac{a}{2^{m-1}} \frac{1}{2i} \int\limits_{-\beta+i\infty}^{-\beta-i\infty} d\xi \frac{\left[\dfrac{a}{(b/2)^{\mu/\nu}} \right]^{2\xi}}{\sin \pi \xi \; \Gamma(2+2\xi)} \frac{N_m(\xi)}{2\nu} \left(\frac{b}{2} \right)^{-\frac{1+\gamma+\mu}{\nu}}$$

$$\times \frac{1}{2i} \int\limits_{-\beta'+i\infty}^{-\beta'-i\infty} d\eta \frac{\left[\dfrac{c}{(b/2)^{\delta/\nu}} \right]^{\eta}}{\sin \pi \eta \; \Gamma(1+\eta)} \Gamma\left(-\sigma + \frac{1+\gamma+\mu+2\mu\xi+\delta\eta+\nu\sigma}{2\nu} \right)$$

$$\times \Gamma\left(\frac{1+\gamma+\mu+2\mu\xi+\delta\eta+\nu\sigma}{2\nu} \right),$$

(8.45)

where $N_m(\xi)$ ($m = 1, 3, 5, \ldots$) is defined by (1.34) in Chapter 1.

8.1.14 103rd *General Formula*

$$N_{103} = \int\limits_0^\infty dx\, x^\gamma\, K_\sigma(bx^\nu)\, \cos^m(ax^\mu)\, e^{-cx^\delta}$$

$$= \frac{1}{2}\left(\frac{b/2}{c^{\nu/\delta}}\right)^\sigma \frac{1}{2i} \int\limits_{-\beta'+i\infty}^{-\beta'-i\infty} d\eta \frac{\left[\dfrac{b/2}{c^{\nu/\delta}}\right]^{2\eta} \Gamma(-\sigma-\eta)}{\sin\pi\eta\,\Gamma(1+\eta)}\, \frac{1}{\delta}\, c^{-\frac{1+\gamma}{\delta}}$$

$$\times\frac{1}{2^{m-1}}\frac{1}{2i}\int\limits_{-\beta+i\infty}^{-\beta-i\infty} d\xi \frac{\left[\dfrac{a}{c^{\nu/\delta}}\right]^{2\xi}}{\sin\pi\xi\,\Gamma(1+2\xi)}\, N'_m(\xi)$$

$$\times\Gamma\left(\frac{1+\gamma+\nu\sigma+2\nu\eta+2\mu\xi}{\delta}\right)$$

$$(8.46)$$

or

$$N_{103} = \frac{1}{2}\left[\frac{b/2}{(a/2)^{\nu/\mu}}\right]^\sigma \frac{\sqrt{\pi}}{2\mu}\frac{1}{2^{m-1}}$$

$$\times\frac{1}{2i}\int\limits_{-\beta'+i\infty}^{-\beta'-i\infty} d\eta \frac{\left[\dfrac{b/2}{(a/2)^{\nu/\mu}}\right]^{2\eta}\Gamma(-\sigma-\eta)}{\sin\pi\eta\,\Gamma(1+\eta)}$$

$$\times\left(\frac{a}{2}\right)^{-\frac{1+\gamma}{\mu}}\frac{1}{2i}\int\limits_{-\beta+i\infty}^{-\beta-i\infty} d\xi \frac{\left[\dfrac{c}{(a/2)^{\delta/\mu}}\right]^\xi}{\sin\pi\xi\,\Gamma(1+\xi)}$$

$$\times N'_q\left(\xi = -\frac{1+\gamma+\nu\delta+2\nu\eta+\delta\xi}{2\mu}\right)\frac{\Gamma\left(\dfrac{1+\gamma+\nu\sigma+2\nu\eta+\delta\xi}{2\mu}\right)}{\Gamma\left(\dfrac{1}{2}-\dfrac{1+\gamma+\nu\sigma+2\nu\eta+\delta\xi}{2\mu}\right)}$$

$$(8.47)$$

or

$$
N_{103} = \frac{1}{2} \left(\frac{b}{2}\right)^{-\frac{1+\gamma}{\nu}} \frac{1}{2^{m-1}} \frac{1}{2i} \int\limits_{-\beta+i\infty}^{-\beta-i\infty} d\xi \frac{\left[\dfrac{a}{(b/2)^{\mu/\nu}}\right]^{2\xi}}{\sin \pi\xi \ \Gamma(1+2\xi)} \frac{N'_m(\xi)}{2\nu}
$$

$$
\times \frac{1}{2i} \int\limits_{-\beta+i\infty}^{-\beta-i\infty} d\eta \frac{\left[\dfrac{c}{(b/2)^{\delta/\nu}}\right]^{\eta}}{\sin \pi\eta \ \Gamma(1+\eta)} \Gamma\left(-\sigma + \frac{1+\gamma+2\mu\xi+\delta\eta+\nu\sigma}{2\nu}\right)
$$

$$
\times \Gamma\left(\frac{1+\gamma+2\mu\xi+\delta\eta+\nu\sigma}{2\nu}\right),
$$

(8.48)

where $N'_m(\xi)$ $(m = 1, 3, 5, \ldots)$ is defined by (1.36) in Chapter 1.

8.1.15 104th *General Formula*

$$
N_{104} = \int\limits_0^\infty dx x^\gamma \ K_\sigma(bx^\nu) \left[\cos^q(ax^\mu) - 1\right] e^{-cx^\delta}
$$

$$
= \frac{1}{2} \left(\frac{b/2}{c^{\nu/\delta}}\right)^\sigma \frac{1}{2i} \int\limits_{-\beta'+i\infty}^{-\beta'-i\infty} d\eta \frac{\left[\dfrac{b/2}{c^{\nu/\delta}}\right]^{2\eta} \Gamma(-\sigma-\eta)}{\sin \pi\eta \ \Gamma(1+\eta)} \frac{1}{\delta} c^{-\frac{1+\gamma}{\delta}}
$$

$$
\times \frac{1}{2^{q-1}} \frac{1}{2i} \int\limits_{\alpha+i\infty}^{\alpha-i\infty} d\xi \frac{\left[\dfrac{2a}{c^{\nu/\delta}}\right]^{2\xi}}{\sin \pi\xi \ \Gamma(1+2\xi)} I'_q(\xi)
$$

$$
\times \Gamma\left(\frac{1+\gamma+\nu\sigma+2\nu\eta+2\mu\xi}{\delta}\right)
$$

(8.49)

or

$$
N_{104} = \frac{1}{2} \left[\frac{b/2}{a^{\nu/\mu}} \right]^{\sigma} \frac{\sqrt{\pi}}{2\mu} \, a^{-\frac{1+\gamma}{\mu}} \frac{1}{2^{q-1}}
$$

$$
\times \frac{1}{2i} \int\limits_{-\beta'+i\infty}^{-\beta'-i\infty} d\eta \frac{\left[\frac{b/2}{(a/2)^{\nu/\mu}} \right]^{2\eta} \Gamma(-\sigma-\eta)}{\sin \pi\eta \, \Gamma(1+\eta)}
$$

$$
\times \frac{1}{2i} \int\limits_{-\beta+i\infty}^{-\beta-i\infty} d\xi \frac{\left[\frac{c}{a^{\delta/\mu}} \right]^{\xi}}{\sin \pi\xi \, \Gamma(1+\xi)}
$$

$$
\times I_q' \left(\xi = -\frac{1+\gamma+\nu\delta+2\nu\eta+\delta\xi}{2\mu} \right) \frac{\Gamma\left(\dfrac{1+\gamma+\nu\sigma+2\nu\eta+\delta\xi}{2\mu} \right)}{\Gamma\left(\dfrac{1}{2} - \dfrac{1+\gamma+\nu\sigma+2\nu\eta+\delta\xi}{2\mu} \right)}
$$

$$(8.50)$$

or

$$
N_{104} = \frac{1}{2} \left(\frac{b}{2} \right)^{-\frac{1+\gamma}{\nu}} \frac{1}{2^{q-1}} \frac{1}{2i} \int\limits_{\alpha+i\infty}^{\alpha-i\infty} d\xi \frac{\left[\frac{2a}{(b/2)^{\mu/\nu}} \right]^{2\xi}}{\sin \pi\xi \, \Gamma(1+2\xi)} \frac{I_q'(\xi)}{2\nu}
$$

$$
\times \frac{1}{2i} \int\limits_{-\beta+i\infty}^{-\beta-i\infty} d\eta \frac{\left[\frac{c}{(b/2)^{\delta/\nu}} \right]^{\eta}}{\sin \pi\eta \, \Gamma(1+\eta)} \Gamma\left(-\sigma + \frac{1+\gamma+2\mu\xi+\delta\eta+\nu\sigma}{2\nu} \right)
$$

$$
\times \Gamma\left(\frac{1+\gamma+2\mu\xi+\delta\eta+\nu\sigma}{2\nu} \right),
$$

$$(8.51)$$

where $I_q'(\xi)$ ($m = 2,\ 4,\ 6, \ldots$) is given by (1.38) in Chapter 1.

8.1.16 *Some Examples of Calculation of Integrals*

(1) The formula (8.46) with $\gamma = -\frac{1}{2}$, $\sigma = 0$, $b = 1$, $a \to 4a$, $\mu = \frac{1}{2}$, $c = -1$, $\delta = 1$, $m = 1$ gives

$$
S_{78} = \int\limits_{0}^{\infty} dx \frac{1}{\sqrt{x}} \, K_0(x) \, \cos(4a\sqrt{x}) \, e^x
$$

$$
= \sqrt{\frac{\pi}{2}} \, e^{a^2} K_0(a^2), \ a > 0.
$$

$$(8.52)$$

(2) The above example with $c = 1$ reads

$$S_{79} = \int_0^\infty dx \frac{1}{\sqrt{x}} K_0(x) \, \cos(4a\sqrt{x}) \, e^{-x}$$

$$= \frac{1}{\sqrt{2}} \pi^{3/2} e^{-a^2} I_0(a^2). \tag{8.53}$$

8.2 Integrals Involving the Struve Function

$$\mathbf{H}_\sigma(x) = \frac{1}{2i} \int_{-\beta+i\infty}^{-\beta-i\infty} d\xi \frac{(x/2)^{2\xi+\sigma+1}}{\sin \pi\xi \, \Gamma\left(\xi+\frac{3}{2}\right) \, \Gamma\left(\xi+\sigma+\frac{3}{2}\right)} \tag{8.54}$$

or

$$\mathbf{L}_\sigma(x) = -i \, e^{-i\sigma \frac{\pi}{2}} \, \mathbf{H}_\sigma\left(x \, e^{i\frac{\pi}{2}}\right)$$

$$= \frac{1}{2i} \int_{-\beta+i\infty}^{-\beta-i\infty} d\xi \frac{(-1)^\xi (x/2)^{2\xi+\sigma+1}}{\sin \pi\xi \, \Gamma\left(\xi+\frac{3}{2}\right) \, \Gamma\left(\xi+\sigma+\frac{3}{2}\right)}, \tag{8.55}$$

where $\mathbf{L}_\sigma(x)$ is called the modified Struve function.

8.2.1 105th *General Formula*

$$N_{105} = \int_0^\infty dx x^\gamma \, \mathbf{H}_\sigma(bx^\nu) = \frac{1}{2\nu} \left(\frac{b}{2}\right)^{-\frac{1+\gamma}{\nu}} \frac{1}{\sin \pi \left(\frac{1+\gamma+\nu\sigma+\nu}{2\nu}\right)}$$

$$\times \frac{1}{\Gamma\left(\frac{3}{2} - \frac{1+\gamma+\nu\sigma+\nu}{2\nu}\right)} \frac{1}{\Gamma\left(\sigma + \frac{3}{2} - \frac{1+\gamma+\nu\sigma+\nu}{2\nu}\right)}.$$

$$\tag{8.56}$$

8.2.2 106th General Formula

$$N_{106} = \int\limits_0^\infty dx \frac{x^\gamma}{\left[p + tx^{\text{æ}}\right]^\lambda} \mathbf{H}_\sigma(bx^\nu)$$

$$= \frac{1}{\text{æ}} \frac{1}{\Gamma(\lambda)} \frac{1}{p^\lambda} \left(\frac{b}{t}\right)^{\frac{\gamma+1+\nu\sigma+\nu}{\text{æ}}}$$

$$\times \frac{1}{2i} \int\limits_{-\beta+i\infty}^{-\beta-i\infty} d\xi \frac{\left(\dfrac{b}{2}\right)^{2\xi+\sigma+1} \left(\dfrac{p}{t}\right)^{2\nu\xi/\text{æ}}}{\sin \pi\xi\, \Gamma\left(\xi + \dfrac{3}{2}\right) \Gamma\left(\xi + \sigma + \dfrac{3}{2}\right)}$$

$$\times \Gamma\left(\frac{1 + \gamma + \nu(\sigma + 2\xi) + \nu}{\text{æ}}\right) \Gamma\left(\lambda - \frac{1 + \gamma + \nu(\sigma + 2\xi) + \nu}{\text{æ}}\right).$$

(8.57)

8.2.3 107th General Formula

$$N_{107} = \int\limits_0^\infty dx\, x^\gamma\, e^{-ax^\mu} \mathbf{H}_\sigma(bx^\nu) = \frac{1}{\mu} a^{-\frac{1+\gamma}{\mu}} \left(\frac{b/2}{a^{\nu/\mu}}\right)^{\sigma+1}$$

$$\times \frac{1}{2i} \int\limits_{-\beta+i\infty}^{-\beta-i\infty} d\xi \frac{\left(\dfrac{b/2}{a^{\nu/\mu}}\right)^{2\xi}}{\sin \pi\xi\, \Gamma\left(\xi + \dfrac{3}{2}\right) \Gamma\left(\xi + \sigma + \dfrac{3}{2}\right)}$$

$$\times \Gamma\left(\frac{1 + \gamma + \nu(\sigma + 2\xi) + \nu}{\mu}\right).$$

(8.58)

8.2.4 108th General Formula

$$N_{108} = \int\limits_0^\infty dx\, x^\gamma\, \mathbf{H}_\sigma(bx^\nu)\, \sin^m(ax^\mu)$$

$$= \frac{1}{2\nu} \left(\frac{b}{2}\right)^{-\frac{1+\gamma+\mu}{\nu}} \frac{a}{2^{m-1}} \frac{1}{2i} \int\limits_{-\beta+i\infty}^{-\beta-i\infty} d\xi\, \frac{\left[\dfrac{a}{(b/2)^{\mu/\nu}}\right]^{2\xi}}{\sin\pi\xi\, \Gamma(2+2\xi)}\, N_m(\xi)$$

$$\times \frac{1}{\sin\pi\left(\dfrac{1+\gamma+\mu+2\mu\xi+\nu\sigma+\nu}{2\nu}\right)} \frac{1}{\Gamma\left(\dfrac{3}{2} - \dfrac{1+\gamma+\nu\sigma+\nu+\mu+2\mu\xi}{2\nu}\right)}$$

$$\times \frac{1}{\Gamma\left(\sigma + \dfrac{3}{2} - \dfrac{1+\gamma+\mu+2\mu\xi+\nu\sigma+\nu}{2\nu}\right)}$$

$$(8.59)$$

or

$$N_{108} = \frac{\sqrt{\pi}}{2\mu} \left(\frac{a}{2}\right)^{-\frac{1+\gamma}{\mu}} \left[\frac{b/2}{(a/2)^{\nu/\mu}}\right]^{\sigma+1} \frac{1}{2^{m-1}}$$

$$\times \frac{1}{2i} \int\limits_{-\beta+i\infty}^{-\beta-i\infty} d\eta\, \frac{\left[\dfrac{b/2}{(a/2)^{\nu/\mu}}\right]^{2\eta}}{\sin\pi\eta\, \Gamma\left(\eta + \dfrac{3}{2}\right) \Gamma\left(\eta + \sigma + \dfrac{3}{2}\right)}$$

$$\times N_m\left(\xi = -\frac{1}{2}\left[1 + \frac{\gamma+1+\nu\sigma+\nu+2\nu\eta}{\mu}\right]\right)$$

$$\times \frac{\Gamma\left(\dfrac{1}{2}\left[1 + \dfrac{1+\gamma+\nu\sigma+\nu+2\nu\eta}{\mu}\right]\right)}{\Gamma\left(1 - \dfrac{1+\gamma+\nu\sigma+\nu+2\nu\eta}{2\mu}\right)},$$

$$(8.60)$$

where $N_m(\xi)$ ($m = 1,\ 3,\ 5, \ldots$) is defined by (1.34) in Chapter 1.

Notice that it is easy to obtain such types of general formulas for integrals with other $\sin^q(ax^\mu)$, $\cos^m(ax^\mu)$, $\cos^q(ax^\mu) - 1$, and $\mathbf{H}(bx^\nu)$-functions.

8.2.5 *Exercises*

(1) From the formula (8.56) where $\gamma = 0$, $\nu = 1$, one gets

$$S_{80} = \int\limits_0^\infty dx\, \mathbf{H}_\sigma(bx) = -\frac{1}{b} \cot\left(\frac{\pi\sigma}{2}\right), \qquad (8.61)$$

where $-2 < \mathrm{Re}\ \sigma < 0,\ b > 0$.

(2) Let $\gamma = -1 - \sigma$ and $b = 1,\ \nu = 1$ be in (8.56), then we have

$$S_{81} = \int_0^\infty dx\, x^{-1-\sigma}\, \mathbf{H}_\sigma(x) = \frac{\pi}{\Gamma(1+\sigma)}\, 2^{-1-\sigma}, \tag{8.62}$$

where $\mathrm{Re}\ \sigma > -\dfrac{3}{2}$.

(3) If we put

$$\gamma = 0,\ \lambda = 1,\ p \to a^2,\ t = 1,\ \text{æ} = 2,\ \sigma = 1,\ \nu = 1$$

in the formula (8.57), then we have

$$S_{82} = \int_0^\infty dx\, \frac{1}{x^2 + a^2}\, \mathbf{H}_1(bx) = \frac{\pi}{2a}\Big[I_1(ab) - \mathbf{L}_1(ab)\Big], \tag{8.63}$$

where $\mathrm{Re}\ a > 0,\ b > 0$.

(4) Assuming $\gamma = 0,\ \mu = 1,\ \sigma = 0,\ \nu = 1$ in (8.58), one gets

$$S_{83} = \int_0^\infty dx\, e^{-ax}\, \mathbf{H}_0(bx) = \frac{2}{\pi}(a^2 + b^2)^{-1/2}\, \ln\left[\frac{\sqrt{a^2 + b^2} + b}{a}\right], \tag{8.64}$$

where $\mathrm{Re}\ a > |\mathrm{Im}\ b|$.

(5) Moreover, a similar calculation reads

$$S_{84} = \int_0^\infty dx\, e^{-ax}\, \mathbf{L}_0(bx) = \frac{2}{\pi}\, \frac{\arcsin\left(\frac{b}{a}\right)}{\sqrt{a^2 + b^2}}, \tag{8.65}$$

where $\mathrm{Re}\ a > |\mathrm{Re}\ b|$.

(6) Let $\gamma = -\sigma,\ \nu = 1,\ m = 1,\ \mu = 1$ be in (8.59) and (8.60), then we have

$$S_{85} = \int_0^\infty dx\, x^{-\sigma}\, \mathbf{H}_\sigma(bx)\, \sin(ax) \tag{8.66}$$

$$= \begin{cases} 0 & \text{if } 0 < b < a,\ \mathrm{Re}\ \sigma > -\dfrac{1}{2} \\[3mm] \sqrt{2}\, 2^{-\sigma}\, b^{-\sigma}\, \dfrac{(b^2 - a^2)^{\sigma - \frac{1}{2}}}{\Gamma\left(\sigma + \frac{1}{2}\right)} & \text{if } 0 < a < b,\ \mathrm{Re}\ \sigma > -\dfrac{1}{2}. \end{cases}$$

(7) Let $\gamma = \frac{1}{2},\ \sigma = \frac{1}{4},\ \nu = 2,\ b \to b^2,\ \mu = 1,\ m = 1$ be in (8.59), then one gets

$$S_{86} = \int_0^\infty dx\, \sqrt{x}\, \mathbf{H}_{\frac{1}{4}}(b^2 x^2)\, \sin(ax)$$

$$= -2^{-\frac{3}{2}}\sqrt{\pi}\, \frac{\sqrt{a}}{b^2}\, N_{\frac{1}{4}}\left(\frac{a^2}{4b^2}\right), \tag{8.67}$$

where $a > 0$.

8.3 Integrals Involving Other Special Functions

By using the Mellin representations for the following special functions, one can derive corresponding general formulas for integrals involving them as the above procedure.

8.3.1 *Hypergeometric Functions, Order (1.1)*

$$
{}_1F_1(\mu;\ \nu;\ x) = \frac{\Gamma(\nu)}{\Gamma(\mu)}\frac{1}{2i}\int\limits_{-\beta+i\infty}^{-\beta-i\infty} d\xi \frac{(-x)^\xi}{\sin\pi\xi\ \Gamma(1+\xi)}\frac{\Gamma(\mu+\xi)}{\Gamma(\nu+\xi)}. \tag{8.68}
$$

8.3.2 *Hypergeometric Function*

$$
{}_2F_1(\mu,\nu;\ \lambda;\ x) = \frac{\Gamma(\lambda)}{\Gamma(\mu)\,\Gamma(\nu)}\frac{1}{2i}\int\limits_{-\beta+i\infty}^{-\beta-i\infty} d\xi \frac{(-x)^\xi}{\sin\pi\xi\ \Gamma(1+\xi)}
$$
$$
\times\ \frac{\Gamma(\mu+\xi)\,\Gamma(\nu+\xi)}{\Gamma(\lambda+\xi)}. \tag{8.69}
$$

8.3.3 *Tomson Function*

$$
\mathrm{ber}(x) = \frac{1}{2i}\int\limits_{-\beta+i\infty}^{-\beta-i\infty} d\xi \frac{(x/2)^{4\xi}}{\sin\pi\xi\ \Gamma^2(1+2\xi)} \tag{8.70}
$$

or

$$
\mathrm{bei}(x) = \frac{1}{2i}\int\limits_{-\beta+i\infty}^{-\beta-i\infty} d\xi \frac{(x/2)^{4\xi+2}}{\sin\pi\xi\ \Gamma^2(2+2\xi)}. \tag{8.71}
$$

8.3.4 *Anger Function*

$$
\mathbf{J}_\nu(x) = \frac{1}{2i}\int\limits_{-\beta+i\infty}^{-\beta-i\infty} d\xi \frac{(x/2)^{2\xi}}{\sin\pi\xi}\left[\frac{\cos(\pi\nu/2)}{\Gamma\left(1+\xi+\frac{1}{2}\nu\right)\Gamma\left(\xi+1-\frac{1}{2}\nu\right)}\right.
$$
$$
\left. +\ \frac{\sin\left(\frac{\pi\nu}{2}\right)\left(\frac{x}{2}\right)}{\Gamma\left(\xi+\frac{3}{2}+\frac{1}{2}\nu\right)\Gamma\left(\xi+\frac{3}{2}-\frac{1}{2}\nu\right)}\right]. \tag{8.72}
$$

8.3.5   Veber Function

$$
\mathbf{E}_\nu(x) = \frac{1}{2i} \int\limits_{-\beta+i\infty}^{-\beta-i\infty} d\xi \frac{(x/2)^{2\xi}}{\sin \pi \xi} \left[\frac{\sin\left(\frac{\pi\nu}{2}\right)}{\Gamma\left(\xi+1+\frac{1}{2}\nu\right)\Gamma\left(\xi+1-\frac{1}{2}\nu\right)} \right.
$$

$$
\left. - \frac{\cos\left(\frac{\pi\nu}{2}\right)\left(\frac{x}{2}\right)}{\Gamma\left(\xi+\frac{3}{2}+\frac{1}{2}\nu\right)\Gamma\left(\xi+\frac{3}{2}-\frac{1}{2}\nu\right)} \right]. \tag{8.73}
$$

8.3.6   Legendre's Function of the Second Kind

$$
Q_\sigma(x) = \frac{\Gamma(\sigma+1)\Gamma\left(\frac{1}{2}\right)}{2^{1+\sigma}\Gamma\left(\sigma+\frac{3}{2}\right)} x^{-1-\sigma} \frac{\Gamma\left(\frac{\sigma+3}{2}\right)}{\Gamma\left(\frac{\sigma+2}{2}\right)\Gamma\left(\frac{1+\sigma}{2}\right)} \tag{8.74}
$$

$$
\times \frac{1}{2i} \int\limits_{-\beta+i\infty}^{-\beta-i\infty} d\xi \frac{\left(-x^{-2}\right)^\xi}{\sin \pi \xi\, \Gamma(1+\xi)} \frac{\Gamma\left(\xi+1+\frac{\sigma}{2}\right)\Gamma\left(\frac{1}{2}+\frac{\sigma}{2}+\xi\right)}{\Gamma\left(\xi+\frac{3}{2}+\frac{\sigma}{2}\right)}.
$$

8.3.7   Complete Elliptic Integral of the First Kind

$$
\mathbf{K}(x) = \frac{\pi}{2} \frac{\Gamma(1)}{\Gamma\left(\frac{1}{2}\right)\Gamma\left(\frac{1}{2}\right)} \frac{1}{2i} \int\limits_{-\beta+i\infty}^{-\beta-i\infty} d\xi \frac{(-x^2)^\xi}{\sin \pi \xi\, \Gamma(1+\xi)}
$$

$$
\times \frac{\Gamma\left(\frac{1}{2}+\xi\right)\Gamma\left(\frac{1}{2}+\xi\right)}{\Gamma(1+\xi)}. \tag{8.75}
$$

8.3.8   Complete Elliptic Integral of the Second Kind

$$
\mathbf{E}(x) = \frac{\pi}{2} \frac{\Gamma(1)}{\Gamma\left(\frac{1}{2}\right)\Gamma\left(-\frac{1}{2}\right)} \frac{1}{2i} \int\limits_{-\beta+i\infty}^{-\beta-i\infty} d\xi \frac{(-x^2)^\xi}{\sin \pi \xi\, \Gamma(1+\xi)}
$$

$$
\times \frac{\Gamma\left(-\frac{1}{2}+\xi\right)\Gamma\left(\frac{1}{2}+\xi\right)}{\Gamma(1+\xi)}. \tag{8.76}
$$

8.3.9   Exponential Integral Functions

1. For $x < 0$,

$$
Ei(x) = -\int\limits_{-x}^{\infty} \frac{d\tau}{\tau} e^{-\tau}
$$

$$
= C + \ln(-x) + \frac{1}{2i} \int\limits_{\alpha+i\infty}^{\alpha-i\infty} d\xi \frac{(-1)^\xi x^\xi}{\sin \pi \xi\, \xi\, \Gamma(1+\xi)}. \tag{8.77}
$$

2. For $x > 0$,

$$Ei(x) = C + \ln(x) + \frac{1}{2i} \int\limits_{\alpha+i\infty}^{\alpha-i\infty} d\xi \frac{(-1)^\xi \, x^\xi}{\sin \pi\xi \, \xi \, \Gamma(1+\xi)}, \qquad (8.78)$$

$(0 < \alpha < 1)$.

8.3.10 *Sine Integral Function*

$$Si(x) = -\frac{\pi}{2} - \frac{1}{2i} \int\limits_{\alpha+i\infty}^{\alpha-i\infty} d\xi \frac{x^{2\xi-1}}{\sin \pi\xi \, (2\xi-1) \, \Gamma(2\xi)}. \qquad (8.79)$$

8.3.11 *Cosine Integral Function*

$$Ci(x) = C + \ln(x) + \frac{1}{2i} \int\limits_{\alpha+i\infty}^{\alpha-i\infty} d\xi \frac{x^{2\xi}}{\sin \pi\xi \, (2\xi) \, \Gamma(1+2\xi)}. \qquad (8.80)$$

8.3.12 *Probability Integral*

$$\Phi(x) = \frac{2}{\sqrt{\pi}} \int\limits_0^x d\tau \, e^{-\tau^2} = -\frac{2}{\sqrt{\pi}} \frac{1}{2i} \int\limits_{\alpha+i\infty}^{\alpha-i\infty} d\xi \frac{x^{2\xi-1}}{\sin \pi\xi \, (2\xi-1) \, \Gamma(\xi)} \qquad (8.81)$$

or

$$\Phi(x) = e^{-x^2} \frac{1}{2i} \int\limits_{-\beta+i\infty}^{-\beta-i\infty} d\xi \frac{(-1)^\xi}{\sin \pi\xi} \frac{x^{2\xi+1}}{\Gamma\left(\frac{3}{2}+\xi\right)}, \qquad (8.82)$$

where

$$\Gamma\left(n+\frac{3}{2}\right) = \frac{\sqrt{\pi}}{2^{n+1}}(2n+1)!!.$$

8.3.13 *Frenel Functions*

$$S(x) = \frac{2}{\sqrt{2\pi}} \int\limits_0^x d\tau \, \sin \tau^2$$

$$= \frac{2}{\sqrt{2\pi}} \frac{1}{2i} \int\limits_{-\beta+i\infty}^{-\beta-i\infty} d\xi \frac{x^{4\xi+3}}{\sin \pi\xi \, (4\xi+3) \, \Gamma(2+2\xi)} \qquad (8.83)$$

and

$$C(x) = \frac{2}{\sqrt{2\pi}} \int\limits_0^x d\tau \, \cos \tau^2$$

$$= \frac{2}{\sqrt{2\pi}} \frac{1}{2i} \int\limits_{-\beta+i\infty}^{-\beta-i\infty} d\xi \frac{x^{4\xi+1}}{\sin \pi\xi \, (4\xi+1) \, \Gamma(1+2\xi)}. \tag{8.84}$$

8.3.14 Incomplete Gamma Function

$$\gamma(\alpha, x) = \int\limits_0^x d\tau \, e^{-\tau} \, \tau^{\alpha-1}$$

$$= \frac{1}{2i} \int\limits_{-\beta+i\infty}^{-\beta-i\infty} d\xi \frac{x^{\alpha+\xi}}{\sin \pi\xi \, \Gamma(1+\xi) \, (\alpha+\xi)} \tag{8.85}$$

and

$$\Gamma(\alpha, x) = \int\limits_x^\infty d\tau \, e^{-\tau} \, \tau^{\alpha-1}$$

$$= \Gamma(\alpha) - \frac{1}{2i} \int\limits_{-\beta+i\infty}^{-\beta-i\infty} d\xi \frac{x^{\alpha+\xi}}{\sin \pi\xi \, \Gamma(1+\xi) \, (\alpha+\xi)}. \tag{8.86}$$

8.3.15 Psi-Functions $\Psi(x)$

$$\Psi(x) = \frac{d}{dx} \ln \Gamma(x)$$

$$= -C - \frac{1}{2i} \int\limits_{-\beta+i\infty}^{-\beta-i\infty} d\xi \frac{(-1)^\xi}{\sin \pi\xi} \left(\frac{1}{x+\xi} - \frac{1}{1+\xi} \right) \tag{8.87}$$

$$= -C - \frac{1}{x} + x \frac{1}{2i} \int\limits_{\alpha+i\infty}^{\alpha-i\infty} d\xi \frac{(-1)^\xi}{\sin \pi\xi} \frac{1}{\xi(x+\xi)}. \tag{8.88}$$

8.3.16 Euler's Constant

$$C = \gamma = 0.5772157\ldots \tag{8.89}$$

8.3.17 Hankel Function

$$H_\sigma^{(1)}(x) = J_\sigma(x) + iN_\sigma(x), \tag{8.90}$$

$$H_\sigma^{(2)}(x) = J_\sigma(x) - iN_\sigma(x), \tag{8.91}$$

$$H_\sigma^{(1)}(x) = \frac{1}{2i} \int\limits_{-\beta+i\infty}^{-\beta-i\infty} d\xi \frac{(x/2)^{2\xi}}{\sin \pi\xi \ \Gamma(1+\xi)} \left\{ \frac{(x/2)^\sigma}{\Gamma(1+\xi+\sigma)} \right.$$
$$\left. + \frac{i}{\sin \pi\sigma} \left[\frac{\cos \pi\sigma \ (x/2)^\sigma}{\Gamma(1+\xi+\sigma)} - \frac{(x/2)^{-\sigma}}{\Gamma(1+\xi-\sigma)} \right] \right\}, \tag{8.92}$$

$$H_\sigma^{(2)}(x) = \frac{1}{2i} \int\limits_{-\beta+i\infty}^{-\beta-i\infty} d\xi \frac{(x/2)^{2\xi}}{\sin \pi\xi \ \Gamma(1+\xi)} \left\{ \frac{(x/2)^\sigma}{\Gamma(1+\xi+\sigma)} \right.$$
$$\left. - \frac{i}{\sin \pi\sigma} \left[\frac{\cos \pi\sigma \ (x/2)^\sigma}{\Gamma(1+\xi+\sigma)} - \frac{(x/2)^{-\sigma}}{\Gamma(1+\xi-\sigma)} \right] \right\}. \tag{8.93}$$

8.3.18 Cylindrical Function of Imaginary Arguments

$$I_\sigma(x) = e^{-\frac{\pi}{2}\sigma i} \ J_\sigma\left(e^{\frac{\pi}{2}i}x\right), \tag{8.94}$$

$$I_\sigma(x) = \frac{1}{2i} \int\limits_{-\beta+i\infty}^{-\beta-i\infty} d\xi \frac{(-1)^\xi \ (x/2)^{2\xi+\sigma}}{\sin \pi\xi \ \Gamma(1+\xi) \ \Gamma(1+\xi+\sigma)}. \tag{8.95}$$

$I_\sigma(x)$ is called the modified Bessel function of the first kind.

8.4 Some Examples

(1)

$$S_{87} = \int\limits_0^\infty dx x^{\sigma-1} \ Ei(-bx) = -\frac{\Gamma(\sigma)}{\sigma \ b^\sigma}, \tag{8.96}$$

where Re $b \geq 0$, Re $\sigma > 0$.

(2)

$$S_{88} = \int\limits_0^\infty dx e^{-ax} \ Ei(-bx) = -\frac{1}{a} \ln\left(1 + \frac{a}{b}\right), \tag{8.97}$$

where Re$(b + a) \geq 0$, $a > 0$.

(3)

$$S_{89} = \int_0^\infty dx \; e^{-ax} \; Ei(bx) = -\frac{1}{a} \ln \left[\frac{a}{b} - 1 \right], \tag{8.98}$$

where $b > 0$, Re $a > 0, a > b$.

(4)

$$S_{90} = \int_0^\infty dx \; \frac{Ci(bx)}{a^2 + b^2} = \frac{\pi}{2a} Ei(-ab), \tag{8.99}$$

where $a, b > 0$.

(5)

$$S_{91} = \int_0^\infty dx \; Si(ax) \; x^{\sigma-1} = -\frac{\Gamma(\sigma)}{\sigma \, a^\sigma} \sin \frac{\pi\sigma}{2}, \tag{8.100}$$

where $a > 0$, $0 < \text{Re } \sigma < 1$.

(6)

$$S_{92} = \int_0^\infty dx \; Ci(ax) \; x^{\sigma-1} = -\frac{\Gamma(\sigma)}{\sigma \, a^\sigma} \cos \frac{\pi\sigma}{2}, \tag{8.101}$$

where $a > 0$, $0 < \text{Re } \sigma < 1$.

(7)

$$S_{93} = \int_0^\infty dx \; Si(bx) \; e^{-ax} = -\frac{1}{a} \arctan \frac{a}{b}, \tag{8.102}$$

Re $a > 0$.

(8)

$$S_{94} = \int_0^\infty dx \; Ci(bx) \; e^{-ax} = -\frac{1}{a} \ln \left(\sqrt{1 + \frac{a^2}{b^2}} \right), \tag{8.103}$$

Re $a > 0$.

(9)

$$S_{95} = \int_0^\infty dx \left[1 - \Phi(bx) \right] x^{2\sigma-1} = \frac{\Gamma\left(\frac{1}{2} + \sigma\right)}{2\sqrt{\pi} \, \sigma \, b^{2\sigma}}, \tag{8.104}$$

where Re $\sigma > 0$, Re $b > 0$.

(10)

$$S_{96} = \int_0^\infty dx \left[1 - \Phi(x) \right] e^{-a^2 x^2} = \frac{\arctan a}{\sqrt{\pi} \, a}, \tag{8.105}$$

where Re $a > 0$.

(11)

$$S_{97} = \int_0^\infty dx \; \Phi(bx) \; e^{-ax^2} = \frac{b}{2a\sqrt{b^2 + a}},$$ (8.106)

where Re $a > -$Re b^2, Re $a > 0$.

(12)

$$S_{98} = \int_0^\infty dx \left[\frac{1}{2} - S(bx)\right] x^{2\sigma-1}$$

$$= \frac{\sqrt{2} \; \Gamma\left(\frac{1}{2} + \sigma\right) \; \sin\left(\frac{2\sigma+1}{4}\pi\right)}{4\sqrt{\pi} \; \sigma \; b^{2\sigma}},$$ (8.107)

where $0 <$ Re $\sigma < \dfrac{3}{2}$, $b > 0$.

(13)

$$S_{99} = \int_0^\infty dx \left[\frac{1}{2} - C(bx)\right] x^{2\sigma-1}$$

$$= \frac{\sqrt{2} \; \Gamma\left(\frac{1}{2} + \sigma\right) \; \cos\left(\frac{2\sigma+1}{4}\pi\right)}{4\sqrt{\pi} \; \sigma \; b^{2\sigma}},$$ (8.108)

where $0 <$ Re $\sigma < \dfrac{3}{2}$, $b > 0$.

(14)

$$S_{100} = \int_0^\infty dx \; S(x) \; \sin b^2 x^2$$

$$= \begin{cases} \dfrac{1}{b}\sqrt{\pi} \; 2^{-\frac{5}{2}} & \text{if} \quad 0 < b^2 < 1 \\[2ex] 0 & \text{if} \quad b^2 > 1. \end{cases}$$ (8.109)

(15)

$$S_{101} = \int_0^\infty dx \; C(x) \; \cos(b^2 x^2)$$

$$= \begin{cases} \dfrac{1}{b}\sqrt{\pi} \; 2^{-\frac{5}{2}} & \text{if} \quad 0 < b^2 < 1 \\[2ex] 0 & \text{if} \quad b^2 > 1. \end{cases}$$ (8.110)

(16)

$$S_{102} = \int_0^\infty dx \; e^{-ax} \; \gamma(b, x) = \frac{1}{a} \; \Gamma(b) \; (1+a)^{-b},$$ (8.111)

where $b > 0$.

(17)

$$S_{103} = \int\limits_0^\infty dx \ e^{-ax} \ \Gamma(b, x) = \frac{1}{a} \ \Gamma(b) \left[1 - \frac{1}{(1+a)^b} \right], \qquad (8.112)$$

where $b > 0$.

(18)

$$S_{104} = \int\limits_0^\infty dx x^{-a} \left[C + \Psi(1+x) \right] = -\pi \ \csc(\pi a) \ \zeta(a), \qquad (8.113)$$

where $1 < \mathrm{Re} \ a < 2$, and $\zeta(x)$ is the Weierstrass zeta function.

(19)

$$S_{105} = \int\limits_0^\infty dx \ Si(ax) \ J_0(bx)$$

$$= \begin{cases} -\dfrac{1}{b} \arcsin\left(\dfrac{b}{a}\right) & \text{if} \quad 0 < b < a \\[4mm] 0 & \text{if} \quad 0 < a < b. \end{cases} \qquad (8.114)$$

(20)

$$S_{106} = \int\limits_0^\infty dx \ Ci(x) \ J_0(2\sqrt{bx}) = \frac{\cos b - 1}{b}. \qquad (8.115)$$

(21)

$$S_{107} = \int\limits_0^\infty dx x \ Si(a^2 x^2) \ J_0(bx) = -\frac{2}{b^2} \ \sin\left(\frac{b^2}{4a^2}\right), \ a > 0. \qquad (8.116)$$

(22)

$$S_{108} = \int\limits_0^\infty dx x \ Ci(a^2 x^2) \ J_0(bx) = \frac{2}{b^2} \left[1 - \cos\left(\frac{b^2}{4a^2}\right) \right], \ a > 0. \qquad (8.117)$$

(23)

$$S_{109} = \int\limits_0^\infty dx \ e^{-ax} \ \mathrm{ber}(2\sqrt{x}) = \frac{1}{a} \ \cos\frac{1}{a}. \qquad (8.118)$$

(24)

$$S_{110} = \int\limits_0^\infty dx\, e^{-ax}\, \mathrm{bei}(2\sqrt{x}) = \frac{1}{a}\, \sin\frac{1}{a}. \tag{8.119}$$

(25)

$$S_{111} = \int\limits_0^\infty dx\, x^{-\sigma-1}\, F(a, b;\ c;\ -x)$$

$$= \frac{\Gamma(a+\sigma)\,\Gamma(b+\sigma)\,\Gamma(c)\,\Gamma(-\sigma)}{\Gamma(a)\,\Gamma(b)\,\Gamma(c+\sigma)}, \tag{8.120}$$

where $c \neq 0,\ -1, -2, \ldots,$ $\mathrm{Re}\,\sigma > 0,\ \mathrm{Re}(a+\sigma) > 0,\ \mathrm{Re}(b+\sigma) > 0.$

Chapter 9

Integrals Involving Two Trigonometric Functions

9.1 109th General Formula

$$N_{109} = \int\limits_0^\infty dx\, x^\gamma\, \sin^{q_1}(ax^\mu)\, \sin^{q_2}(bx^\nu)$$

$$= \frac{\sqrt{\pi}}{2\mu}\, \frac{1}{2^{q_1-1}}\, a^{-\frac{1+\gamma}{\mu}}\, \frac{1}{2^{q_2-1}}\, \frac{1}{2i} \int\limits_{\alpha+i\infty}^{\alpha-i\infty} d\xi \frac{\left[\dfrac{2b}{a^{\nu/\mu}}\right]^{2\xi}}{\sin \pi\xi\, \Gamma(1+2\xi)}$$

$$\times I_{q_2}(\xi)\, I_{q_1}\left(\xi = -\frac{1+\gamma+2\xi\nu}{2\mu}\right) \frac{\Gamma\left(\dfrac{1+\gamma+2\xi\nu}{2\mu}\right)}{\Gamma\left(\dfrac{1}{2} - \dfrac{1+\gamma+2\xi\nu}{2\mu}\right)} \tag{9.1}$$

or

$$N_{109} = \frac{\sqrt{\pi}}{2\nu}\, \frac{1}{2^{q_2-1}}\, b^{-\frac{1+\gamma}{\nu}}\, \frac{1}{2^{q_1-1}}\, \frac{1}{2i} \int\limits_{\alpha+i\infty}^{\alpha-i\infty} d\xi \frac{\left[\dfrac{2a}{b^{\mu/\nu}}\right]^{2\xi}}{\sin \pi\xi\, \Gamma(1+2\xi)}$$

$$\times I_{q_1}(\xi)\, I_{q_2}\left(\xi = -\frac{1+\gamma+2\xi\mu}{2\nu}\right) \frac{\Gamma\left(\dfrac{1+\gamma+2\xi\mu}{2\nu}\right)}{\Gamma\left(\dfrac{1}{2} - \dfrac{1+\gamma+2\xi\mu}{2\nu}\right)}, \tag{9.2}$$

where $I_{q_2}(\xi)$ and $I_{q_1}(\xi)$ (q_1, $q_2 = 2$, 4, 6,...) are given by (1.32) in Chapter 1.

213

9.2   110th General Formula

$$N_{110} = \int\limits_0^\infty dx\, x^\gamma \, \sin^{q_1}(ax^\mu)\, \sin^m(bx^\nu)$$

$$= \frac{\sqrt{\pi}}{2\mu}\, \frac{1}{2^{q_1-1}}\, a^{-\frac{1+\gamma+\nu}{\mu}}\, \frac{b}{2^{m-1}}\, \frac{1}{2i} \int\limits_{-\beta+i\infty}^{-\beta-i\infty} d\xi \frac{\left[\dfrac{b}{a^{\nu/\mu}}\right]^{2\xi}}{\sin\pi\xi\, \Gamma(2+2\xi)}$$

$$\times N_m(\xi)\, I_{q_1}\left(\xi = -\frac{1+\gamma+\nu+2\xi\nu}{2\mu}\right) \frac{\Gamma\left(\dfrac{1+\gamma+\nu+2\xi\nu}{2\mu}\right)}{\Gamma\left(\dfrac{1}{2} - \dfrac{1+\gamma+\nu+2\xi\nu}{2\mu}\right)}$$

(9.3)

or

$$N_{110} = \frac{\sqrt{\pi}}{2\nu}\, \frac{1}{2^{m-1}}\, \left(\frac{b}{2}\right)^{-\frac{1+\gamma}{\nu}}\, \frac{1}{2^{q_1-1}}\, \frac{1}{2i} \int\limits_{\alpha+i\infty}^{\alpha-i\infty} d\xi \frac{\left[\dfrac{2a}{(b/2)^{\mu/\nu}}\right]^{2\xi}}{\sin\pi\xi\, \Gamma(1+2\xi)}$$

$$\times I_{q_1}(\xi)\, N_m\left(\xi = -\frac{1}{2}\left[1 + \frac{1+\gamma+2\xi\mu}{\nu}\right]\right) \frac{\Gamma\left(\dfrac{1}{2}\left[1 + \dfrac{1+\gamma+2\xi\mu}{\nu}\right]\right)}{\Gamma\left(1 - \dfrac{1+\gamma+2\xi\mu}{2\nu}\right)},$$

(9.4)

where $N_m(\xi)$ $(m = 1, 3, 5, \ldots)$ and $I_{q_1}(q_1 = 2, 4, 6, \ldots)$ are defined by (1.34) and (1.32), respectively.

9.3   111$^\text{th}$ General Formula

$$N_{111} = \int\limits_0^\infty dx x^\gamma \, \sin^{q_1}(ax^\mu) \, \cos^m(bx^\nu)$$

$$= \frac{\sqrt{\pi}}{2\mu} \frac{1}{2^{q_1-1}} a^{-\frac{1+\gamma}{\mu}} \frac{1}{2^{m-1}} \frac{1}{2i} \int\limits_{-\beta+i\infty}^{-\beta-i\infty} d\xi \frac{\left[\dfrac{b}{a^{\nu/\mu}}\right]^{2\xi}}{\sin \pi\xi \, \Gamma(1+2\xi)}$$

$$\times N_m'(\xi) \, I_{q_1}\left(\xi = -\frac{1+\gamma+2\xi\nu}{2\mu}\right) \frac{\Gamma\left(\dfrac{1+\gamma+2\xi\nu}{2\mu}\right)}{\Gamma\left(\dfrac{1}{2} - \dfrac{1+\gamma+2\xi\nu}{2\mu}\right)}$$

$$(9.5)$$

or

$$N_{111} = \frac{\sqrt{\pi}}{2\nu} \frac{1}{2^{m-1}} \left(\frac{b}{2}\right)^{-\frac{1+\gamma}{\nu}} \frac{1}{2^{q_1-1}} \frac{1}{2i} \int\limits_{\alpha+i\infty}^{\alpha-i\infty} d\xi \frac{\left[\dfrac{2a}{(b/2)^{\mu/\nu}}\right]^{2\xi}}{\sin \pi\xi \, \Gamma(1+2\xi)}$$

$$\times I_{q_1}(\xi) \, N_m'\left(\xi = -\frac{1+\gamma+2\xi\mu}{2\nu}\right) \frac{\Gamma\left(\dfrac{1+\gamma+2\xi\mu}{2\nu}\right)}{\Gamma\left(\dfrac{1}{2} - \dfrac{1+\gamma+2\xi\mu}{2\nu}\right)},$$

$$(9.6)$$

where $N_m'(\xi)$ ($m = 1, 3, 5, \ldots$) and I_{q_1} ($q_1 = 2, 4, 6, \ldots$) are defined by (1.36) and (1.32) in Chapter 1, respectively.

9.4 112th General Formula

$$N_{112} = \int_0^\infty dx\, x^\gamma\, \sin^{q_1}(ax^\mu) \left[\cos^{q_2}(bx^\nu) - 1 \right]$$

$$= \frac{\sqrt{\pi}}{2\mu} \frac{1}{2^{q_1-1}}\, a^{-\frac{1+\gamma}{\mu}} \frac{1}{2^{q_2-1}} \frac{1}{2i} \int\limits_{\alpha+i\infty}^{\alpha-i\infty} d\xi \frac{\left[\dfrac{2b}{a^{\nu/\mu}} \right]^{2\xi}}{\sin \pi\xi\, \Gamma(1+2\xi)}$$

$$\times I'_{q_2}(\xi)\, I_{q_1}\left(\xi = -\frac{1+\gamma+2\xi\nu}{2\mu} \right) \frac{\Gamma\left(\dfrac{1+\gamma+2\xi\nu}{2\mu} \right)}{\Gamma\left(\dfrac{1}{2} - \dfrac{1+\gamma+2\xi\nu}{2\mu} \right)}$$

(9.7)

or

$$N_{112} = \frac{\sqrt{\pi}}{2\nu} \frac{1}{2^{q_2-1}}\, b^{-\frac{1+\gamma}{\nu}} \frac{1}{2^{q_1-1}} \frac{1}{2i} \int\limits_{\alpha+i\infty}^{\alpha-i\infty} d\xi \frac{\left[\dfrac{2a}{b^{\mu/\nu}} \right]^{2\xi}}{\sin \pi\xi\, \Gamma(1+2\xi)}$$

$$\times I_{q_1}(\xi)\, I'_{q_2}\left(\xi = -\frac{1+\gamma+2\xi\mu}{2\nu} \right) \frac{\Gamma\left(\dfrac{1+\gamma+2\xi\mu}{2\nu} \right)}{\Gamma\left(\dfrac{1}{2} - \dfrac{1+\gamma+2\xi\mu}{2\nu} \right)},$$

(9.8)

where $I'_{q_2}(\xi)$ ($q_2 = 2,\ 4,\ 6,\dots$) and $I_{q_1}(\xi)$ ($q_1 = 2,\ 4,\ 6,\dots$) are defined from formulas (1.38) and (1.32) in Chapter 1.

9.5   113th General Formula

$$N_{113} = \int\limits_{0}^{\infty} dx\, x^{\gamma}\, \sin^{m_1}(ax^{\mu})\, \sin^{m_2}(bx^{\nu})$$

$$= \frac{\sqrt{\pi}}{2\mu}\, \frac{1}{2^{m_1-1}}\, \left(\frac{a}{2}\right)^{-\frac{1+\gamma+\nu}{\mu}}\, \frac{b}{2^{m_2-1}}\, \frac{1}{2i}\, \int\limits_{-\beta+i\infty}^{-\beta-i\infty} d\xi\, \frac{\left[\dfrac{b}{(a/2)^{\nu/\mu}}\right]^{2\xi}}{\sin \pi\xi\, \Gamma(2+2\xi)}\, N_{m_2}(\xi)$$

$$\times N_{m_1}\left(\xi = -\frac{1}{2}\left[1 + \frac{1+\gamma+\nu+2\xi\nu}{\mu}\right]\right) \frac{\Gamma\left(\dfrac{1}{2}\left[1 + \dfrac{1+\gamma+\nu+2\xi\nu}{\mu}\right]\right)}{\Gamma\left(1 - \dfrac{1+\gamma+\nu+2\xi\nu}{2\mu}\right)}$$

(9.9)

or

$$N_{113} = \frac{\sqrt{\pi}}{2\nu}\, \frac{1}{2^{m_2-1}}\, \left(\frac{b}{2}\right)^{-\frac{1+\gamma+\mu}{\nu}}\, \frac{a}{2^{m_1-1}}\, \frac{1}{2i}\, \int\limits_{-\beta+i\infty}^{-\beta-i\infty} d\xi\, \frac{\left[\dfrac{a}{(b/2)^{\mu/\nu}}\right]^{2\xi}}{\sin \pi\xi\, \Gamma(2+2\xi)}$$

$$\times N_{m_1}(\xi)\, N_{m_2}\left(\xi = -\frac{1}{2}\left[1 + \frac{1+\gamma+\mu+2\xi\mu}{\nu}\right]\right)$$

$$\times \frac{\Gamma\left(\dfrac{1}{2}\left(1 + \dfrac{1+\gamma+\mu+2\xi\mu}{\nu}\right]\right)}{\Gamma\left(1 - \dfrac{1+\gamma+\mu+2\xi\mu}{2\nu}\right)},$$

(9.10)

where $N_{m_2}(\xi)$ and $N_{m_1}(\xi)$ (m_1, $m_2 = 1$, 3, 5, 7, ...) are given by formula (1.34) in Chapter 1.

9.6   114th General Formula

$$N_{114} = \int\limits_0^\infty dx\, x^\gamma\, \sin^{m_1}(ax^\mu)\, \cos^{m_2}(bx^\nu)$$

$$= \frac{\sqrt{\pi}}{2\mu}\, \frac{1}{2^{m_1-1}}\, \left(\frac{a}{2}\right)^{-\frac{1+\gamma}{\mu}}\, \frac{1}{2^{m_2-1}}\, \frac{1}{2i}\, \int\limits_{-\beta+i\infty}^{-\beta-i\infty} d\xi\, \frac{\left[\dfrac{b}{(a/2)^{\nu/\mu}}\right]^{2\xi}}{\sin \pi\xi\, \Gamma(1+2\xi)}\, N'_{m_2}(\xi)$$

$$\times N_{m_1}\left(\xi = -\frac{1}{2}\left[1 + \frac{1+\gamma+2\xi\nu}{\mu}\right]\right)\, \frac{\Gamma\left(\dfrac{1}{2}\left[1 + \dfrac{1+\gamma+2\xi\nu}{\mu}\right]\right)}{\Gamma\left(1 - \dfrac{1+\gamma+2\xi\nu}{2\mu}\right)}$$

$$(9.11)$$

or

$$N_{114} = \frac{\sqrt{\pi}}{2\nu}\, \frac{1}{2^{m_2-1}}\, \left(\frac{b}{2}\right)^{-\frac{1+\gamma+\mu}{\nu}}\, \frac{a}{2^{m_1-1}}\, \frac{1}{2i}\, \int\limits_{-\beta+i\infty}^{-\beta-i\infty} d\xi\, \frac{\left[\dfrac{a}{(b/2)^{\mu/\nu}}\right]^{2\xi}}{\sin \pi\xi\, \Gamma(2+2\xi)}$$

$$\times N_{m_1}(\xi)\, N'_{m_2}\left(\xi = -\frac{1+\gamma+\mu+2\xi\mu}{2\nu}\right)\, \frac{\Gamma\left(\dfrac{1+\gamma+\mu+2\xi\mu}{2\nu}\right)}{\Gamma\left(\dfrac{1}{2} - \dfrac{1+\gamma+\mu+2\xi\mu}{2\nu}\right)},$$

$$(9.12)$$

where $N_{m_1}(\xi)$ and $N'_{m_2}(\xi)$ (m_1, $m_2 = 1,\ 3,\ 5,\ 7,\dots$) are defined by (1.34) and (1.36) in Chapter 1, respectively.

9.7   115th General Formula

$$N_{115} = \int\limits_0^\infty dx x^\gamma \, \sin^{m_1}(ax^\mu) \left[\cos^q(bx^\nu) - 1\right]$$

$$= \frac{\sqrt{\pi}}{2\mu} \frac{1}{2^{m_1-1}} \left(\frac{a}{2}\right)^{-\frac{1+\gamma}{\mu}} \frac{1}{2^{q-1}} \frac{1}{2i} \int\limits_{\alpha+i\infty}^{\alpha-i\infty} d\xi \frac{\left[\dfrac{2b}{(a/2)^{\nu/\mu}}\right]^{2\xi}}{\sin \pi\xi \, \Gamma(1+2\xi)} \, I'_q(\xi)$$

$$\times N_{m_1}\left(\xi = -\frac{1}{2}\left[1 + \frac{1+\gamma+2\xi\nu}{\mu}\right]\right) \frac{\Gamma\left(\dfrac{1}{2}\left[1 + \dfrac{1+\gamma+2\xi\nu}{\mu}\right]\right)}{\Gamma\left(1 - \dfrac{1+\gamma+2\xi\nu}{2\mu}\right)}$$

(9.13)

or

$$N_{115} = \frac{\sqrt{\pi}}{2\nu} \frac{1}{2^{q-1}} b^{-\frac{1+\gamma+\mu}{\nu}} \frac{a}{2^{m_1-1}} \frac{1}{2i} \int\limits_{-\beta+i\infty}^{-\beta-i\infty} d\xi \frac{\left[\dfrac{a}{b^{\mu/\nu}}\right]^{2\xi}}{\sin \pi\xi \, \Gamma(2+2\xi)}$$

$$\times N_{m_1}(\xi) \, I'_q\left(\xi = -\frac{1+\gamma+\mu+2\xi\mu}{2\nu}\right) \frac{\Gamma\left(\dfrac{1+\gamma+\mu+2\xi\mu}{2\nu}\right)}{\Gamma\left(\dfrac{1}{2} - \dfrac{1+\gamma+\mu+2\xi\mu}{2\nu}\right)},$$

(9.14)

where $I'_q(\xi)$ and $N_{m_1}(\xi)$ $(q = 2, 4, 6, \ldots)$, $(m_1 = 1, 3, 5, \ldots)$ are determined by (1.38) and (1.34) in Chapter 1.

9.8   116th General Formula

$$N_{116} = \int_0^\infty dx\, x^\gamma\, \cos^{m_1}(ax^\mu)\, \cos^{m_2}(bx^\nu)$$

$$= \frac{\sqrt{\pi}}{2\mu}\, \frac{1}{2^{m_1-1}}\, \left(\frac{a}{2}\right)^{-\frac{1+\gamma}{\mu}}\, \frac{1}{2^{m_2-1}}\, \frac{1}{2i}\, \int_{-\beta+i\infty}^{-\beta-i\infty} d\xi\, \frac{\left[\dfrac{b}{(a/2)^{\nu/\mu}}\right]^{2\xi}}{\sin\pi\xi\,\, \Gamma(1+2\xi)}$$

$$\times N'_{m_2}(\xi)\, N'_{m_1}\left(\xi = -\frac{1+\gamma+2\xi\nu}{2\mu}\right)\, \frac{\Gamma\left(\dfrac{1+\gamma+2\xi\nu}{2\mu}\right)}{\Gamma\left(\dfrac{1}{2} - \dfrac{1+\gamma+2\xi\nu}{2\mu}\right)}$$

(9.15)

or

$$N_{116} = \frac{\sqrt{\pi}}{2\nu}\, \frac{1}{2^{m_2-1}}\, \left(\frac{b}{2}\right)^{-\frac{1+\gamma}{\nu}}\, \frac{1}{2^{m_1-1}}\, \frac{1}{2i}\, \int_{-\beta+i\infty}^{-\beta-i\infty} d\xi\, \frac{\left[\dfrac{a}{(b/2)^{\mu/\nu}}\right]^{2\xi}}{\sin\pi\xi\,\, \Gamma(1+2\xi)}$$

$$\times N'_{m_1}(\xi)\, N'_{m_2}\left(\xi = -\frac{1+\gamma+2\xi\mu}{2\nu}\right)\, \frac{\Gamma\left(\dfrac{1+\gamma+2\xi\mu}{2\nu}\right)}{\Gamma\left(\dfrac{1}{2} - \dfrac{1+\gamma+2\xi\mu}{2\nu}\right)},$$

(9.16)

where $N'_{m_2}(\xi)$ and $N'_{m_1}(\xi)$ (m_2, $m_1 = 1, 3, 5, 7, \ldots$) are derived from formula (1.36) in Chapter 1.

9.9 117th General Formula

$$N_{117} = \int\limits_0^\infty dx\, x^\gamma\, \cos^{m_1}(ax^\mu)\, \left[\cos^q(bx^\nu) - 1\right]$$

$$= \frac{\sqrt{\pi}}{2\mu}\, \frac{1}{2^{m_1-1}}\, \left(\frac{a}{2}\right)^{-\frac{1+\gamma}{\mu}}\, \frac{1}{2^{q-1}}\, \frac{1}{2i} \int\limits_{\alpha+i\infty}^{\alpha-i\infty} d\xi\, \frac{\left[\dfrac{2b}{(a/2)^{\nu/\mu}}\right]^{2\xi}}{\sin\pi\xi\, \Gamma(1+2\xi)}$$

$$\times I'_q(\xi)\, N'_{m_1}\left(\xi = -\frac{1+\gamma+2\xi\nu}{2\mu}\right)\, \frac{\Gamma\left(\dfrac{1+\gamma+2\xi\nu}{2\mu}\right)}{\Gamma\left(\dfrac{1}{2} - \dfrac{1+\gamma+2\xi\nu}{2\mu}\right)}$$

(9.17)

or

$$N_{117} = \frac{\sqrt{\pi}}{2\nu}\, \frac{1}{2^{q-1}}\, b^{-\frac{1+\gamma}{\nu}}\, \frac{1}{2^{m_1-1}}\, \frac{1}{2i} \int\limits_{-\beta+i\infty}^{-\beta-i\infty} d\xi\, \frac{\left[\dfrac{a}{b^{\mu/\nu}}\right]^{2\xi}}{\sin\pi\xi\, \Gamma(1+2\xi)}$$

$$\times N'_{m_1}(\xi)\, I'_q\left(\xi = -\frac{1+\gamma+2\xi\mu}{2\nu}\right)\, \frac{\Gamma\left(\dfrac{1+\gamma+2\xi\mu}{2\nu}\right)}{\Gamma\left(\dfrac{1}{2} - \dfrac{1+\gamma+2\xi\mu}{2\nu}\right)},$$

(9.18)

where $N'_{m_1}(\xi)$ ($m_1 = 1,\, 3,\, 5,\, 7, \ldots$) and $I'_q(\xi)$ ($q = 2,\, 4,\, 6, \ldots$) are given by (1.36) and (1.38) in Chapter 1.

9.10   118th General Formula

$$N_{118} = \int_0^\infty dx\, x^\gamma \left[\cos^{q_1}(ax^\mu) - 1\right] \left[\cos^{q_2}(bx^\nu) - 1\right]$$

$$= \frac{\sqrt{\pi}}{2\mu} \frac{1}{2^{q_1-1}} \, a^{-\frac{1+\gamma}{\mu}} \frac{1}{2^{q_2-1}} \frac{1}{2i} \int_{\alpha+i\infty}^{\alpha-i\infty} d\xi \frac{\left[\dfrac{2b}{a^{\nu/\mu}}\right]^{2\xi}}{\sin \pi\xi\, \Gamma(1+2\xi)}$$

$$\times I'_{q_2}(\xi)\, I'_{q_1}\left(\xi = -\frac{1+\gamma+2\xi\nu}{2\mu}\right) \frac{\Gamma\left(\dfrac{1+\gamma+2\xi\nu}{2\mu}\right)}{\Gamma\left(\dfrac{1}{2} - \dfrac{1+\gamma+2\xi\nu}{2\mu}\right)}$$

(9.19)

or

$$N_{118} = \frac{\sqrt{\pi}}{2\nu} \frac{1}{2^{q_2-1}} \, b^{-\frac{1+\gamma}{\nu}} \frac{1}{2^{q_1-1}} \frac{1}{2i} \int_{\alpha+i\infty}^{\alpha-i\infty} d\xi \frac{\left[\dfrac{2a}{b^{\mu/\nu}}\right]^{2\xi}}{\sin \pi\xi\, \Gamma(1+2\xi)}$$

$$\times I'_{q_1}(\xi)\, I'_{q_2}\left(\xi = -\frac{1+\gamma+2\xi\mu}{2\nu}\right) \frac{\Gamma\left(\dfrac{1+\gamma+2\xi\mu}{2\nu}\right)}{\Gamma\left(\dfrac{1}{2} - \dfrac{1+\gamma+2\xi\mu}{2\nu}\right)},$$

(9.20)

where $I'_{q_1}(\xi)$ and $I'_{q_2}(\xi)$ ($q_1,\ q_2 = 2,\ 4,\ 6, \ldots$) are expressed by (1.38) in Chapter 1.

9.11 Exercises

By means of formulas (9.1)-(9.20) calculate following integrals:

$$n_1 = \int_0^\infty dx \; \sin(ax^2) \; \sin(2bx), \qquad (9.21)$$

$$n_2 = \int_0^\infty dx \; \sin(ax^2) \; \cos(2bx), \qquad (9.22)$$

$$n_3 = \int_0^\infty dx \; \cos(ax^2) \; \cos(2bx), \qquad (9.23)$$

$$n_4 = \int_0^\infty dx \; \sin(ax^2) \; \cos(bx^2), \qquad (9.24)$$

$$n_5 = \int_0^\infty dx \; \sin\left(\frac{a^2}{x}\right) \; \sin(bx), \qquad (9.25)$$

$$n_6 = \int_0^\infty dx \; \sin\left(\frac{a^2}{x^2}\right) \; \sin(b^2x^2), \qquad (9.26)$$

$$n_7 = \int_0^\infty dx \; \sin\left(\frac{a^2}{x^2}\right) \; \cos(b^2x^2), \qquad (9.27)$$

$$n_8 = \int_0^\infty dx \; \cos\left(\frac{a^2}{x^2}\right) \; \sin(b^2x^2), \qquad (9.28)$$

$$n_9 = \int_0^\infty dx \; \cos\left(\frac{a^2}{x^2}\right) \; \cos(b^2x^2), \qquad (9.29)$$

$$n_{10} = \int_0^\infty dx \frac{\sin(ax) \; \sin(bx)}{x}, \qquad (9.30)$$

$$n_{11} = \int_0^\infty dx \frac{\sin(ax) \; \cos(bx)}{x}, \qquad (9.31)$$

$$n_{12} = \int_0^\infty dx \frac{\sin(ax)\ \sin(bx)}{x^2}, \tag{9.32}$$

$$n_{13} = \int_0^\infty dx\, x^{\gamma-1}\ \sin(ax)\ \cos(bx), \tag{9.33}$$

$$n_{14} = \int_0^\infty dx\, x^{\gamma-1}\ \cos(ax)\ \cos(bx), \tag{9.34}$$

$$n_{15} = \int_0^\infty dx \frac{\sin(ax)\ \sin(bx)}{x^2}, \tag{9.35}$$

$$n_{16} = \int_0^\infty dx \frac{\sin^2(ax)\ \sin(bx)}{x}, \tag{9.36}$$

$$n_{17} = \int_0^\infty dx \frac{\sin^2(ax)\ \cos(bx)}{x}, \tag{9.37}$$

$$n_{18} = \int_0^\infty dx \frac{\sin^2(ax)\ \cos(2bx)}{x^2}, \tag{9.38}$$

$$n_{19} = \int_0^\infty dx \frac{\sin(2ax)\ \cos^2(bx)}{x}, \tag{9.39}$$

$$n_{20} = \int_0^\infty dx \frac{\sin^2(ax)\ \sin^2(bx)}{x^2}, \tag{9.40}$$

$$n_{21} = \int_0^\infty dx \frac{\sin^2(ax)\ \sin^2(bx)}{x^4}, \tag{9.41}$$

$$n_{22} = \int_0^\infty dx \frac{\sin^2(ax)\ \cos^2(bx)}{x^2}, \tag{9.42}$$

$$n_{23} = \int_0^\infty dx \frac{\sin^3(ax)\ \sin(3bx)}{x^4}, \tag{9.43}$$

$$n_{24} = \int_0^\infty dx \frac{\sin^3(ax)\ \cos(bx)}{x}, \tag{9.44}$$

$$n_{25} = \int_0^\infty dx \frac{\sin^3(ax)\ \cos(bx)}{x^3}, \tag{9.45}$$

$$n_{26} = \int_0^\infty dx \frac{\sin^3(ax)\ \sin(bx)}{x^4}, \tag{9.46}$$

$$n_{27} = \int_0^\infty dx \frac{\sin^3(ax)\ \sin^2(bx)}{x}, \tag{9.47}$$

$$n_{28} = \int_0^\infty dx x\ \sin(ax^2)\ \sin(2bx), \tag{9.48}$$

$$n_{29} = \int_0^\infty dx x\ \cos(ax^2)\ \sin(2bx), \tag{9.49}$$

$$n_{30} = \int_0^\infty \frac{dx}{x}\ \sin\left(\frac{b}{x}\right)\ \sin(ax), \tag{9.50}$$

$$n_{31} = \int_0^\infty \frac{dx}{x}\ \cos\left(\frac{b}{x}\right)\ \cos(ax), \tag{9.51}$$

where $n_{12} = n_{15}$.

9.12   Answers

$$n_1 = \sqrt{\frac{\pi}{2a}}\left\{\cos\frac{b^2}{a}\ C\left(\frac{b}{\sqrt{a}}\right) + \sin\frac{b^2}{a}\ S\left(\frac{b}{\sqrt{a}}\right)\right\},$$

$$n_2 = \frac{1}{2}\sqrt{\frac{\pi}{2a}}\left\{\cos\frac{b^2}{a} - \sin\frac{b^2}{a}\right\} = \frac{1}{2}\sqrt{\frac{\pi}{a}}\cos\left(\frac{b^2}{a} + \frac{\pi}{4}\right)$$

in these formulas $a > 0$, and $b > 0$,

$$n_3 = \frac{1}{2}\sqrt{\frac{\pi}{2a}}\left\{\cos\frac{b^2}{a} + \sin\frac{b^2}{a}\right\},$$

$$n_4 = \frac{1}{4}\sqrt{\frac{\pi}{2}} \begin{cases} \dfrac{1}{\sqrt{a+b}} + \dfrac{1}{\sqrt{a-b}} & \text{if} \quad a > b > 0 \\[3mm] \dfrac{1}{\sqrt{a+b}} - \dfrac{1}{\sqrt{b-a}} & \text{if} \quad b > a > 0, \end{cases}$$

$$n_5 = \frac{a\pi}{2\sqrt{b}}\, J_1(2a\sqrt{b}),$$

$$n_6 = \frac{1}{4b}\sqrt{\pi/2}\left[\sin(2ab) - \cos(2ab) + e^{-2ab}\right],$$

$$n_7 = \frac{1}{4b}\sqrt{\pi/2}\left[\sin(2ab) + \cos(2ab) + e^{-2ab}\right],$$

$$n_8 = \frac{1}{4b}\sqrt{\pi/2}\left[\sin(2ab) + \cos(2ab) + e^{-2ab}\right],$$

$$n_9 = \frac{1}{4b}\sqrt{\pi/2}\left[\cos(2ab) - \sin(2ab) + e^{-2ab}\right],$$

$$n_{10} = \frac{1}{4}\ln\left(\frac{a+b}{a-b}\right)^2, \quad a > 0,\ b > 0,\ a \neq b,$$

$$n_{11} = \begin{cases} \dfrac{\pi}{2} & \text{if} \quad a > b \geq 0 \\[3mm] \dfrac{\pi}{4} & \text{if} \quad a = b > 0 \\[3mm] 0 & \text{if} \quad b > a \geq 0, \end{cases}$$

$$n_{12} = \begin{cases} \dfrac{a\pi}{2} & \text{if} \quad 0 < a \leq b \\[3mm] \dfrac{b\pi}{2} & \text{if} \quad 0 < b \leq a, \end{cases}$$

$$n_{13} = \frac{1}{2}\,\sin\frac{\gamma\pi}{2}\,\Gamma(\gamma)\left[(a+b)^{-\gamma} + |a-b|^{-\gamma}\,\mathrm{sign}(a-b)\right],$$

where $a > 0,\ b > 0,\ |\mathrm{Re}\,\gamma| < 1$,

$$n_{14} = \frac{1}{2}\,\cos\frac{\gamma\pi}{2}\,\Gamma(\gamma)\left[(a+b)^{-\gamma} + |a-b|^{-\gamma}\right],$$

where $a > 0,\ b > 0,\ 0 < |\mathrm{Re}\,\gamma| < 1$,

$$n_{15} = \begin{cases} \dfrac{a\pi}{2} & \text{if} \quad a \le b \\[2ex] \dfrac{b\pi}{2} & \text{if} \quad b \le a, \end{cases}$$

$$n_{16} = \begin{cases} \dfrac{\pi}{4} & \text{if} \quad 0 < b < 2a \\[2ex] \dfrac{\pi}{8} & \text{if} \quad b = 2a \\[2ex] 0 & \text{if} \quad b > 2a, \end{cases}$$

$$n_{17} = \frac{1}{4} \ln \frac{4a^2 - b^2}{b^2},$$

$$n_{18} = \begin{cases} \dfrac{\pi}{2}(a - b) & \text{if} \quad b < a \\[2ex] 0 & \text{if} \quad b > a, \end{cases}$$

$$n_{19} = \begin{cases} \dfrac{\pi}{2} & \text{if} \quad a > b \\[2ex] \dfrac{3\pi}{8} & \text{if} \quad a = b \\[2ex] \dfrac{\pi}{4} & \text{if} \quad a < b, \end{cases}$$

$$n_{20} = \begin{cases} \dfrac{\pi}{4}a & \text{if} \quad 0 \le a \le b \\[2ex] \dfrac{\pi}{4}b & \text{if} \quad 0 \le b \le a, \end{cases}$$

$$n_{21} = \begin{cases} \dfrac{\pi a^2}{6}(3b - a) & \text{if} \quad 0 \le a \le b \\[2ex] \dfrac{\pi b^2}{6}(3a - b) & \text{if} \quad 0 \le b \le a, \end{cases}$$

$$n_{22} = \begin{cases} \dfrac{2a - b}{4}\pi & \text{if} \quad a \ge b > 0 \\[2ex] \dfrac{a}{4}\pi & \text{if} \quad 0 < a < b, \end{cases}$$

$$n_{23} = \begin{cases} \dfrac{a^3}{2}\pi & \text{if} \quad b > a \\[3mm] \dfrac{\pi}{16}\left[8a^3 - 9(a-b)^3\right] & \text{if} \quad 0 \le 3b \le 3a \\[3mm] \dfrac{9b\pi}{8}(a^2 - b^2) & \text{if} \quad 3b \le a, \end{cases}$$

$$n_{24} = \begin{cases} 0 & \text{if} \quad b > 3a \\[3mm] \dfrac{-\pi}{16} & \text{if} \quad b = 3a \\[3mm] \dfrac{-\pi}{8} & \text{if} \quad 3a > b > a \\[3mm] \dfrac{\pi}{16} & \text{if} \quad b = a \\[3mm] \dfrac{\pi}{4} & \text{if} \quad a > b, \quad a > 0, \quad b > 0, \end{cases}$$

$$n_{25} = \begin{cases} \dfrac{\pi}{8}(3a^2 - b^2) & \text{if} \quad b < a \\[3mm] \dfrac{\pi}{4}b^2 & \text{if} \quad b = a \\[3mm] \dfrac{\pi}{16}(3a - b)^2 & \text{if} \quad a < b < 3a \\[3mm] 0 & \text{if} \quad 3a < b, \quad a > 0, \quad b > 0, \end{cases}$$

$$n_{26} = \begin{cases} \dfrac{b\pi}{24}(9a^2 - b^2) & \text{if} \quad 0 < b \le a \\[3mm] \dfrac{\pi}{48}\left[24a^3 - (3a - b)^3\right] & \text{if} \quad 0 < a \le b \le 3a \\[3mm] \dfrac{\pi}{2}a^3 & \text{if} \quad 0 < 3a \le b, \end{cases}$$

$$n_{27} = \begin{cases} \dfrac{\pi}{8} & \text{if} \quad 2b > 3a \\[2ex] \dfrac{5\pi}{32} & \text{if} \quad 2b = 3a \\[2ex] \dfrac{3\pi}{16} & \text{if} \quad 3a > 2b > a \\[2ex] \dfrac{3\pi}{32} & \text{if} \quad 2b = a \\[2ex] 0 & \text{if} \quad a > 2b, \quad a > 0, \quad b > 0, \end{cases}$$

$$n_{28} = \frac{b}{2a}\sqrt{\frac{\pi}{2a}}\left(\sin\frac{b^2}{a} + \cos\frac{b^2}{a}\right), \quad a > 0, \quad b > 0,$$

$$n_{29} = \frac{b}{2a}\sqrt{\frac{\pi}{2a}}\left(\sin\frac{b^2}{a} - \cos\frac{b^2}{a}\right), \quad a > 0, \quad b > 0,$$

$$n_{30} = \frac{\pi}{2} N_0(2\sqrt{ab}) + K_0(2\sqrt{ab}), \quad a > 0, \quad b > 0,$$

$$n_{31} = -\frac{\pi}{2} N_0(2\sqrt{ab}) + K_0(2\sqrt{ab}), \quad a > 0, \quad b > 0.$$

9.13 An Example of a Solution

Let $\gamma = 0$, $m_1 = m_2 = 1$, $a \to a^2$, $\mu = -1$, $\nu = 1$ be in the main formula (9.10). Then we have

$$n_5 = \int_0^\infty dx \, \sin\left(\frac{a^2}{x}\right) \sin(bx)$$

$$= \frac{\sqrt{\pi}}{2}a^2 \frac{1}{2i} \int_{-\beta+i\infty}^{-\beta-i\infty} d\xi \frac{\left[\frac{ba^2}{2}\right]^{2\xi}}{\sin\pi\xi \, \Gamma(2+2\xi)} \frac{\Gamma\left(\frac{1}{2}(1-2\xi)\right)}{\Gamma(1+\xi)}, \tag{9.52}$$

where

$$\Gamma\left(\frac{1}{2} - \xi\right) = \frac{\pi}{\cos\pi\xi \, \Gamma\left(\frac{1}{2} + \xi\right)},$$

$$\Gamma(1+\xi)\,\Gamma\left(\frac{1}{2} + \xi\right) = \frac{\sqrt{\pi}}{2^{2\xi}}\,\Gamma(1+2\xi).$$

Therefore going to the integration variable $2\xi = x$, one gets

$$n_5 = \frac{\pi a^2}{2} \frac{1}{2i} \int_{-\beta+i\infty}^{-\beta-i\infty} dx \frac{(a\sqrt{b})^{2x}}{\sin\pi x \, \Gamma(1+x)\,\Gamma(2+x)}. \tag{9.53}$$

Taking into account the definition of the Bessel function $J_1(x)$, we have

$$n_5 = \frac{\pi a}{2\sqrt{b}} \, J_1(2a\sqrt{b})$$

as it should be.

Chapter 10

Derivation of Universal Formulas for Calculation of Fractional Derivatives and Inverse Operators

10.1 Introduction

Recently, fractional derivatives have played an important role in mathematical methods and their physical and chemical applications (for example, see: Dzrbashan, 1966, Samko, Kilbas and Marichev, 1993 and Hilfer, 2000). Many attempts (Zabodal, Vilhena and Livotto, 2001, Dattoli, Quanttromini and Torre, 1999, Turmetov and Umarov, 1993) (where earlier references concerning this problem are cited) have been devoted to the problem of definition of fractional derivatives.

The most usual definition for fractional derivatives consists in a natural extension of the integer-order derivative operators (Dattoli, Quanttromini and Torre, 1999).

Let us consider the polynomial expansion for an arbitrary function

$$F(x) = \sum_{k=0}^{\infty} a_k x^k. \tag{10.1}$$

Then the m-th derivative of $F(x)$ is given by

$$\frac{d^m}{dx^m} F(x) = \sum_{k=0}^{\infty} a_k \frac{k!}{(k-m)!} x^{k-m}, \ 0 \le m \le k. \tag{10.2}$$

Therefore its extension is

$$\left(\frac{d}{dx}\right)^\nu F(x) = \sum_{k=0}^{\infty} a_k \frac{k!}{\Gamma(k-\nu+1)} (x-x_0)^{k-\nu}, \ 0 \le \nu \le k \tag{10.3}$$

where x_0 is called the lower differintegration limit, plays some role of the lower limit of integration. The fractional derivative of a function does not depend upon the lower limit only when ν is a non-negative integer.

Riemann-Liouville definition for fractional derivative is

$$\left(\frac{d}{dx}\right)^\nu F(x) = \frac{1}{\Gamma(-\nu)} \int_0^x dt F(t)(x-t)^{-1-\nu}, \tag{10.4}$$

Re $\nu < 0$, where the lower differintegration limit was taken as zero.

In this Chapter, by using the formulas obtained in previous chapters, we study fractional derivatives and inverse operators by means of infinite integer-order differentials. This allows us to derive some universal formulas by taking fractional derivatives and calculation of inverse operators for wide classes of functions.

10.2   Derivation of General Formulas for Taking Fractional Derivatives

10.2.1   The First General Formula

According to the formula (5.2) in Chapter 5, we have

$$D_1 = \left(\frac{d}{dx}\right)^{-\frac{1}{\nu}} F(x) = \frac{\nu}{\Gamma\left(\frac{1}{\nu}\right)} \int_0^\infty dt\; e^{-t^\nu \frac{d}{dx}}\; F(x).$$

(10.5)

10.2.2   Consequences of the First General Formula

Let

$$F(x) = \sin x.$$

(10.6)

Then the direct calculations give

$$\left(\frac{d}{dx}\right)^{-\frac{1}{\nu}} \sin x = \frac{\nu}{\Gamma\left(\frac{1}{\nu}\right)} \left[\sin x \int_0^\infty dt \cos t^\nu \right.$$
$$\left. - \cos x \int_0^\infty dt \sin t^\nu \right],$$

(10.7)

where we have used the decomposition

$$\exp\left[-t^\nu \frac{d}{dx}\right] = 1 - t^\nu \frac{d}{dx} + \frac{1}{2!}t^{2\nu}\frac{d^2}{dx^2} - \cdots .$$

(10.8)

Further, taking into account the following standard integrals (see formulas (2.57) and (2.32), in the previous Chapter 2, Gradshteyn and Ryzhik, 1980)

$$\vartheta_1 = \int_0^\infty dt\; \cos t^\nu = \frac{1}{\nu}\Gamma\left(\frac{1}{\nu}\right)\cos\frac{\pi}{2\nu},$$

$$\vartheta_2 = \int_0^\infty dt\; \sin t^\nu = \frac{1}{\nu}\Gamma\left(\frac{1}{\nu}\right)\sin\frac{\pi}{2\nu}$$

we obtain a nice general formula

$$\left(\frac{d}{dx}\right)^{-\frac{1}{\nu}} \sin x = \sin x\; \cos\frac{\pi}{2\nu} - \cos x\; \sin\frac{\pi}{2\nu}$$
$$= \sin\left(x - \frac{\pi}{2\nu}\right).$$

(10.9)

Let

$$F(x) = \cos x,$$

then a similar calculation as above reads

$$\left(\frac{d}{dx}\right)^{-\frac{1}{\nu}} \cos x = \cos x \, \cos \frac{\pi}{2\nu} + \sin x \, \sin \frac{\pi}{2\nu}$$

$$= \cos \left(x - \frac{\pi}{2\nu}\right). \tag{10.10}$$

Examples:

1) $\nu = -1,$ 2) $\nu = 1,$

$$\frac{d}{dx} \sin x = \cos x, \qquad \left(\frac{d}{dx}\right)^{-1} \sin x = -\cos x,$$

$$\frac{d}{dx} \cos x = -\sin x, \qquad \left(\frac{d}{dx}\right)^{-1} \cos x = \sin x$$

as it should be.

From these simple formulas, we have unexpected properties:

$$\left(\frac{d}{dx}\right)\left[\left(\frac{d}{dx}\right)^{-1} \sin x\right] = \left(\frac{d}{dx}\right)^{-1}\left[\frac{d}{dx} \sin x\right] \equiv \sin x \tag{10.11}$$

and

$$\left(\frac{d}{dx}\right)\left[\left(\frac{d}{dx}\right)^{-1} \cos x\right] = \left(\frac{d}{dx}\right)^{-1}\left[\frac{d}{dx} \cos x\right] \equiv \cos x. \tag{10.12}$$

Thus, in our scheme, one can define the **inverse operation** with respect to the usual differential one. Moreover, these two operations are commutative:

$$\frac{d}{dx}\left(\frac{d}{dx}\right)^{-1} \equiv \left(\frac{d}{dx}\right)^{-1}\left(\frac{d}{dx}\right). \tag{10.13}$$

In general case, for any q_1 and $-q_1$, we have

$$\left(\frac{d}{dx}\right)^{q_1}\left(\frac{d}{dx}\right)^{-q_1} \equiv \left(\frac{d}{dx}\right)^{-q_1}\left(\frac{d}{dx}\right)^{q_1} = \left(\frac{d}{dx}\right)^{0}.$$

Now we continue to consider these examples.

3) $\nu = -2,$

$$\left(\frac{d}{dx}\right)^{1/2} \sin x = \sin x \, \cos \frac{\pi}{4} + \cos x \, \sin \frac{\pi}{4}$$

$$= \frac{1}{\sqrt{2}}(\sin x + \cos x), \tag{10.14}$$

$$\left(\frac{d}{dx}\right)^{1/2} \cos x = \frac{1}{\sqrt{2}}(\cos x - \sin x). \tag{10.15}$$

4) $\nu = 2$,

$$\left(\frac{d}{dx}\right)^{-\frac{1}{2}} \sin x = \frac{1}{\sqrt{2}}(\sin x - \cos x), \tag{10.16}$$

$$\left(\frac{d}{dx}\right)^{-\frac{1}{2}} \cos x = \frac{1}{\sqrt{2}}(\cos x + \sin x). \tag{10.17}$$

From these two formulas (10.16) and (10.17), one gets

$$\frac{d}{dx}\left[\left(\frac{d}{dx}\right)^{-1/2} \sin x\right] = \frac{1}{\sqrt{2}}\frac{d}{dx}(\sin x - \cos x)$$

$$= \frac{1}{\sqrt{2}}(\cos x + \sin x) \tag{10.18}$$

and

$$\frac{d}{dx}\left[\left(\frac{d}{dx}\right)^{-1/2} \cos x\right] = \frac{1}{\sqrt{2}}\frac{d}{dx}(\cos x + \sin x)$$

$$= \frac{1}{\sqrt{2}}(-\sin x + \cos x). \tag{10.19}$$

These two identities mean that

$$\boxed{\frac{d}{dx}\left(\frac{d}{dx}\right)^{-1/2} \equiv \left(\frac{d}{dx}\right)^{1/2}.} \tag{10.20}$$

Moreover, we have

A

$$\left(\frac{d}{dx}\right)^{1/2}\left[\left(\frac{d}{dx}\right)^{-1/2} \sin x\right] = \left(\frac{d}{dx}\right)^{1/2}\left\{\frac{1}{\sqrt{2}}(\sin x - \cos x)\right\}$$

$$= \frac{1}{\sqrt{2}}\left\{\frac{1}{\sqrt{2}}(\sin x + \cos x) - \frac{1}{\sqrt{2}}(\cos x - \sin x)\right\} = \sin x. \tag{10.21}$$

B

$$\left(\frac{d}{dx}\right)^{-1/2}\left[\left(\frac{d}{dx}\right)^{1/2} \sin x\right] = \left(\frac{d}{dx}\right)^{-1/2}\left\{\frac{1}{\sqrt{2}}(\sin x + \cos x)\right\}$$

$$= \frac{1}{\sqrt{2}}\left\{\frac{1}{\sqrt{2}}(\sin x - \cos x) + \frac{1}{\sqrt{2}}(\cos x + \sin x)\right\} = \sin x. \tag{10.22}$$

It means that two operations are commutative:

$$\left(\frac{d}{dx}\right)^{1/2}\left[\left(\frac{d}{dx}\right)^{-1/2}\right] = \left(\frac{d}{dx}\right)^{-1/2}\left[\left(\frac{d}{dx}\right)^{1/2}\right] = \left(\frac{d}{dx}\right)^{0}. \tag{10.23}$$

Similar formulas are valid for cosine-functions. It is easily seen that

$$\left(\frac{d}{dx}\right)^{-\frac{1}{2}}\left[\left(\frac{d}{dx}\right)^{-1/2}\sin x\right] = \left(\frac{d}{dx}\right)^{-1/2}\left[\frac{1}{\sqrt{2}}(\sin x - \cos x)\right]$$

$$= \frac{1}{\sqrt{2}}\left\{\frac{1}{\sqrt{2}}(\sin x - \cos x) - \frac{1}{\sqrt{2}}(\cos x + \sin x)\right\} = -\cos x. \tag{10.24}$$

Similarly

$$\left(\frac{d}{dx}\right)^{-\frac{1}{2}}\left[\left(\frac{d}{dx}\right)^{-1/2}\cos x\right] = \sin x. \tag{10.25}$$

The last two formulas mean that

$$\left(\frac{d}{dx}\right)^{-\frac{1}{2}}\left[\left(\frac{d}{dx}\right)^{-1/2}\sin x\right] = \left(\frac{d}{dx}\right)^{-1}\sin x = -\cos x \tag{10.26}$$

and

$$\left(\frac{d}{dx}\right)^{-\frac{1}{2}}\left[\left(\frac{d}{dx}\right)^{-1/2}\cos x\right] = \left(\frac{d}{dx}\right)^{-1}\cos x = \sin x \tag{10.27}$$

as it should be. Let us continue with further examples:

C   Let $\nu = -\dfrac{2}{3}$,

$$\left(\frac{d}{dx}\right)^{\frac{3}{2}}\sin x = \sin x \, \cos\frac{3\pi}{4} + \cos x \, \sin\frac{3\pi}{4}$$

$$= \frac{1}{\sqrt{2}}(\cos x - \sin x), \tag{10.28}$$

and therefore

$$\left(\frac{d}{dx}\right)^{\frac{3}{2}}\sin x = \left(\frac{d}{dx}\right)^{1/2}\frac{d}{dx}\sin x = \left(\frac{d}{dx}\right)^{1/2}\cos x$$

$$= \frac{1}{\sqrt{2}}(\cos x - \sin x), \tag{10.29}$$

or

$$\left(\frac{d}{dx}\right)^{\frac{3}{2}}\sin x = \frac{d}{dx}\left[\left(\frac{d}{dx}\right)^{1/2}\sin x\right]$$

$$= \frac{1}{\sqrt{2}}\frac{d}{dx}(\sin x + \cos x) = \frac{1}{\sqrt{2}}(\cos x - \sin x). \tag{10.30}$$

Similar identities hold for cosine-functions:

D $\nu = -\dfrac{2}{3}$,

$$\left(\frac{d}{dx}\right)^{\frac{3}{2}}\cos x = \cos x \, \cos\frac{3\pi}{4} - \sin x \, \sin\frac{3\pi}{4}$$

$$= \frac{1}{\sqrt{2}}(-\cos x - \sin x), \tag{10.31}$$

$$\left(\frac{d}{dx}\right)^{\frac{3}{2}}\cos x = \left(\frac{d}{dx}\right)^{1/2}\left[\frac{d}{dx}\cos x\right] = -\left(\frac{d}{dx}\right)^{1/2}\sin x$$

$$= -\frac{1}{\sqrt{2}}(\sin x + \cos x),\qquad(10.32)$$

or

$$\left(\frac{d}{dx}\right)^{\frac{3}{2}}\cos x = \frac{d}{dx}\left(\frac{d}{dx}\right)^{1/2}\cos x = \frac{1}{\sqrt{2}}\frac{d}{dx}(\cos x - \sin x)$$

$$= \frac{1}{\sqrt{2}}(-\sin x - \cos x).\qquad(10.33)$$

E   Let $\nu = \frac{2}{3}$, then

$$\left(\frac{d}{dx}\right)^{-\frac{3}{2}}\sin x = \sin x\,\cos\frac{3\pi}{4} - \cos x\,\sin\frac{3\pi}{4}$$

$$= -\frac{1}{\sqrt{2}}(\sin x + \cos x),\qquad(10.34)$$

or

$$\left(\frac{d}{dx}\right)^{-\frac{3}{2}}\sin x = \left(\frac{d}{dx}\right)^{-1/2}\left[\left(\frac{d}{dx}\right)^{-1}\sin x\right] = -\left(\frac{d}{dx}\right)^{-1/2}\cos x$$

$$= -\frac{1}{\sqrt{2}}(\cos x + \sin x).\qquad(10.35)$$

Moreover

$$\left(\frac{d}{dx}\right)^{-\frac{3}{2}}\sin x = \left(\frac{d}{dx}\right)^{-1}\left[\left(\frac{d}{dx}\right)^{-1/2}\sin x\right]$$

$$= \frac{1}{\sqrt{2}}\left(\frac{d}{dx}\right)^{-1}[\sin x - \cos x]$$

$$= -\frac{1}{\sqrt{2}}(\cos x + \sin x).\qquad(10.36)$$

F   Similar formulas hold for cosine-functions:

$$\left(\frac{d}{dx}\right)^{-\frac{3}{2}}\cos x = \cos x\,\cos\frac{3\pi}{4} + \sin x\,\sin\frac{3\pi}{4}$$

$$= \frac{1}{\sqrt{2}}(-\cos x + \sin x),\qquad(10.37)$$

or

$$\left(\frac{d}{dx}\right)^{-\frac{3}{2}}\cos x = \left(\frac{d}{dx}\right)^{-1/2}\left[\left(\frac{d}{dx}\right)^{-1}\cos x\right] = \left(\frac{d}{dx}\right)^{-1/2}\sin x$$

$$= \frac{1}{\sqrt{2}}(\sin x - \cos x),\qquad(10.38)$$

$$\left(\frac{d}{dx}\right)^{-\frac{3}{2}}\cos x = \left(\frac{d}{dx}\right)^{-1}\left[\left(\frac{d}{dx}\right)^{-1/2}\cos x\right]$$

$$= \left(\frac{d}{dx}\right)^{-1}\left[\frac{1}{\sqrt{2}}(\cos x + \sin x)\right]$$

$$= \frac{1}{\sqrt{2}}(\sin x - \cos x). \tag{10.39}$$

G   Let $\nu = -4$, then we have

$$\left(\frac{d}{dx}\right)^{1/4}\sin x = \sin x \,\cos\frac{\pi}{8} + \cos x \,\sin\frac{\pi}{8} \tag{10.40}$$

and using the trigonometric formulas $\cos^2\frac{\pi}{8} - \sin^2\frac{\pi}{8} = \cos\frac{\pi}{4}$, $2\sin\frac{\pi}{8}\times\cos\frac{\pi}{8} = \sin\frac{\pi}{4}$, one gets

$$\left(\frac{d}{dx}\right)^{\frac{1}{4}}\left[\left(\frac{d}{dx}\right)^{\frac{1}{4}}\sin x\right] = \left(\frac{d}{dx}\right)^{\frac{1}{4}}\left[\sin x \,\cos\frac{\pi}{8} + \cos x \,\sin\frac{\pi}{8}\right]$$

$$= \frac{1}{\sqrt{2}}(\sin x + \cos x). \tag{10.41}$$

It means that

$$\boxed{\left(\frac{d}{dx}\right)^{\frac{1}{4}}\left(\frac{d}{dx}\right)^{\frac{1}{4}}\sin x \equiv \left(\frac{d}{dx}\right)^{1/2}\sin x.} \tag{10.42}$$

Similar formulas for cosine-functions read

$$\left(\frac{d}{dx}\right)^{\frac{1}{4}}\cos x = \cos x \,\cos\frac{\pi}{8} - \sin x \,\sin\frac{\pi}{8} \tag{10.43}$$

and therefore

$$\left(\frac{d}{dx}\right)^{\frac{1}{4}}\left[\left(\frac{d}{dx}\right)^{\frac{1}{4}}\cos x\right] = \left(\frac{d}{dx}\right)^{\frac{1}{4}}\left[\cos x \,\cos\frac{\pi}{8} - \sin x \,\sin\frac{\pi}{8}\right]$$

$$= \frac{1}{\sqrt{2}}(\cos x - \sin x). \tag{10.44}$$

It means that

$$\boxed{\left(\frac{d}{dx}\right)^{\frac{1}{4}}\left(\frac{d}{dx}\right)^{\frac{1}{4}}\cos x \equiv \left(\frac{d}{dx}\right)^{1/2}\cos x} \tag{10.45}$$

as it should be.

H Finally, we consider yet one case, when $\nu = 4$. Then

$$\left(\frac{d}{dx}\right)^{-1/4}\sin x = \sin x \,\cos\frac{\pi}{8} - \cos x \,\sin\frac{\pi}{8}. \tag{10.46}$$

Here a simple calculation gives

$$\left(\frac{d}{dx}\right)^{-\frac{1}{4}}\left[\left(\frac{d}{dx}\right)^{-\frac{1}{4}}\sin x\right] = \left(\frac{d}{dx}\right)^{-\frac{1}{4}}\left[\sin x\,\cos\frac{\pi}{8} - \cos x\,\sin\frac{\pi}{8}\right]$$

$$= \frac{1}{\sqrt{2}}(\sin x - \cos x). \tag{10.47}$$

It means that

$$\left(\frac{d}{dx}\right)^{-\frac{1}{4}}\left(\frac{d}{dx}\right)^{-\frac{1}{4}}\sin x \equiv \left(\frac{d}{dx}\right)^{-\frac{1}{2}}\sin x. \tag{10.48}$$

For the cosine-function, we have

$$\left(\frac{d}{dx}\right)^{-\frac{1}{4}}\cos x = \cos x\,\cos\frac{\pi}{8} + \sin x\,\sin\frac{\pi}{8} \tag{10.49}$$

and therefore

$$\left(\frac{d}{dx}\right)^{-\frac{1}{4}}\left[\left(\frac{d}{dx}\right)^{-\frac{1}{4}}\cos x\right] = \left(\frac{d}{dx}\right)^{-\frac{1}{4}}\left[\cos x\,\cos\frac{\pi}{8} + \sin x\,\sin\frac{\pi}{8}\right]$$

$$= \frac{1}{\sqrt{2}}(\cos x + \sin x). \tag{10.50}$$

It means that

$$\left(\frac{d}{dx}\right)^{-\frac{1}{4}}\left(\frac{d}{dx}\right)^{-\frac{1}{4}}\cos x \equiv \left(\frac{d}{dx}\right)^{-\frac{1}{2}}\cos x. \tag{10.51}$$

10.2.3 *Additivity Properties of the Fractional Derivatives*

From definitions (10.9), (10.10) and above concrete examples, we see that the fractional derivatives defined by the formula (10.5) for sine- and cosine-functions obey properties of **additivity**, i.e., for any q_1- and q_2-orders, we have

$$\left(\frac{d}{dx}\right)^{q_1}\left[\left(\frac{d}{dx}\right)^{q_2}\left(\begin{matrix}\sin x\\ \cos x\end{matrix}\right)\right] \equiv \left(\frac{d}{dx}\right)^{q_1+q_2}\left(\begin{matrix}\sin x\\ \cos x\end{matrix}\right). \tag{10.52}$$

10.2.4 *Commutativity Properties of the Fractional Derivatives*

Fractional derivatives defined by the formula (10.5) for sine- and cosine-functions possess also properties of commutativity, i.e., for any q_1- and q_2-orders, we obtain the identity:

$$\left(\frac{d}{dx}\right)^{q_1}\left[\left(\frac{d}{dx}\right)^{q_2}\left(\begin{array}{c}\sin x\\ \cos x\end{array}\right)\right] \equiv \left(\frac{d}{dx}\right)^{q_2}\left[\left(\frac{d}{dx}\right)^{q_1}\left(\begin{array}{c}\sin x\\ \cos x\end{array}\right)\right]. \qquad (10.53)$$

10.2.5 *Standard Case*

Usual rules for integer-order derivatives are also given by the formula (10.5), when $\nu = -\dfrac{1}{n}$, $n = 1,\ 2,\ 3,\ \ldots$. Thus

$$\left(\frac{d}{dx}\right)^2\sin x = \sin x \cos \pi + \cos x \sin \pi = -\sin x,$$

$$\left(\frac{d}{dx}\right)^3\sin x = \sin x \cos \frac{3\pi}{2} + \cos x \sin \frac{3\pi}{2} = -\cos x,$$

$$\left(\frac{d}{dx}\right)^4\sin x = \sin x \cos 2\pi + \cos x \sin 2\pi = \sin x,$$

$$\frac{d^2}{dx^2}\cos x = \cos x \cos \pi - \sin x \sin \pi = -\cos x,$$

$$\frac{d^3}{dx^3}\cos x = \cos x \cos \frac{3\pi}{2} - \sin x \sin \frac{3\pi}{2} = \sin x,$$

$$\frac{d^4}{dx^4}\cos x = \cos x \cos 2\pi - \sin x \sin 2\pi = \cos x,$$

$$\ldots\ldots\ldots\ldots\ldots\ldots\ldots\ldots\ldots$$

10.2.6 *Fractional Derivatives for* $\sin ax$, $\cos ax$, e^{ax} *and* a^x*-Functions*

From the formula (10.5), it follows directly

$$\left(\frac{d}{dx}\right)^{-\frac{1}{\nu}}\sin ax = a^{-\frac{1}{\nu}}\sin\left(ax - \frac{\pi}{2\nu}\right), \qquad (10.54)$$

$$\left(\frac{d}{dx}\right)^{-\frac{1}{\nu}}\cos ax = a^{-\frac{1}{\nu}}\cos\left(ax - \frac{\pi}{2\nu}\right), \qquad (10.55)$$

$$\left(\frac{d}{dx}\right)^{-\frac{1}{\nu}} e^{ax} = a^{-\frac{1}{\nu}} e^{ax}, \tag{10.56}$$

$$\left(\frac{d}{dx}\right)^{-\frac{1}{\nu}} a^x = (\ln a)^{-\frac{1}{\nu}} a^x, \ a > 1. \tag{10.57}$$

10.3 Derivation of General Formulas for Calculation of Fractional Derivatives for Some Infinite Differentiable Functions

10.3.1 The Second General Formula

By using the formula (2.32) in Chapter 2, one gets

$$D_2 = \left(\frac{d}{dx}\right)^{-\frac{1}{\nu}} F(x) = \frac{\nu}{\Gamma\left(\frac{1}{\nu}\right)\sin\frac{\pi}{2\nu}} \int_0^\infty dt \sin\left(t^\nu \frac{d}{dx}\right) F(x), \tag{10.58}$$

where

$$\sin t^\nu \frac{d}{dx} = t^\nu \frac{d}{dx} - \frac{1}{3!}\left(t^\nu \frac{d}{dx}\right)^3 + \frac{1}{5!}\left(t^\nu \frac{d}{dx}\right)^5 - \cdots . \tag{10.59}$$

This decomposition is important for obtaining concrete results of explicit differentiation.

10.3.2 Fractional Derivatives for the Functions: $F(x) = e^{-ax}$, e^{ax} and a^x

It is obvious that from the second general formula, it follows directly

$$\left(\frac{d}{dx}\right)^{-\frac{1}{\nu}} e^{-ax} = -a^{-\frac{1}{\nu}} e^{-ax}, \tag{10.60}$$

$$\left(\frac{d}{dx}\right)^{-\frac{1}{\nu}} e^{ax} = a^{-\frac{1}{\nu}} e^{ax}, \tag{10.61}$$

$$\left(\frac{d}{dx}\right)^{-\frac{1}{\nu}} a^x = (\ln a)^{-\frac{1}{\nu}} a^x, \ a > 1. \tag{10.62}$$

It is natural, that formulas (10.56) and (10.61), (10.57) and (10.62) arise from different general formulas (10.5) and (10.58) coincide with each other, and that is as it should be.

10.3.3 *Fractional Derivatives for the Hyperbolic Functions* $F(x) =$ sinh ax, cosh ax

Due to simple properties of usual derivatives for these functions, for example,

$$(\sinh ax)' = a \cosh ax, \qquad (\cosh ax)' = a \sinh ax,$$
$$(\sinh ax)''' = a^3 \cosh ax, \qquad (\cosh ax)''' = a^3 \sinh ax,$$

etc., one can obtain the following nice formulas:

$$\left(\frac{d}{dx}\right)^{-\frac{1}{\nu}} \sinh ax = a^{-\frac{1}{\nu}} \cosh ax, \tag{10.63}$$

$$\left(\frac{d}{dx}\right)^{-\frac{1}{\nu}} \cosh ax = a^{-\frac{1}{\nu}} \sinh ax. \tag{10.64}$$

10.3.4 *Fractional Derivatives for the* $1/x$, $1/x^2$, $\ln x$, $\sqrt{x}$, $1/\sqrt{x}$*-Functions*

Let $F(x) = \dfrac{1}{x}$, then putting its derivatives

$$F' = -\frac{1}{x^2}, \quad F''' = -\frac{6}{x^4}, \quad F^{\mathrm{V}} = -\frac{120}{x^6}, \quad F^{\mathrm{VII}} = -\frac{7!}{x^8}, \quad \cdots$$

in the decomposition (10.59), we obtain the series

$$-\frac{1}{x^2}\left[t^\nu - \frac{t^{3\nu}}{x^2} + \frac{t^{5\nu}}{x^4} - \frac{(t^\nu)^7}{x^6} + \cdots\right] = -\frac{1}{x^2}\sum_{n=0}^{\infty}\frac{(-1)^n t^{2n\nu+\nu}}{x^{2n}}.$$

Thus, from the formula (10.58), one gets

$$\left(\frac{d}{dx}\right)^{-\frac{1}{\nu}}\frac{1}{x} = -\frac{N(\nu)}{x^2}\frac{1}{2i}\int_{-\beta+i\infty}^{-\beta-i\infty} d\xi\frac{1}{\sin\pi\xi}x^{-2\xi}\lim_{\epsilon\to 0}\int_{\epsilon}^{\infty}dt\, t^{2\nu\xi+\nu}$$

$$= \frac{N(\nu)}{x^2}\lim_{\epsilon\to 0}\frac{1}{2i}\int_{-\beta+i\infty}^{-\beta-i\infty} d\xi\frac{1}{\sin\pi\xi}x^{-2\xi}\frac{\epsilon^{2\nu\xi+\nu+1}}{2\nu\xi+\nu+1}, \tag{10.65}$$

where $N(\nu) = \dfrac{\nu}{\Gamma(1/\nu)}\dfrac{1}{\sin(\pi/2\nu)}$. Further, we calculate the residue at the point

$$\xi = -\frac{\nu+1}{2\nu}$$

and obtain

$$\left(\frac{d}{dx}\right)^{-\frac{1}{\nu}}\frac{1}{x} = -\frac{\pi}{\cos\frac{\pi}{2\nu}}\frac{1}{2\nu}x^{\frac{1}{\nu}-1}N(\nu). \tag{10.66}$$

After some elementary transformations, one gets

$$\left(\frac{d}{dx}\right)^{-\frac{1}{\nu}}\left(\frac{1}{x}\right) = -\Gamma\left(1 - \frac{1}{\nu}\right)x^{\frac{1}{\nu}-1}. \tag{10.67}$$

Let $F(x) = \ln x$, then inserting its derivatives

$$F' = \frac{1}{x}, \quad F''' = \frac{2}{x^3}, \quad F^{\text{V}} = \frac{24}{x^5}, \quad F^{\text{VII}} = \frac{24 \cdot 5 \cdot 6}{x^7}, \quad \cdots$$

into the series (10.59), we have

$$\left(\frac{d}{dx}\right)^{-\frac{1}{\nu}} \ln x = N(\nu) \lim_{\epsilon \to 0} \frac{1}{2i} \int_{-\beta+i\infty}^{-\beta-i\infty} d\xi \frac{1}{\sin \pi\xi \,(2\xi + 1)} x^{-2\xi-1}$$
$$\times \int_{\epsilon}^{\infty} dt \, t^{2\nu\xi+\nu}.$$

Similar calculation as above reads

$$\left(\frac{d}{dx}\right)^{-\frac{1}{\nu}} \ln x = N(\nu)\left[-\frac{\pi}{2\cos\frac{\pi}{2\nu}}\right]x^{\frac{1}{\nu}}$$

or

$$\left(\frac{d}{dx}\right)^{-\frac{1}{\nu}} \ln x = \Gamma\left(-\frac{1}{\nu}\right)x^{\frac{1}{\nu}}. \tag{10.68}$$

Let $F(x) = \frac{1}{x^2}$, then taking into account its derivatives

$$F' = -\frac{2}{x^3}, \quad F''' = -\frac{24}{x^5}, \quad F^{\text{V}} = -\frac{24 \cdot 5 \cdot 6}{x^7}, \quad \cdots$$

and series (10.58), one gets

$$\left(\frac{d}{dx}\right)^{-\frac{1}{\nu}}\left(\frac{1}{x^2}\right) = -N(\nu)\frac{2}{x^3} \lim_{\epsilon \to 0} \frac{1}{2i} \int_{-\beta+i\infty}^{-\beta-i\infty} d\xi \frac{x^{-2\xi}}{\sin \pi\xi}(1 + \xi)$$
$$\times \int_{\epsilon}^{\infty} dt \, t^{\nu+2\nu\xi} = -\frac{N(\nu)}{x^3}\frac{(\nu-1)}{2\nu^2}\frac{\pi \, x^{\frac{1+\nu}{\nu}}}{\sin \pi \left(\frac{1+\nu}{2\nu}\right)}.$$

Finally, we have

$$\left(\frac{d}{dx}\right)^{-\frac{1}{\nu}}\left(\frac{1}{x^2}\right) = -\Gamma\left(2 - \frac{1}{\nu}\right)x^{\frac{1}{\nu}-2}. \tag{10.69}$$

Let

$$F(x) = \frac{1}{\sqrt{x}},$$

where

$$F' = -\frac{1}{2}x^{-\frac{3}{2}}, \quad F''' = -\frac{3 \cdot 5}{2^3}x^{-\frac{7}{2}}, \quad F^{\mathrm{V}} = -\frac{3 \cdot 5 \cdot 7 \cdot 9}{2^5}x^{-\frac{11}{2}},$$

$$F^{\mathrm{VII}} = -\frac{3 \cdot 5 \cdot 7 \cdot 9 \cdot 11 \cdot 13}{2^7}x^{-\frac{15}{2}}, \cdots.$$

For this case, series (10.58) takes the form

$$-\sum_{n=0}^{\infty} \frac{(-1)^n}{(2n+1)!} \frac{(t^{\nu})^{2n+1}}{2^{2n+1}} x^{-\frac{4n+3}{2}} (4n+1)!!.$$

Writing this series in the form of the Mellin representation and calculating the resulting integral by means of residue, one gets

$$\left(\frac{d}{dx}\right)^{-\frac{1}{\nu}} \frac{1}{\sqrt{x}} = -2^{\frac{1}{\nu}} x^{\frac{2-\nu}{2\nu}} \left[-\frac{2+2\nu}{\nu} + 1\right]!!. \tag{10.70}$$

Let

$$F(x) = \sqrt{x},$$

where

$$F' = \frac{1}{2}\frac{1}{\sqrt{x}}, \quad F''' = \frac{3}{2^3}x^{-\frac{5}{2}}, \quad F^{\mathrm{V}} = \frac{3 \cdot 5 \cdot 7}{2^5}x^{-\frac{9}{2}},$$

$$F^{\mathrm{VII}} = \frac{3 \cdot 5 \cdot 7 \cdot 9 \cdot 11}{2^7}x^{-\frac{13}{2}}, \cdots.$$

Then the series (10.58) in this case takes the form

$$\sum_{n=0}^{\infty} \frac{(-1)^n}{(2n+1)!} \frac{x^{-\frac{4n+1}{2}}}{2^{2n+1}} (t^{\nu})^{2n+1} (4n-1)!!.$$

Similar calculation reads

$$\left(\frac{d}{dx}\right)^{-\frac{1}{\nu}} \sqrt{x} = 2^{\frac{1}{\nu}} x^{\frac{2+\nu}{2\nu}} \left[-\frac{2+2\nu}{\nu} - 1\right]!!. \tag{10.71}$$

Here by definitions:

$$(-1)!! = 1, \quad (-5)!! = \frac{1}{3}, \quad (-3)!! = -1, \quad 0!! = 1$$

and so on.

10.3.5　Some Examples

By using the above formulas (10.67), (10.68), (10.69), (10.70) and (10.71), one can obtain particular interesting fractional derivatives, like

1) $\left(\dfrac{d}{dx}\right)^{-1/2}\dfrac{1}{x}=-\dfrac{\sqrt{\pi}}{\sqrt{x}},\quad \left(\dfrac{d}{dx}\right)^{\frac{3}{2}}\left(\dfrac{1}{x}\right)=-\dfrac{3}{4}\sqrt{\pi}\,x^{-\frac{5}{2}},$

2) $\left(\dfrac{d}{dx}\right)^{-\frac{1}{3}}\dfrac{1}{x}=-\dfrac{2\pi}{\sqrt{3}\,\Gamma\left(\dfrac{1}{3}\right)}x^{-\frac{2}{3}},$

3) $\dfrac{d}{dx}\left[\left(\dfrac{d}{dx}\right)^{-1/2}\left(\dfrac{1}{x}\right)\right]=\dfrac{\sqrt{\pi}}{2}x^{-\frac{3}{2}},\quad \left(\dfrac{d}{dx}\right)^{-1/2}\left(-\dfrac{1}{x^2}\right)=\dfrac{\sqrt{\pi}}{2}x^{-\frac{3}{2}},$

it means that

$$\frac{d}{dx}\left[\left(\frac{d}{dx}\right)^{-1/2}\right]\left(\frac{1}{x}\right)\equiv\left(\frac{d}{dx}\right)^{-1/2}\frac{d}{dx}\left(\frac{1}{x}\right),\qquad(10.72)$$

4) $\left(\dfrac{d}{dx}\right)^{-1}\ln x=\infty,\quad \left(\dfrac{d}{dx}\right)^{-1}\left(\dfrac{1}{x}\right)=\infty,$

5) $\left(\dfrac{d}{dx}\right)^{-1}\dfrac{1}{x^2}=-\dfrac{1}{x},$

6) $\left(\dfrac{d}{dx}\right)^{-1/2}\ln x=-2\sqrt{\pi x},$

7) $\left(\dfrac{d}{dx}\right)^{1/2}\ln x=\sqrt{\dfrac{\pi}{x}},$

8) $\left(\dfrac{d}{dx}\right)^{3/2}\ln x=\dfrac{\sqrt{\pi}}{2}x^{-\frac{3}{2}},$

9) $\dfrac{d}{dx}\left(\dfrac{d}{dx}\right)^{-1/2}\ln x=-\sqrt{\dfrac{\pi}{x}}\equiv-\left(\dfrac{d}{dx}\right)^{1/2}\ln x,$

10) $\dfrac{d}{dx}\left(\dfrac{d}{dx}\right)^{-1/2}\ln x=\left(\dfrac{d}{dx}\right)^{-1/2}\left(\dfrac{1}{x}\right)=-\sqrt{\dfrac{\pi}{x}},$

11) $\left[\dfrac{d}{dx}\left(\dfrac{d}{dx}\right)^{-1/2}+\left(\dfrac{d}{dx}\right)^{1/2}\right]\ln x=0,$

12) $\left[\dfrac{d}{dx}\left(\dfrac{d}{dx}\right)^{-1/2}+\left(\dfrac{d}{dx}\right)^{1/2}\right]\left(\dfrac{1}{x}\right)=0,$

13) $\left[\left(\dfrac{d}{dx}\right)^{\frac{3}{2}}-\dfrac{d^2}{dx^2}\left(\dfrac{d}{dx}\right)^{-1/2}\right]\left(\dfrac{1}{x}\right)=0,$

14) $\left[\dfrac{d}{dx} \left(\dfrac{d}{dx} \right)^{1/2} - \left(\dfrac{d}{dx} \right)^{1/2} \dfrac{d}{dx} \right] \left(\dfrac{1}{x} \right) = 0,$

15) $\left[\dfrac{d}{dx} \left(\dfrac{d}{dx} \right)^{1/2} - \left(\dfrac{d}{dx} \right)^{1/2} \dfrac{d}{dx} \right] \ln x = 0,$

16) $\left(\dfrac{d}{dx} \right)^{1/2} \left(\dfrac{1}{x} \right) = -\dfrac{\sqrt{\pi}}{2} x^{-\frac{3}{2}},$

17) $\left(\dfrac{d}{dx} \right)^{-1/2} \left(\dfrac{1}{x^2} \right) = -\dfrac{\sqrt{\pi}}{2} x^{-\frac{3}{2}},$

18) $\left(\dfrac{d}{dx} \right)^{1/2} \left(\dfrac{1}{x^2} \right) = -\dfrac{3}{4} \sqrt{\pi} x^{-\frac{5}{2}},$

19) $\left(\dfrac{d}{dx} \right)^{\frac{3}{2}} \left(\dfrac{1}{x^2} \right) = -\dfrac{15}{8} \sqrt{\pi} x^{-\frac{7}{2}},$

20) $\left(\dfrac{d}{dx} \right)^{\frac{3}{2}} \sqrt{x} = 2^{-\frac{3}{2}} x^{-1},$

21) $\left(\dfrac{d}{dx} \right)^{1/2} \dfrac{1}{\sqrt{x}} = -\dfrac{1}{\sqrt{2}} x^{-1},$

22) $\left(\dfrac{d}{dx} \right)^{1/2} \sqrt{x} = \dfrac{1}{\sqrt{2}},$

23) $\left(\dfrac{d}{dx} \right)^{-1/2} \sqrt{x} = -\sqrt{2}\, x.$

10.3.6 Usual Case, When $-\dfrac{1}{\nu} = n$

Thus, generalization of above formulas for standard integer-order derivatives gives

$$\left(\frac{d}{dx} \right)^n \ln x = -(-1)^n \Gamma(n) x^{-n}, \tag{10.73}$$

$$\left(\frac{d}{dx} \right)^n \left(\frac{1}{x} \right) = (-1)^n \Gamma(n+1) x^{-n-1}, \tag{10.74}$$

$$\left(\frac{d}{dx} \right)^n \left(\frac{1}{x^2} \right) = (-1)^n \Gamma(2+n) x^{-n-2}, \tag{10.75}$$

$$\left(\frac{d}{dx} \right)^n \sqrt{x} = -(-1)^n 2^{-n} x^{\frac{1}{2}-n} (2n-3)!!, \tag{10.76}$$

$$\left(\frac{d}{dx} \right)^n \frac{1}{\sqrt{x}} = (-1)^n 2^{-n} x^{-n-\frac{1}{2}} (2n-1)!!. \tag{10.77}$$

Here in order to obtain these standard formulas, we have to put a sign variable factor $(-1)^n$ by hand.

10.4   Representation for Inverse Derivatives in the Form of Integer-Order Differentials

The physical application point of view is usually used in the following inhomogeneous equation

$$Lu = J, \tag{10.78}$$

where $J = J(x)$ is some external source or current and

$$L = \left(\frac{\partial}{\partial x}\right)^2 + W(x) \tag{10.79}$$

is the given operator. Here $W(x)$ is a known function. The problem is to find a solution of this inhomogeneous equation (10.78) in the form

$$u(x) = u_0(x) + \frac{1}{L}J(x)$$

$$= u_0(x) + \int dx' \, G(x, x')J(x'). \tag{10.80}$$

Here $u_0(x)$ is a solution of homogeneous equation. Then the Green function is given by the well-known expression

$$G(x, x') = \frac{1}{L}\delta(x - x'). \tag{10.81}$$

The main problem in physics is that it is necessary to find the Green function i.e., to define the action of the inverse operator (see Efimov, 2008):

$$\frac{1}{L}J = L^{-1}J = \ ? \tag{10.82}$$

In general, for this purpose, we consider here some representations of inverse operators by means of usual integer-order differentials.

10.4.1   *The Third General Formula*

$$\frac{1}{\left[\frac{d^2}{dx^2} + b^2\right]^{\sigma + \frac{1}{2}}}F(x) = \frac{\sqrt{\pi}}{2^\sigma \, \Gamma\left(\sigma + \frac{1}{2}\right)}$$

$$\times \int_0^\infty dt \, t^\sigma \, \frac{J_\sigma(bt)}{b^\sigma} e^{-t\frac{d}{dx}} F(x) \tag{10.83}$$

or

$$\frac{1}{\left[p^2 + \frac{d^2}{dx^2}\right]^{\sigma + \frac{1}{2}}}F(x) = \frac{\sqrt{\pi}}{2^\sigma \, \Gamma\left(\sigma + \frac{1}{2}\right)}$$

$$\times \int_0^\infty dt \, t^\sigma e^{-pt} \frac{J_\sigma\left(t\frac{d}{dx}\right)}{\left(\frac{d}{dx}\right)^\sigma} F(x). \tag{10.84}$$

Let us consider the case $\sigma = 0$ in (10.83) and $F(x) = \sin ax, \cos ax$. Then

$$\frac{1}{\sqrt{b^2 + \frac{d^2}{dx^2}}} \sin ax = \int_0^\infty dt\, J_0(bt) e^{-t\frac{d}{dx}} \sin ax, \qquad (10.85)$$

where

$$e^{-t\frac{d}{dx}} \sin ax = \sin ax \, \cos at - \cos ax \, \sin at.$$

Therefore

$$\frac{1}{\sqrt{b^2 + \frac{d^2}{dx^2}}} \sin ax = \begin{cases} -\frac{1}{\sqrt{a^2-b^2}} \cos ax & \text{if} \quad 0 < b < a \\ \frac{1}{\sqrt{b^2-a^2}} \sin ax & \text{if} \quad 0 < a < b. \end{cases} \qquad (10.86)$$

Here we have used the integrals:

$$l_1 = \int_0^\infty dt\, J_0(bt) \sin at = \begin{cases} 0 & \text{if} \quad 0 < a < b \\ \frac{1}{\sqrt{a^2-b^2}} & \text{if} \quad 0 < b < a \end{cases}$$

and

$$l_2 = \int_0^\infty dt\, J_0(bt) \cos at = \begin{cases} \frac{1}{\sqrt{b^2-a^2}} & \text{if} \quad 0 < a < b \\ \infty & \text{if} \quad a = b \\ 0 & \text{if} \quad 0 < b < a. \end{cases}$$

Similar calculation for $F(x) = \cos ax$ reads

$$\frac{1}{[b^2 + \frac{d^2}{dx^2}]^{1/2}} \cos ax = \cos ax \, l_2 + \sin ax \, l_1,$$

where functions l_1 and l_2 are given by the above formulas.

Thus

$$\frac{1}{\sqrt{b^2 + \frac{d^2}{dx^2}}} \cos ax = \begin{cases} \frac{1}{\sqrt{b^2-a^2}} \cos ax & \text{if} \quad 0 < a < b \\ \frac{1}{\sqrt{a^2-b^2}} \sin ax & \text{if} \quad 0 < b < a. \end{cases}$$

Let $F(x) = x$ and $\sigma = 0$ in the formula (10.84). Then

$$\frac{1}{[p^2 + \frac{d^2}{dx^2}]^{1/2}} x = x \int_0^\infty dt\, e^{-pt} = \frac{x}{p}, \qquad (10.87)$$

where we have used the following series

$$J_0(z) = \sum_{k=0}^\infty \frac{(-1)^k}{2^{2k}} \frac{z^{2k}}{(k!)^2} = 1 - \frac{1}{4}z^2 + \frac{1}{64}z^4 - \cdots .$$

The above case with $F(x) = x^2$ reads

$$\frac{1}{[p^2 + \frac{d^2}{dx^2}]^{1/2}} x^2 = \frac{x^2}{p} - \frac{1}{p^3}. \qquad (10.88)$$

If $F(x) = x^3$. Then

$$\frac{1}{\left[p^2 + \frac{d^2}{dx^2}\right]^{1/2}} x^3 = \frac{x^3}{p} - \frac{3x}{p^3}. \tag{10.89}$$

If $F(x) = x^4$. Then

$$\frac{1}{\left[p^2 + \frac{d^2}{dx^2}\right]^{1/2}} x^4 = \frac{x^4}{p} - \frac{6x^2}{p^3} + \frac{9}{p^5} \tag{10.90}$$

and etc.

Notice that similar calculations hold for any σ in (10.84), for example, let $\sigma = 1$, then we have

$$\frac{1}{\left[p^2 + \frac{d^2}{dx^2}\right]^{3/2}} F(x) = \int_0^\infty dt\, t\, e^{-pt} \frac{J_1\left(t\frac{d}{dx}\right)}{\left(\frac{d}{dx}\right)} F(x). \tag{10.91}$$

It is important to notice that here the function

$$J_1\left(t\frac{d}{dx}\right) \Big/ (d/dx)$$

is an entire analytic function over its independent variable (or differential).

1) Let $F(x) = x$ in (10.91). Then

$$\frac{1}{\left[p^2 + \frac{d^2}{dx^2}\right]^{3/2}} x = \int_0^\infty dt\, t\, e^{-pt} \frac{t}{2}[1 - \cdots]x = \frac{x}{p^3}. \tag{10.92}$$

2) Let $F(x) = x^2$ for this case, then

$$\frac{1}{\left[p^2 + \frac{d^2}{dx^2}\right]^{3/2}} x^2 = \int_0^\infty dt\, t\, e^{-pt} \frac{t}{2}\left[1 - \frac{t^2}{8}\frac{d^2}{dx^2}\right] x^2$$

$$= \frac{x^2}{p^3} - \frac{4!}{16p^5} = \frac{x^2}{p^3} - \frac{3}{2}\frac{1}{p^5}. \tag{10.93}$$

3) For $F(x) = x^3$, we have

$$\frac{1}{\left[p^2 + \frac{d^2}{dx^2}\right]^{3/2}} x^3 = \int_0^\infty dt\, t\, e^{-pt} \frac{t}{2}\left[1 - \frac{t^2}{8}\frac{d^2}{dx^2}\right] x^3$$

$$= \frac{x^3}{p^3} - \frac{9x}{p^5} \tag{10.94}$$

and etc.

10.4.2 4th *General Formula*

$$\left[1 - \frac{\frac{d}{dx}}{\sqrt{\frac{d^2}{dx^2} + b^2}}\right] F(x) = \int_0^\infty dt\, bJ_1(bt)e^{-t\frac{d}{dx}} F(x) \qquad (10.95)$$

or

$$\left[1 - \frac{p}{\sqrt{p^2 + \frac{d^2}{dx^2}}}\right] F(x) = \int_0^\infty dt e^{-pt} J\left(t\frac{d}{dx}\right)\left(\frac{d}{dx}\right) F(x). \qquad (10.96)$$

Let $F(x) = \cos ax$ in (10.95), then

$$\left[1 - \frac{\frac{d}{dx}}{\sqrt{b^2 + \frac{d^2}{dx^2}}}\right] \cos ax = \frac{b}{\sqrt{b^2 - a^2}}\left[\cos ax\ \cos\left(\arcsin\frac{a}{b}\right)\right.$$

$$\left. + \sin ax\ \sin\left(\arcsin\frac{a}{b}\right)\right], \qquad (10.97)$$

if $a < b$, or

$$\left[1 - \frac{\frac{d}{dx}}{\sqrt{b^2 + \frac{d^2}{dx^2}}}\right] \cos ax = \frac{b^2}{\sqrt{a^2 - b^2}(a + \sqrt{a^2 - b^2})} \cos ax, \qquad (10.98)$$

if $a > b$.

Let us consider the case when $F(x) = \sin ax$ in (10.95). Then

$$\left[1 - \frac{\frac{d}{dx}}{\sqrt{b^2 + \frac{d^2}{dx^2}}}\right] \sin ax = \frac{b}{\sqrt{b^2 - a^2}}\left[\sin ax\ \cos\left(\arcsin\frac{a}{b}\right)\right.$$

$$\left. - \cos ax\ \sin\left(\arcsin\frac{a}{b}\right)\right], \qquad (10.99)$$

if $a < b$, or

$$\left[1 - \frac{\frac{d}{dx}}{\sqrt{b^2 + \frac{d^2}{dx^2}}}\right] \sin ax = \frac{b^2}{\sqrt{a^2 - b^2}(a + \sqrt{a^2 - b^2})} \sin ax, \qquad (10.100)$$

if $a > b$.

10.4.3   5th *General Formula*

$$\frac{1}{\left[p^2 + \frac{d^2}{dx^2}\right]^{5/2}} F(x) = \frac{1}{3p} \int_0^\infty dt\, t^2 e^{-pt} \frac{J_1\left(t\frac{d}{dx}\right)}{\frac{d}{dx}} F(x),$$

(10.101)

$$\frac{\frac{d}{dx}}{\left[\frac{d^2}{dx^2} + b^2\right]^{5/2}} F(x) = \frac{1}{3b} \int_0^\infty dt\, t^2 J_1(bt) e^{-t\frac{d}{dx}} F(x).$$

(10.102)

Consider the formula (10.101) with $F_1(x) = x^3$ and $F_2(x) = x^6$. Thus

$$\frac{1}{\left[p^2 + \frac{d^2}{dx^2}\right]^{5/2}} \binom{x^3}{x^6} = \frac{1}{3p} \int_0^\infty dt\, t^2 e^{-pt} \frac{t}{2} \left[1 - \frac{1}{8}t^2 \frac{d^2}{dx^2}\right.$$

$$\left. + \frac{1}{2!\Gamma(4)} \frac{t^4}{2^4} \frac{d^4}{dx^4} - \frac{1}{3!} \frac{1}{\Gamma(5)} \frac{t^6}{2^6} \frac{d^6}{dx^6}\right] \binom{x^3}{x^6}.$$

After some elementary calculations, we have

$$\frac{1}{\left[p^2 + \frac{d^2}{dx^2}\right]^{5/2}} \binom{x^3}{x^6} = \begin{pmatrix} \dfrac{x^3}{p^5} - \dfrac{15x}{p^7} \\[2mm] \dfrac{x^6}{p^5} - \dfrac{150x^4}{p^7} + \dfrac{1575x^2}{p^9} - \dfrac{4725}{p^{11}} \end{pmatrix}.$$

(10.103)

The integral (10.102) with functions

$$F(x) = \begin{cases} a^x \\ e^{ax} \end{cases}$$

gives elementary results:

$$\frac{d/dx}{\left[\frac{d^2}{dx^2} + b^2\right]^{5/2}} \binom{a^x}{e^{ax}} = \begin{pmatrix} \dfrac{a^x \ln a}{[\ln^2 a + b^2]^{5/2}} \\[2mm] \dfrac{a e^{ax}}{[a^2 + b^2]^{5/2}} \end{pmatrix}.$$

(10.104)

10.4.4   6th *General Formula*

$$\left[\sqrt{\frac{d^2}{dx^2} + b^2} - \frac{d}{dx}\right]^\nu F(x) = \nu b^\nu \int_0^\infty dt\, \frac{J_\nu(bt)}{t} e^{-t\frac{d}{dx}} F(x)$$

(10.105)

or

$$\left[\sqrt{p^2 + \frac{d^2}{dx^2}} - p\right]^\nu F(x) = \nu \int_0^\infty dt\, e^{-pt} \frac{J_\nu\left(t\frac{d}{dx}\right)}{t} \left(\frac{d}{dx}\right)^\nu F(x).$$

(10.106)

The formula (10.105) with $F(x) = \sin ax$ gives

$$\left[\sqrt{\frac{d^2}{dx^2} + b^2} - \frac{d}{dx}\right]^\nu \sin ax = \nu b^\nu \left[\sin ax \int_0^\infty dt \frac{J_\nu(bt)}{t} \cos at\right.$$

$$\left. - \cos ax \int_0^\infty dt \frac{J_\nu(bt)}{t} \sin at\right],$$

where

$$i_1 = \int_0^\infty \frac{dt}{t} J_\nu(bt) \sin at$$

$$= \begin{cases} \dfrac{\sin\left(\nu \arcsin \frac{a}{b}\right)}{\nu} & \text{if } 0 < a < b \\ & \qquad \text{Re } \nu > -1 \\ \dfrac{b^\nu \sin\left(\frac{\pi\nu}{2}\right)\left[a - \sqrt{a^2-b^2}\right]^{-\nu}}{\nu} & \text{if } 0 < b < a \end{cases}$$

(10.107)

and

$$i_2 = \int_0^\infty \frac{dt}{t} J_\nu(bt) \cos at$$

$$= \begin{cases} \dfrac{\cos\left(\nu \arcsin \frac{a}{b}\right)}{\nu} & \text{if } 0 < a < b \\ & \qquad \text{Re } \nu > 0 \\ \dfrac{b^\nu \cos\left(\frac{\pi\nu}{2}\right)\left[a - \sqrt{a^2-b^2}\right]^{-\nu}}{\nu} & \text{if } 0 < b < a. \end{cases}$$

(10.108)

For $F(x) = \cos ax$, similar formulas are obtained, that is

$$\left[\sqrt{\frac{d^2}{dx^2} + b^2} - \frac{d}{dx}\right]^\nu \cos ax = \nu b^\nu [\cos ax\, i_2 + \sin ax\, i_1],$$

(10.109)

where functions i_1 and i_2 are given by formulas (10.107) and (10.108), respectively.

Notice that for concrete integer values of ν in (10.106), one can easily obtain expressions for $F(x) = x^n$-functions as above.

10.4.5 7th General Formula

$$\frac{1}{\sqrt{\frac{d^2}{dx^2} - b^2}} F(x) = \int_0^\infty dt\, I_0(bt)\, e^{-t\frac{d}{dx}}\, F(x)$$

(10.110)

or

$$\frac{1}{\sqrt{p^2 - \frac{d^2}{dx^2}}} F(x) = \int_0^\infty dt e^{-pt} I_0\left(t\frac{d}{dx}\right) F(x). \tag{10.111}$$

By using the formula (10.110), it is easy to obtain the following expressions for the functions:

$$F(x) = e^{ax} \quad \text{and} \quad F(x) = a^x,$$

that is

$$\frac{1}{\sqrt{\frac{d^2}{dx^2} - b^2}} e^{ax} = \frac{1}{\sqrt{a^2 - b^2}} e^{ax} \quad \text{if } a > b, \tag{10.112}$$

and

$$\frac{1}{\sqrt{\frac{d^2}{dx^2} - b^2}} a^x = \frac{1}{\sqrt{\ln^2 a - b^2}} a^x \quad \text{if } \ln a > b, \ a > 1. \tag{10.113}$$

Let $F(x) = x, x^2, x^5$ be in (10.111), then we have

$$\frac{1}{\sqrt{p^2 - \frac{d^2}{dx^2}}} x = \int_0^\infty dt e^{-pt} \left[1 + \frac{1}{4}t^2 \frac{d^2}{dx^2} + \frac{1}{64}t^4 \frac{d^4}{dx^4}\right] x = \frac{x}{p}, \tag{10.114}$$

$$\frac{1}{\sqrt{p^2 - \frac{d^2}{dx^2}}} x^2 = \frac{x^2}{p} + \frac{1}{p^3}, \tag{10.115}$$

$$\frac{1}{\sqrt{p^2 - \frac{d^2}{dx^2}}} x^5 = \frac{x^5}{p} + \frac{10}{p^3} x^3 + \frac{45x}{p^5} \tag{10.116}$$

so on.

10.4.6   8th *General Formula*

$$\frac{1}{b + \frac{d}{dx}} F(x) = \sqrt{\frac{2b}{\pi}} \int_0^\infty dt\sqrt{t} K_{\pm 1/2}(bt) e^{-t\frac{d}{dx}} F(x) \tag{10.117}$$

or

$$\frac{1}{p + \frac{d}{dx}} F(x) = \sqrt{\frac{2}{\pi}} \int_0^\infty dt\sqrt{t} e^{-pt} \sqrt{\frac{d}{dx}} K_{\pm 1/2}\left(t\frac{d}{dx}\right) F(x), \tag{10.118}$$

where

$$K_{\pm 1/2}(z) = \sqrt{\frac{\pi}{2z}} e^{-z}.$$

Thus, the formulas (10.117) and (10.118) take the following simple forms:

$$\frac{1}{b + \frac{d}{dx}} F(x) = \int_0^\infty dt\, e^{-bt}\, e^{-t\frac{d}{dx}}\, F(x) \tag{10.119}$$

and

$$\frac{1}{p + \frac{d}{dx}} F(x) = \int_0^\infty dt\, e^{-pt}\, e^{-t\frac{d}{dx}}\, F(x). \tag{10.120}$$

The formula (10.119) or (10.120) with the functions $F(x) = \sin ax$ and $F(x) = \cos ax$ gives nice results:

$$\frac{1}{b + \frac{d}{dx}} \sin ax = \sin ax\, \lambda_1 - \cos ax\, \lambda_2 \tag{10.121}$$

and

$$\frac{1}{b + \frac{d}{dx}} \cos ax = \cos ax\, \lambda_1 + \sin ax\, \lambda_2, \tag{10.122}$$

where λ_1 and λ_2 are given by the expressions:

$$\lambda_1 = \int_0^\infty dt\, e^{-bt} \cos at = \frac{1}{\sqrt{a^2 + b^2}} \cos\left(\arctan\frac{a}{b}\right)$$

and

$$\lambda_2 = \int_0^\infty dt\, e^{-bt} \sin at = \frac{1}{\sqrt{a^2 + b^2}} \sin\left(\arctan\frac{a}{b}\right).$$

For completeness, we calculate the following inverse operators:

$$1)\quad \frac{1}{b + \frac{d}{dx}}\, x = \frac{x}{b}, \tag{10.123}$$

$$2)\quad \frac{1}{b + \frac{d}{dx}}\, x^2 = \frac{x^2}{b} - \frac{2x}{b^2} + \frac{2}{b^3}, \tag{10.124}$$

$$3)\quad \frac{1}{b + \frac{d}{dx}}\, x^3 = \frac{x^3}{b} - \frac{3x^2}{b^2} + \frac{6x}{b^3} - \frac{6}{b^4} \tag{10.125}$$

so on.

10.4.7 9th *General Formula*

$$\frac{1}{\left[\frac{d^2}{dx^2} + p^2\right]^{1/2}} F(x) = \frac{2}{\pi} \int_0^\infty dt\, K_0(pt) \cos\left(t\frac{d}{dx}\right) F(x). \qquad (10.126)$$

It is natural that this formula with functions $F(x) = x, x^2, x^3, x^4$ etc. gives identical results as before in (10.84) with the condition $\sigma = 0$. Direct calculations give

1) $\dfrac{1}{\left[\frac{d^2}{dx^2} + p^2\right]^{1/2}} x = \dfrac{2}{\pi} \displaystyle\int_0^\infty dt\, K_0(pt)$

$$= \frac{2}{\pi}\frac{x}{p} \int_0^\infty dy\, K_0(y) = \frac{2}{\pi} x \frac{1}{p}\frac{\pi}{2} = \frac{x}{p},$$

2) $\dfrac{1}{\left[\frac{d^2}{dx^2} + p^2\right]^{1/2}} x^2 = \dfrac{2}{\pi} \displaystyle\int_0^\infty dt\, K_0(pt)\left(x^2 - t^2\right) = \dfrac{x^2}{p} - \dfrac{1}{p^3},$

3) $\dfrac{1}{\left[\frac{d^2}{dx^2} + p^2\right]^{1/2}} x^3 = \dfrac{2}{\pi} \displaystyle\int_0^\infty dt\, K_0(pt)\left(x^3 - \frac{1}{2}t^2\, 6x\right) = \dfrac{x^3}{p} - \dfrac{3x}{p^3},$

4) $\dfrac{1}{\left[\frac{d^2}{dx^2} + p^2\right]^{1/2}} x^4 = \dfrac{2}{\pi} \displaystyle\int_0^\infty dt\, K_0(pt)\left(x^4 - \frac{t^2}{2}12t^2 + \frac{1}{4!}t^4\, 24\right)$

$$= \frac{x^4}{p} - \frac{6x^2}{p^3} + \frac{9}{p^5},$$

where we have used the following formulas (Wheelon and Robacker, 1954):

$$\int_0^\infty dx\, K_0(x) = \frac{\pi}{2},$$

$$\int_0^\infty dx\, x K_0(x) = 1,$$

$$\int_0^\infty dx\, x^\nu K_\nu(x) = 2^{\nu-1}\sqrt{\pi}\,\Gamma\left(\nu + \frac{1}{2}\right),$$

and

$$\int_0^\infty dx\, x^{\mu-1} K_\nu(x) = 2^{\mu-2}\Gamma\left(\frac{\nu+\mu}{2}\right)\Gamma\left(\frac{\mu-\nu}{2}\right),$$
$$\text{Re } \mu > |\text{Re } \nu|.$$

All these results coincide with the ones obtained from (10.84), where $\sigma = 0$.

The following two formulas give the same results

$$\frac{1}{\left[\frac{d^2}{dx^2} + p^2\right]^{3/2}} F(x) = \begin{cases} \int_0^\infty dt\, te^{-pt} \dfrac{J_1\left(t\frac{d}{dx}\right)}{\left(\frac{d}{dx}\right)} F(x) \\[4mm] \frac{2}{\pi} \int_0^\infty dt\, tK_0(pt) \dfrac{\sin\left(t\frac{d}{dx}\right)}{\left(\frac{d}{dx}\right)} F(x) \end{cases}$$

for functions $F(x) = x, x^2, x^3$ and etc.

10.4.8 10th *General Formula*

$$\frac{1}{\left[p^2 + \frac{d^2}{dx^2}\right]^{\nu + \frac{3}{2}}} F(x) = \frac{1}{\sqrt{\pi}(2p)^\nu \Gamma\left(\frac{3}{2} + \nu\right)}$$

$$\times \int_0^\infty dt\, t^{1+\nu} K_\nu(pt) \frac{\sin\left(t\frac{d}{dx}\right)}{\frac{d}{dx}} F(x). \tag{10.127}$$

10.4.9 11th *General Formula*

$$\frac{1}{\left[p^2 + \frac{d^2}{dx^2}\right]^{\mu + \frac{1}{2}}} F(x) = \frac{2}{\sqrt{\pi}(2p)^\mu \Gamma\left(\mu + \frac{1}{2}\right)}$$

$$\times \int_0^\infty dt\, t^\mu K_\mu(pt) \cos\left(t\frac{d}{dx}\right) F(x). \tag{10.128}$$

In these formulas (10.127) and (10.128), we have to take

$$\operatorname{Re} \nu > -\frac{3}{2}, \quad \operatorname{Re} \mu > -\frac{1}{2}, \quad p > 0.$$

In conclusion, notice that in quantum field theory, Green functions for particles are given by the following equations:

1) for photon

$$\Box G_1(x) = \delta^4(x), \tag{10.129}$$

2) for Klein-Gordon or scalar particle

$$(\Box + m^2)G_2(x) = \delta^4(x). \tag{10.130}$$

Moreover, in the square root operator formalism it is used in the equation (Namsrai, 1998):

$$\sqrt{\Box + m^2} G_3(x) = \delta^4(x),$$ (10.131)

where

$$\Box = \left(i\gamma^\nu \frac{\partial}{\partial x^\nu} \right)^2 = -\frac{1}{c^2} \frac{\partial^2}{\partial t^2} + \frac{\vec{\partial}^2}{\partial \vec{x}^2}$$

and γ^ν are the Dirac γ-matrices.

Sometimes, it is used in the integral representation for inverse operator

$$\frac{1}{\Box} = \int_0^\infty dt\, e^{-\Box t}.$$ (10.132)

Instead of which in our above scheme, one can use the following inverse operator representations:

$$\frac{1}{\sqrt{\Box + m^2}} = \int_0^\infty dt\, e^{-mt} J_0 \left[t \left(i\gamma^\nu \frac{\partial}{\partial x^\nu} \right) \right]$$ (10.133)

or

$$\frac{1}{\sqrt{\Box + m^2}} = \int_0^\infty dt\, J_0(mt) \exp\left[-t \left(i\gamma^\nu \frac{\partial}{\partial x^\nu} \right) \right]$$ (10.134)

and so on. Such type of representations for inverse operators allows us to work easily. This problem will be considered separately in another place.

10.5   Fractional Integrals

By definition of fractional derivatives, we have

$$\frac{d^\rho}{dx^\rho} f(x) = F(x, \rho).$$ (10.135)

After taking integral from both sides of this equation, one gets

$$\int d^\rho f(x) = \int \rho x^{\rho-1}\, dx\, F(x, \rho).$$

Sometimes it is interesting to consider the following identity

$$\lambda(x) \frac{d^\rho}{dx^\rho} f(x) = \lambda(x)\, F(x, \rho)$$

and the general integral form:

$$\int \lambda(x)\, d^\rho f(x) = \rho \int x^{\rho-1} \lambda(x)\, F(x, \rho) dx.$$ (10.136)

10.5.1 *Fractional Integrals for Sine and Cosine Functions*

Let us consider some examples, since

$$\frac{d^\rho}{dx^\rho} \sin(ax) = a^\rho \sin\left(ax + \frac{\pi}{2}\rho\right),$$

where $\rho = -\dfrac{1}{\nu}$ in the previous formulas and therefore

$$
\int d^\rho \sin(ax) = a^\rho \rho \left[\cos\frac{\pi}{2}\rho \int dx x^{\rho-1} \sin(ax) \right.
$$
$$
\left. + \sin\frac{\pi}{2}\rho \int dx x^{\rho-1} \cos(ax) \right].
$$

(10.137)

In particular,

$$
L_1 = \int_0^\infty d^\rho \sin(ax) = \rho\, a^\rho \left[\cos\frac{\pi}{2}\rho \int_0^\infty dx x^{\rho-1} \sin(ax) \right.
$$
$$
\left. + \sin\frac{\pi}{2}\rho \int_0^\infty dx x^{\rho-1} \cos(ax) \right],
$$

where

$$
\int_0^\infty dx x^{\rho-1} \sin(ax) = \frac{\Gamma(\rho)}{a^\rho} \sin\frac{\pi}{2}\rho,\ (0 < |\mathrm{Re}\,\rho| < 1);
$$

and

$$
\int_0^\infty dx x^{\rho-1} \cos(ax) = \frac{\Gamma(\rho)}{a^\rho} \cos\frac{\pi}{2}\rho,\ (0 < \mathrm{Re}\,\rho < 1),\ a > 0.
$$

Thus,

$$
Q_1 = \int_0^\infty d^\rho \sin(ax) = \Gamma(1+\rho) \sin(\pi\rho).
$$

(10.138)

Similar calculation for the cosine-function takes the forms:

$$\frac{d^\rho}{dx^\rho} \cos(ax) = a^\rho \cos\left(ax + \frac{\pi}{2}\rho\right),$$

$$
\int d^\rho \cos(ax) = \rho\, a^\rho \left[\cos\frac{\pi}{2}\rho \int dx x^{\rho-1} \cos(ax) \right.
$$
$$
\left. - \sin\frac{\pi}{2}\rho \int dx x^{\rho-1} \sin(ax) \right].
$$

(10.139)

So that

$$Q_2 = \int_0^\infty d^\rho \, \cos(ax) = \Gamma(1+\rho) \, \cos(\pi\rho).$$

(10.140)

Next, making use of formulas (10.137), (10.138), (10.139) and (10.140), one can obtain more interesting particular fractional integrals.

1. Let $\rho = 1$, then

$$\int_0^\infty d\sin(ax) = a \int_0^\infty dx \, \cos(ax) = 0$$

and

$$\int_0^\infty d\cos(ax) = -a \int_0^\infty dx \, \sin(ax) = -1,$$

i.e.,

$$\int_0^\infty dx \, \sin(ax) = \frac{1}{a}, \quad a > 0.$$

2. Let $\rho = -1$, then

$$a^{-1} \int_0^\infty dx^{-1} \, \sin\left(ax - \frac{\pi}{2}\right) = -a^{-1} \int_0^\infty dx^{-1} \, \cos(ax) = -\pi.$$

Here we obtain a very interesting definition for π-number

$$\pi = \frac{1}{a} \int_0^\infty dx^{-1} \, \cos(ax),$$

(10.141)

moreover,

$$\frac{1}{a} \int_0^\infty dx^{-1} \, \sin(ax) = -\infty.$$

(10.142)

3. Let $\rho = \dfrac{1}{3}$, then

$$\int_0^\infty d^{1/3} \, \sin(ax) = \Gamma\left(1 + \frac{1}{3}\right) \sin(60°) = \frac{\sqrt{3}}{2} \frac{1}{3} \Gamma\left(\frac{1}{3}\right),$$

on the other hand

$$a^{1/3} \int_0^\infty dx^{1/3} \, \sin\left(ax + \frac{\pi}{6}\right) = \frac{1}{2} a^{1/3} \int_0^\infty dx^{1/3} \left[\sqrt{3} \, \sin(ax) + \cos(ax)\right].$$

For the cosine-function, we have similar expressions:

$$\int_0^\infty d^{1/3} \cos(ax) = \frac{1}{6} \Gamma\left(\frac{1}{3}\right)$$

and

$$a^{1/3} \int_0^\infty dx^{1/3} \left[\sqrt{3} \, \cos(ax) - \sin(ax)\right] = \frac{1}{3} \Gamma\left(\frac{1}{3}\right).$$

Thus, we have the following system of equations

$$a^{1/3} \int_0^\infty dx^{1/3} \left[\sqrt{3} \, \sin(ax) + \cos(ax)\right] = \frac{\sqrt{3}}{3} \Gamma\left(\frac{1}{3}\right),$$

$$a^{1/3} \int_0^\infty dx^{1/3} \left[\sqrt{3} \, \cos(ax) - \sin(ax)\right] = \frac{1}{3} \Gamma\left(\frac{1}{3}\right).$$

(10.143)

The system of equation (10.143) gives us interesting results:

$$\int_0^\infty dx^{1/3} \, \sin(ax) = a^{-1/3} \frac{1}{6} \Gamma\left(\frac{1}{3}\right),$$

$$\int_0^\infty dx^{1/3} \, \cos(ax) = a^{-1/3} \frac{\sqrt{3}}{6} \Gamma\left(\frac{1}{3}\right).$$

(10.144)

4. Let $\rho = -\dfrac{1}{3}$, then

$$\int_0^\infty d^{-1/3} \cos(ax) = a^{-1/3} \int_0^\infty \cos\left(ax - \frac{\pi}{6}\right) dx^{-1/3},$$

where

$$\cos\left(ax - \frac{\pi}{6}\right) = \cos(ax) \, \cos\frac{\pi}{6} + \sin(ax) \, \sin\frac{\pi}{6}$$

$$= \frac{\sqrt{3}}{2} \cos(ax) + \sin(ax) \, \frac{1}{2},$$

moreover,

$$\int_0^\infty d^{-1/3} \cos(ax) = \Gamma\left(\frac{2}{3}\right) \frac{1}{2}$$

and therefore

$$a^{-1/3} \int_0^\infty dx^{-1/3} \left[\sqrt{3}\cos(ax) + \sin(ax)\right] = \Gamma\left(\frac{2}{3}\right). \qquad (10.145)$$

For the sine-function, we have

$$\int_0^\infty d^{-1/3} \sin(ax) = a^{-1/3} \int_0^\infty dx^{-1/3} \sin\left(ax - \frac{\pi}{6}\right),$$

where

$$\sin\left(ax - \frac{\pi}{6}\right) = \sin(ax)\,\cos\frac{\pi}{6} - \cos(ax)\,\sin\frac{\pi}{6}$$

$$= \frac{\sqrt{3}}{2}\sin(ax) - \frac{1}{2}\cos(ax).$$

Thus

$$a^{-1/3} \int_0^\infty dx^{-1/3} \left[\sqrt{3}\sin(ax) - \cos(ax)\right] = -\Gamma\left(\frac{2}{3}\right)\sqrt{3}. \qquad (10.146)$$

From equations (10.145) and (10.146), we have

$$\int_0^\infty dx^{-1/3} \sin(ax) = -\frac{1}{2}\Gamma\left(\frac{2}{3}\right) a^{1/3},$$

$$\int_0^\infty dx^{-1/3} \cos(ax) = \frac{\sqrt{3}}{2} a^{1/3}\,\Gamma\left(\frac{2}{3}\right). \qquad (10.147)$$

5. Let $\rho = \frac{1}{2}$, then

$$\int_0^\infty d^{1/2}\sin(ax) = \Gamma\left(1 + \frac{1}{2}\right) = \frac{1}{2}\sqrt{\pi}$$

or

$$a^{1/2}\int_0^\infty dx^{1/2}\sin\left(ax + \frac{\pi}{4}\right) = \frac{1}{\sqrt{2}}\,a^{1/2}\int_0^\infty dx^{1/2}\left[\sin(ax) + \cos(ax)\right].$$

For the cosine-function, we have equation

$$\frac{1}{\sqrt{2}}\, a^{1/2} \int\limits_0^\infty dx^{1/2}\, [\cos(ax) - \sin(ax)] = 0.$$

Finally, we have

$$\int\limits_0^\infty dx^{1/2}\, \cos(ax) = \frac{\sqrt{2}}{4}\, \sqrt{\pi}\, a^{-1/2} \qquad (10.148)$$

and

$$\int\limits_0^\infty dx^{1/2}\, \sin(ax) = \frac{\sqrt{2}}{4}\, \sqrt{\pi}\, a^{-1/2}. \qquad (10.149)$$

6. Let $\rho = -\dfrac{1}{2}$, then

$$\int\limits_0^\infty d^{-1/2}\, \sin(ax) = -\sqrt{\pi},$$

$$a^{-1/2} \int\limits_0^\infty dx^{-1/2}\, \sin\left(ax - \frac{\pi}{4}\right) = \frac{1}{\sqrt{2}}\, a^{-1/2} \int\limits_0^\infty dx^{-1/2}\, [\sin(ax) - \cos(ax)]\,.$$

Similar equation holds for the cosine-function:

$$\int\limits_0^\infty d^{-1/2}\, \cos(ax) = 0$$

and

$$a^{-1/2} \int\limits_0^\infty dx^{-1/2}\, \cos\left(ax - \frac{\pi}{4}\right) = \frac{1}{\sqrt{2}}\, a^{-1/2} \int\limits_0^\infty dx^{-1/2}\, [\cos(ax) + \sin(ax)]\,.$$

Thus,

$$\int\limits_0^\infty dx^{-1/2}\, \sin(ax) = -\frac{\sqrt{2}}{2}\, \sqrt{\pi}\, a^{1/2},$$

$$\int\limits_0^\infty dx^{-1/2}\, \cos(ax) = \frac{\sqrt{2}}{2}\, \sqrt{\pi}\, a^{1/2}. \qquad (10.150)$$

7. Let $\rho = \dfrac{1}{4}$, then

$$\int_0^\infty d^{1/4} \cos(ax) = \int_0^\infty d^{1/4} \sin(ax) = \frac{1}{4} \Gamma\left(\frac{1}{4}\right) \frac{1}{\sqrt{2}} = \lambda$$

or

$$\lambda = a^{1/4} \int_0^\infty dx^{1/4} \left[\sin(ax) \, \cos\frac{\pi}{8} + \cos(ax) \, \sin\frac{\pi}{8}\right]$$

and

$$\lambda = a^{1/4} \int_0^\infty dx^{1/4} \left[\cos(ax) \, \cos\frac{\pi}{8} - \sin(ax) \, \sin\frac{\pi}{8}\right].$$

From these equations, we have

$$\int_0^\infty dx^{1/4} \sin(ax) = \frac{1}{4} a^{-1/4} \Gamma\left(\frac{1}{4}\right) \frac{1}{\sqrt{2}} \left(\cos\frac{\pi}{8} - \sin\frac{\pi}{8}\right),$$

$$\int_0^\infty dx^{1/4} \cos(ax) = \frac{1}{4} a^{-1/4} \Gamma\left(\frac{1}{4}\right) \frac{1}{\sqrt{2}} \left(\cos\frac{\pi}{8} + \sin\frac{\pi}{8}\right).$$

$$(10.151)$$

8. Let $\rho = -\dfrac{1}{4}$, then

$$L_1 = \int_0^\infty d^{-1/4} \sin(ax) = -\Gamma\left(\frac{3}{4}\right) \frac{1}{\sqrt{2}},$$

$$L_2 = \int_0^\infty d^{-1/4} \cos(ax) = \Gamma\left(\frac{3}{4}\right) \frac{1}{\sqrt{2}}$$

and

$$L_1 = a^{-1/4} \int_0^\infty dx^{-1/4} \left[\sin(ax) \, \cos\frac{\pi}{8} - \cos(ax) \, \sin\frac{\pi}{8}\right],$$

and

$$L_2 = a^{-1/4} \int_0^\infty dx^{-1/4} \left[\cos(ax) \, \cos\frac{\pi}{8} + \sin(ax) \, \sin\frac{\pi}{8}\right].$$

Therefore:

$$\int_0^\infty dx^{-1/4} \sin(ax) = a^{1/4} \, \Gamma\left(\frac{3}{4}\right) \frac{1}{\sqrt{2}} \left(\sin\frac{\pi}{8} - \cos\frac{\pi}{8}\right),$$

$$\int_0^\infty dx^{-1/4} \cos(ax) = a^{1/4} \, \Gamma\left(\frac{3}{4}\right) \frac{1}{\sqrt{2}} \left(\cos\frac{\pi}{8} + \sin\frac{\pi}{8}\right).$$

(10.152)

It turns out that fractional integrals are reduced to usual ones. For example, by using direct calculations, one gets

$$a^{1/4} \int_0^\infty dx^{1/4} \, \sin(ax) = a^{1/4} \frac{1}{4} \int_0^\infty dx\, x^{-\frac{3}{4}} \, \sin(ax)$$

$$= \frac{1}{4} \, \Gamma\left(\frac{1}{4}\right) \sin\frac{\pi}{8}$$

(10.153)

and

$$a^{1/4} \int_0^\infty dx^{1/4} \, \cos(ax) = a^{1/4} \frac{1}{4} \int_0^\infty dx\, x^{-\frac{3}{4}} \, \cos(ax)$$

$$= \frac{1}{4} \, \Gamma\left(\frac{1}{4}\right) \cos\frac{\pi}{8}.$$

(10.154)

On the other hand, in accordance with the standard trigonometric relations:

$$\cos x \pm \sin x = \sqrt{2} \, \sin\left(\frac{\pi}{4} \pm x\right) = \sqrt{2} \, \cos\left(\frac{\pi}{4} \mp x\right),$$

these formulas (10.153) and (10.154) are easily obtained from the equations (10.151), where

$$\cos\frac{\pi}{8} - \sin\frac{\pi}{8} = \sqrt{2} \, \sin\left(\frac{\pi}{4} - \frac{\pi}{8}\right) = \sqrt{2} \, \sin\frac{\pi}{8}$$

and

$$\cos\frac{\pi}{8} + \sin\frac{\pi}{8} = \sqrt{2} \, \cos\left(\frac{\pi}{4} - \frac{\pi}{8}\right) = \sqrt{2} \, \cos\frac{\pi}{8}.$$

It means that formulas (10.9) and (10.10) are **true definitions** of fractional derivatives for sine- and cosine-functions.

10.5.2 *Fractional Integrals for Infinite Differentiable Functions*

Let $f(x) = 1/x$, then its fractional integral is given by

$$\int d^\rho \left(\frac{1}{x}\right) = \rho \int x^{\rho-1} \, F_1\left(x, \rho\right) dx,$$

(10.155)

where

$$F_1(x, \rho) = -\Gamma(1 + \rho)\, x^{-\rho-1}.$$

Let us consider particular cases of (10.155). That is in accordance with the formula (10.67) in Chapter 10:

$$\int_a^\infty d^\rho\left(\frac{1}{x}\right) = \rho \int_a^\infty dx\, x^{\rho-1}\, \left[-\Gamma(1+\rho)\right]\, x^{-\rho-1}$$

$$= -\rho\,\Gamma(1+\rho)\,\frac{1}{a}. \tag{10.156}$$

Let us calculate a very simple version of this expression:

$$9. \quad \int_a^\infty d^{1/2}\left(\frac{1}{x}\right) = -\frac{\sqrt{\pi}}{4}\frac{1}{a}, \tag{10.157}$$

$$10. \quad \int_a^\infty d^{1/4}\left(\frac{1}{x}\right) = -\frac{1}{16}\Gamma\left(\frac{1}{4}\right)\frac{1}{a}, \tag{10.158}$$

$$11. \quad \int_a^\infty d^{1/10}\left(\frac{1}{x}\right) = -\frac{1}{100}\Gamma\left(\frac{1}{10}\right)\frac{1}{a} \tag{10.159}$$

so on.

Similarly, let $f(x) = \dfrac{1}{x^2}$, then

$$\int_a^\infty d^\rho\left(\frac{1}{x^2}\right) = \rho \int_a^\infty dx\, x^{\rho-1}\, \left[-\Gamma(2+\rho)\right]\, x^{-\rho-2}$$

$$= -\rho\,\Gamma(2+\rho)\,\frac{1}{2}\frac{1}{a^2}. \tag{10.160}$$

From this formula we have

$$12. \quad \int_a^\infty d^{1/2}\left(\frac{1}{x^2}\right) = -\frac{3\sqrt{\pi}}{16a^2}, \tag{10.161}$$

$$13. \quad \int_a^\infty d^{1/3}\left(\frac{1}{x^2}\right) = -\frac{2}{27}\Gamma\left(\frac{1}{3}\right)\frac{1}{a^2}, \tag{10.162}$$

$$14. \quad \int_a^\infty x^{1/5}\left(\frac{1}{x^2}\right) = -\frac{3}{125}\Gamma\left(\frac{1}{5}\right)\frac{1}{a^2}, \tag{10.163}$$

$$15. \quad \int_a^\infty d\left(\frac{1}{x^2}\right) = -\frac{1}{a^2}, \tag{10.164}$$

$$16. \quad \int_a^\infty d^{-1}\left(\frac{1}{x^2}\right) = \frac{1}{2}\frac{1}{a^2} \tag{10.165}$$

etc.

Fractional integral from the exponential function $f(x) = e^{-ax}$ is easy to calculate. For example:

$$\int d^{\rho} \, e^{-ax} = \rho \int dx \, x^{\rho-1} \left(-a^{\rho} \, e^{-ax} \right)$$

$$= -\rho \, a^{\rho} \int dx x^{\rho-1} \, e^{-ax}.$$

In this particular case, it takes the form

$$\int_{0}^{\infty} d^{\rho} \, e^{-ax} = -\rho \, a^{\rho} \int_{0}^{\infty} dx \, x^{\rho-1} \, e^{-ax}$$

$$= -\rho \, a^{\rho} \, a^{-\rho} \Gamma \left(\rho \right) = -\Gamma \left(1 + \rho \right).$$

Thus

$$\int_{0}^{\infty} d^{1/2} \, e^{-ax} = -\frac{1}{2} \sqrt{\pi},$$

$$\int_{0}^{\infty} d^{-1/2} \, e^{-ax} = -\sqrt{\pi},$$

$$\int_{0}^{\infty} d^{1/3} \, e^{-ax} = -\frac{1}{3} \Gamma \left(\frac{1}{3} \right),$$

$$\int_{0}^{\infty} d^{1/4} \, e^{-ax} = -\frac{1}{4} \Gamma \left(\frac{1}{4} \right),$$

$$\int_{0}^{\infty} d^{-1/4} \, e^{-ax} = -\frac{\pi\sqrt{2}}{\Gamma \left(\frac{1}{4} \right)}$$

and so on. From the previous formulas, we see that fractional integrals have pure symbolic character and are reduced to usual ones directly.

Notice that we think in future the fractional differential and integral calculus play a vital role in analytic calculation methods in mathematics.

10.6   Final Table for Taking Fractional Derivatives of Some Elementary Functions

$$\left(\frac{d}{dx}\right)^{\rho} C = 0,$$

$$\left(\frac{d}{dx}\right)^{\rho} \sin(ax) = a^{\rho} \sin\left(ax + \frac{\pi}{2}\rho\right),$$

$$\left(\frac{d}{dx}\right)^{\rho} \cos(ax) = a^{\rho} \cos\left(ax + \frac{\pi}{2}\rho\right),$$

$$\left(\frac{d}{dx}\right)^{\rho} e^{ax} = a^{\rho} e^{ax}, \qquad \left(\frac{d}{dx}\right)^{\rho} e^{-ax} = -a^{\rho} e^{-ax},$$

$$\left(\frac{d}{dx}\right)^{\rho} a^{x} = (\ln a)^{\rho} a^{x}, \quad a > 1, \qquad \left(\frac{d}{dx}\right)^{\rho} \left(\frac{1}{x}\right) = -\Gamma\left(1 + \rho\right) x^{-\rho-1},$$

$$\left(\frac{d}{dx}\right)^{\rho} \sinh(ax) = a^{\rho} \cosh(ax), \qquad \left(\frac{d}{dx}\right)^{\rho} \ln x = \Gamma\left(\rho\right) x^{-\rho},$$

$$\left(\frac{d}{dx}\right)^{\rho} \cosh(ax) = a^{\rho} \sinh(ax), \qquad \left(\frac{d}{dx}\right)^{\rho} \left(\frac{1}{x^2}\right) = -\Gamma\left(2 + \rho\right) x^{-\rho-2},$$

$$\left(\frac{d}{dx}\right)^{\rho} \left(\frac{1}{\sqrt{x}}\right) = -2^{-\rho} \qquad \left(\frac{d}{dx}\right)^{\rho} \sqrt{x} = 2^{-\rho}$$
$$\times x^{-\rho-1/2}\,[2\rho - 1]!!, \qquad \times x^{-\rho+1/2}\,(2\rho - 3)!!,$$

$$\left(\frac{d}{dx}\right)^{\rho} \left[\sin^{2n}(ax)\right] = \frac{1}{2^{2n-1}} \sum_{k=0}^{n-1} (-1)^{n-k} \left(\begin{array}{c} 2n \\ k \end{array}\right) 2^{\rho}(n-k)^{\rho}\, a^{\rho}$$
$$\times \cos\left[2(n-k)ax + \frac{\pi}{2}\rho\right],$$

$$\left(\frac{d}{dx}\right)^{\rho} \left[\sin^{2n-1}(ax)\right] = \frac{1}{2^{2n-2}} \sum_{k=0}^{n-1} (-1)^{n+k-1} \left(\begin{array}{c} 2n\text{-}1 \\ k \end{array}\right) (2n - 2k - 1)^{\rho}$$
$$\times a^{\rho} \sin\left[(2n - 2k - 1)ax + \frac{\pi}{2}\rho\right],$$

$$\left(\frac{d}{dx}\right)^{\rho} \left[\cos^{2n}(ax)\right] = \frac{1}{2^{2n-1}} \sum_{k=0}^{n-1} \left(\begin{array}{c} 2n \\ k \end{array}\right) 2^{\rho}(n-k)^{\rho}\, a^{\rho}$$
$$\times \cos\left[2(n-k)ax + \frac{\pi}{2}\rho\right],$$

$$\left(\frac{d}{dx}\right)^{\rho} \left[\cos^{2n-1}(ax)\right] = \frac{1}{2^{2n-2}} \sum_{k=0}^{n-1} \left(\begin{array}{c} 2n\text{-}1 \\ k \end{array}\right) (2n - 2k - 1)^{\rho}\, a^{\rho}$$
$$\times \cos\left[(2n - 2k - 1)ax + \frac{\pi}{2}\rho\right],$$

where $n = 1, 2, 3, \ldots$.

In particular,

$$\left(\frac{d}{dx}\right)^{-1} f(x) = \int dx \ f(x),$$

$$\left(\frac{d}{dx}\right)^{-1} \sin x = -\cos x = \int dx \ \sin x,$$

$$\left(\frac{d}{dx}\right)^{-1} \cos x = \sin x = \int dx \ \cos x,$$

$$\left(\frac{d}{dx}\right)^{-1} \frac{1}{x^2} = -\frac{1}{x} = \int dx \ x^{-2},$$

$$\left(\frac{d}{dx}\right)^{-1} x^n = \frac{x^{n+1}}{n+1} + C = \int dx \ x^n, \quad n \geq 0,$$

$$\left(\frac{d}{dx}\right)^{-2} x^n = \frac{x^{n+2}}{(n+1)(n+2)} + C = \int dx \int_0^x dy \, y^n,$$

$$\left(\frac{d}{dx}\right)^{-1} 1 = x = \int dx,$$

$$\left(\frac{d}{dx}\right)^{-1/2} 1 = \sqrt{x} = \int dx^{1/2},$$

$$\left(\frac{d}{dx}\right)^{-1/2} x = \frac{1}{3} x^{3/2},$$

$$\left(\frac{d}{dx}\right)^{-1/2} x^2 = \frac{1}{5} x^{5/2},$$

$$\left(\frac{d}{dx}\right)^{-1/2} x^3 = \frac{1}{7} x^{7/2},$$

$$\left(\frac{d}{dx}\right)^{1/2} const. = 0,$$

$$\left(\frac{d}{dx}\right)^{1/2} x = \sqrt{x},$$

$$\left(\frac{d}{dx}\right)^{1/2} x^2 = \frac{2}{3} x^{3/2},$$

$$\left(\frac{d}{dx}\right)^{1/2} x^3 = \frac{3}{5} x^{5/2},$$

$$\left(\frac{d}{dx}\right)^{1/2} x^4 = \frac{4}{7} x^{7/2},$$

and so on. These last formulas arise from the following integral representations:

$$\int_0^1 dt\, t^\mu (1-x)^\mu \sin\left(2t\frac{d}{dx}\right) \cdot F(x)$$

$$= \frac{\sqrt{\pi}}{2^{\mu+1/2}} \Gamma(1+\mu) \left[\left(\frac{d}{dx}\right)^{-\mu-1/2} \sin\frac{d}{dx}\, J_{\mu+1/2}\left(\frac{d}{dx}\right)\right] F(x),$$

$$\int_0^1 dt\, t^\mu (1-t)^\mu \cos\left(2t\frac{d}{dx}\right) \cdot F(x)$$

$$= \frac{\sqrt{\pi}}{2^{\mu+1/2}} \Gamma(1+\mu) \left[\left(\frac{d}{dx}\right)^{-\mu-1/2} \cos\frac{d}{dx}\, J_{\mu+1/2}\left(\frac{d}{dx}\right)\right] F(x),$$

$\operatorname{Re}\mu > -1,$

$$\int_0^1 dt\, t^{\nu+1}(1-t^2)^\mu J_\nu\left(t\frac{d}{dx}\right) \cdot F(x)$$

$$= 2^\mu \Gamma(1+\mu) \left[\left(\frac{d}{dx}\right)^{-(1+\mu)} J_{\nu+\mu+1}\left(\frac{d}{dx}\right)\right] F(x),$$

$\operatorname{Re}\nu > -1, \quad \operatorname{Re}\mu > -1,$

and

$$\int_0^1 dt\, t^{\nu+1}(1-t^2)^{-\nu-\frac{1}{2}} J_\nu\left(t\frac{d}{dx}\right) \cdot F(x)$$

$$= 2^{-\nu} \frac{1}{\sqrt{\pi}} \Gamma\left(\frac{1}{2} - \nu\right) \left[\left(\frac{d}{dx}\right)^{\nu-1} \sin\frac{d}{dx}\right] F(x),$$

$|\operatorname{Re}\nu| < \frac{1}{2},$

$$\int_0^1 dt\, t^\nu (1-t^2)^{\nu-1/2} J_\nu\left(t\frac{d}{dx}\right) \cdot F(x)$$

$$= \sqrt{\pi}\, 2^{\nu-1} \Gamma\left(\nu + \frac{1}{2}\right) \left[\left(\frac{d}{dx}\right)^{-\nu} J_\nu^2\left(\frac{1}{2}\frac{d}{dx}\right)\right] F(x),$$

where $\operatorname{Re}\nu > -\frac{1}{2}.$

Appendix

Tables of the Definitions for Fractional Derivatives and Inverse Operators

A.1 Taking Fractional Derivatives

1. $\left(\dfrac{d}{dx}\right)^{-n-\frac{1}{2}} F(x) = \dfrac{1}{\sqrt{\pi}\left(\frac{1}{2}\cdot\frac{3}{2}\cdots\frac{2n-1}{2}\right)} \displaystyle\int_0^\infty dt\, t^{n-\frac{1}{2}} e^{-t\,\frac{d}{dx}}\, F(x),$

   where $n = 0, 1, \ldots$.

2. $\left(\dfrac{d}{dx}\right)^{-\nu} F(x) = \dfrac{1}{\Gamma(\nu)} \displaystyle\int_0^\infty dt\, t^{\nu-1}\, e^{-t\,\frac{d}{dx}}\, F(x).$

3. $\left[\dfrac{\ln\left(\frac{d}{dx}\right)}{\frac{d}{dx}}\right] F(x) = \displaystyle\int_{-\infty}^\infty \dfrac{dt\, e^t}{\left[\frac{d}{dt} + e^t\right]^2}\, F(x).$

4. $\left(\dfrac{d}{dx}\right)^{-\frac{\nu}{p}} F(x) = \dfrac{|p|}{\Gamma\left(\frac{\nu}{p}\right)} \displaystyle\int_0^\infty dt\, t^{\nu-1}\, e^{-t^p\,\frac{d}{dx}}\, F(x),$

   where $\operatorname{Re}\nu > 0$.

5. $\left(\dfrac{d}{dx}\right)^{-\nu} F(x) = \dfrac{1}{\Gamma(\nu)\sin\frac{\pi\nu}{2}} \displaystyle\int_0^\infty dt\, t^{\nu-1}\, \sin\left(t\,\dfrac{d}{dx}\right) F(x),$

   where $0 < \operatorname{Re}\nu \leq 1$.

6. $\left(\dfrac{d}{dx}\right)^{-\nu} F(x) = \dfrac{1}{\Gamma(\nu)\cos\frac{\pi\nu}{2}} \displaystyle\int_0^\infty dt\, t^{\nu-1}\, \cos\left(t\,\dfrac{d}{dx}\right) F(x),$

   where $0 < \operatorname{Re}\nu < 1$.

7. $\left(\dfrac{d}{dx}\right)^{-1-\nu} F(x)$

   $= \dfrac{1}{\Gamma(1+\nu)\cos\left(b + \frac{\pi\nu}{2}\right)} \displaystyle\int_0^\infty dt\, t^{\nu}\, \sin\left(t\,\dfrac{d}{dx} + b\right) F(x),$

   where $-1 < \nu < 0$.

8. $\left(\dfrac{d}{dx}\right)^{-1-\nu} F(x)$

$$= -\frac{1}{\Gamma(1+\nu)\sin\left(b+\frac{\pi\nu}{2}\right)} \int_0^\infty dt\, t^\nu \, \cos\left(t\,\frac{d}{dx}+b\right) F(x),$$

where $-1 < \nu < 0$.

9. $\ln\left[\dfrac{b}{\frac{d}{dx}}\right] F(x) = \displaystyle\int_0^\infty dt\,\frac{\cos\left(t\,\frac{d}{dx}\right) - \cos(bt)}{t}\, F(x).$

10. $\ln\left[\dfrac{\frac{d}{dx}}{a}\right] F(x) = \displaystyle\int_0^\infty dt\,\frac{\cos(at) - \cos\left(t\,\frac{d}{dx}\right)}{t}\, F(x).$

11. $\left\{\dfrac{d}{dx}\,\ln\left[\dfrac{1}{b}\dfrac{d}{dx}\right]\right\} F(x) = \dfrac{1}{b}\displaystyle\int_0^\infty dt\,\frac{\frac{d}{dx}\sin(bt) - b\sin\left(t\,\frac{d}{dx}\right)}{t^2}\, F(x).$

12. $\left\{\dfrac{d}{dx}\,\ln\left[\dfrac{a}{\frac{d}{dx}}\right]\right\} F(x) = \dfrac{1}{a}\displaystyle\int_0^\infty dt\,\frac{a\sin\left(t\,\frac{d}{dx}\right) - \frac{d}{dx}\sin(at)}{t^2}\, F(x).$

13. $\ln\left[\dfrac{1}{b}\dfrac{d}{dx}\right] F(x) = 2\displaystyle\int_0^\infty dt\,\frac{\sin^2\left(t\,\frac{d}{dx}\right) - \sin^2(bt)}{t}\, F(x).$

14. $\ln\left[\dfrac{a}{\frac{d}{dx}}\right] F(x) = 2\displaystyle\int_0^\infty dt\,\frac{\sin^2(at) - \sin^2\left(t\,\frac{d}{dx}\right)}{t}\, F(x).$

15. $\left(\dfrac{d}{dx}\right)^{-\nu} F(x) = -\dfrac{2^{1+\nu}}{\Gamma(\nu)\cos\frac{\pi\nu}{2}}\displaystyle\int_0^\infty dt\, t^{\nu-1}\, \sin^2\left(t\,\frac{d}{dx}\right) F(x),$

where $-2 < \operatorname{Re}\nu < 0$.

16. $\cosh^{-2}\left(\dfrac{\pi}{\beta}\dfrac{d}{dx}\right) F(x) = \dfrac{4\beta^2}{\pi^2}\displaystyle\int_0^\infty dt\,\frac{\cos\left(2t\,\frac{d}{dx}\right)}{\sinh(\beta t)}\, F(x).$

17. $\left(\dfrac{d}{dx}\right)^{-\nu} F(x) = \dfrac{\nu\sin(\pi\nu)}{\pi}\displaystyle\int_0^\infty dt\, t^{\nu-1}\, \ln\left(1+t\,\frac{d}{dx}\right) F(x),$

where $-1 < \operatorname{Re}\nu < 0$.

18. $\left(\dfrac{d}{dx}\right)^{-1-\mu} F(x) = \dfrac{\Gamma\left(\frac{1}{2}+\frac{1}{2}\nu-\frac{1}{2}\mu\right)}{2^\mu\,\Gamma\left(\frac{1}{2}+\frac{1}{2}\nu+\frac{1}{2}\mu\right)}\displaystyle\int_0^\infty dt\, t^\mu\, J_\nu\left(t\,\frac{d}{dx}\right) F(x),$

where $-\operatorname{Re}\nu-1 < \operatorname{Re}\mu < \dfrac{1}{2}.$

19. $\left(\dfrac{d}{dx}\right)^{-1-\mu} F(x) = \dfrac{\Gamma\left(\frac{1}{2}+\frac{1}{2}\nu-\frac{1}{2}\mu\right)}{2^\mu\,\Gamma\left(\frac{1}{2}+\frac{1}{2}\nu+\frac{1}{2}\mu\right)}\dfrac{1}{\cot\left[\frac{1}{2}(\nu+1-\mu)\pi\right]}$

$$\times \int\limits_0^\infty dt\, t^\mu\, N_\nu \left(t\, \frac{d}{dx} \right) F(x),$$

where $|\mathrm{Re}\,\nu| - 1 < \mu < \dfrac{1}{2}$.

20. $\left(\dfrac{d}{dx} \right)^{-1-\mu} F(x) = \dfrac{1}{2^{\mu-1}\, \Gamma\left(\frac{1+\mu+\nu}{2} \right)\, \Gamma\left(\frac{1+\mu-\nu}{2} \right)}$

$$\times \int\limits_0^\infty dt\, t^\mu\, K_\nu \left(t\, \frac{d}{dx} \right) F(x),$$

where $\mathrm{Re}(\mu + 1 \pm \nu) > 0$.

21. $\left(\dfrac{d}{dx} \right)^{-1-q+\nu} F(x) = \dfrac{2^{\nu-q}\Gamma\left(\nu - \frac{1}{2}q + \frac{1}{2} \right)}{\Gamma\left(\frac{1}{2}q + \frac{1}{2} \right)} \int\limits_0^\infty dt\, \dfrac{J_\nu\left(t\, \frac{d}{dx} \right)}{t^{\nu-q}}\, F(x),$

where $-1 < \mathrm{Re}\, q < \mathrm{Re}\, \nu - \dfrac{1}{2}$.

22. $\left[\sin \dfrac{d}{dx} \left(\dfrac{d}{dx} \right)^{\nu-1} \right] F(x) = \dfrac{\sqrt{\pi}\, 2^\nu}{\Gamma\left(\frac{1}{2} - \nu \right)}$

$$\times \int\limits_0^1 dt\, t^{\nu+1}(1 - t^2)^{-\nu-\frac{1}{2}}\, J_\nu \left(t\, \frac{d}{dx} \right) F(x),$$

where $|\mathrm{Re}\,\nu| < \dfrac{1}{2}$.

23. $\left[\cos \dfrac{d}{dx} \left(\dfrac{d}{dx} \right)^{-1-\nu} \right] F(x) = \dfrac{\sqrt{\pi}\, 2^\nu}{\Gamma\left(\frac{1}{2} + \nu \right)}$

$$\times \int\limits_1^\infty dt\, t^{-\nu+1}(t^2 - 1)^{\nu-\frac{1}{2}}\, J_\nu \left(t\, \frac{d}{dx} \right) F(x),$$

where $|\mathrm{Re}\,\nu| < \dfrac{1}{2}$.

A.2 Calculation of Inverse Operators

24. $\left(\dfrac{d}{dx} \right)^{-\nu} F(x) = \dfrac{1}{2^{\nu-1}\, \Gamma(\nu)}\, \dfrac{z^\mu}{J_\mu(\alpha z)} \int\limits_0^\infty dt\, t^{\nu-1}\, \dfrac{J_\mu\left[\alpha\sqrt{t^2 + z^2} \right]}{\sqrt{(t^2 + z^2)^\mu}}$

$$\times J_\nu \left(t\, \frac{d}{dx} \right) F(x),$$

where $\mathrm{Re}(\mu + 2) > \mathrm{Re}\, \nu > 0,\ \alpha > 0$.

25. $\left(\dfrac{d}{dx}\right)^{-1-\mu} \left[J_{\nu-\mu-1}\left(z\,\dfrac{d}{dx}\right)\right] F(x) = \dfrac{z^{\nu-\mu-1}}{2^{\mu}\,\Gamma(1+\mu)}$

$$\times \int_0^{\infty} dt\, t^{1+2\mu} \frac{J_{\nu}\left(\sqrt{t^2+z^2}\,\frac{d}{dt}\right)}{\sqrt{(t^2+z^2)^{\nu}}} F(x),$$

where $\operatorname{Re}\left[\dfrac{1}{2}\nu - \dfrac{1}{4}\right] > \operatorname{Re}\mu > -1.$

26. $\dfrac{1}{\left(\frac{d}{dx}\right)^{\nu}\left[b^2 + \frac{d^2}{dx^2}\right]} F(x) = \dfrac{1}{b^{\nu}} \int_0^{\infty} dt\, J_{\nu}(bt)\, K_{\nu}\left(t\,\dfrac{d}{dx}\right) F(x)$

or

27. $\dfrac{\left(\frac{d}{dx}\right)^{\nu}}{\left[\frac{d^2}{dx^2}+a^2\right]} F(x) = a^{\nu} \int_0^{\infty} dt\, K_{\nu}(at)\, J_{\nu}\left(t\,\dfrac{d}{dx}\right) F(x),$

where $\operatorname{Re}\nu > -1.$

28. $\dfrac{\left[\sqrt{\frac{d^2}{dx^2}+\beta^2} - \frac{d}{dx}\right]^{\nu}}{\sqrt{\frac{d^2}{dx^2}+\beta^2}} F(x) = \beta^{\nu} \int_0^{\infty} dt\, J_{\nu}(\beta t)\, e^{-t\frac{d}{dx}} F(x)$

or

29. $\left(\dfrac{d}{dx}\right)^{-\nu} \dfrac{\left[\sqrt{\alpha^2 + \frac{d^2}{dx^2}} - \alpha\right]^{\nu}}{\sqrt{\alpha^2 + \frac{d^2}{dx^2}}} F(x) = \int_0^{\infty} dt\, e^{-\alpha t} J_{\nu}\left(t\,\dfrac{d}{dx}\right) F(x),$

where $\operatorname{Re}\nu > -1.$

30. $\dfrac{\left(\frac{d}{dx}\right)^{\nu}}{\sqrt{\alpha^2 - \frac{d^2}{dx^2}}} \dfrac{1}{\left(\alpha^2 + \sqrt{\alpha^2 - \frac{d^2}{dx^2}}\right)^{\nu}} F(x)$

$$= \int_0^{\infty} dt\, e^{-\alpha t} I_{\nu}\left(t\,\dfrac{d}{dx}\right) F(x)$$

or

31. $\dfrac{1}{\sqrt{\frac{d^2}{dx^2} - \beta^2}} \dfrac{1}{\left(\frac{d^2}{dx^2} + \sqrt{\frac{d^2}{dx^2} - \beta^2}\right)^{\nu}} F(x)$

$$= \dfrac{1}{\beta^{\nu}} \int_0^{\infty} dt\, I_{\nu}(t\beta)\, e^{-t\frac{d}{dx}} F(x).$$

32. $\dfrac{\arccos\left(\frac{1}{\beta}\frac{d}{dx}\right)}{\sqrt{\beta^2 - \frac{d^2}{dx^2}}} F(x) = \int_0^{\infty} dt\, K_0(\beta t)\, e^{-t\frac{d}{dx}} F(x).$

33. $\dfrac{1}{\left[\frac{d^2}{dx^2}+\beta^2\right]^{\nu+\frac{1}{2}}}F(x)$

$$=\frac{\sqrt{\pi}}{(2\beta)^{\nu}\,\Gamma\left(\nu+\frac{1}{2}\right)}\int\limits_{0}^{\infty}dt\,J_{\nu}(\beta t)\,e^{-t\frac{d}{dx}}\,F(x).$$

34. $\left(\dfrac{d}{dx}\right)^{\nu}\dfrac{1}{\left[\alpha^2+\frac{d^2}{dx^2}\right]^{\nu+\frac{1}{2}}}F(x)$

$$=\frac{\sqrt{\pi}}{2^{\nu}\,\Gamma\left(\nu+\frac{1}{2}\right)}\int\limits_{0}^{\infty}dt\,e^{-\alpha t}\,J_{\nu}\left(t\frac{d}{dx}\right)F(x),$$

where Re $\nu>-\dfrac{1}{2}$.

35. $\dfrac{\frac{d}{dx}}{\left[\frac{d^2}{dx^2}+\beta^2\right]^{\nu+\frac{3}{2}}}F(x)$

$$=\frac{\sqrt{\pi}}{2(2\beta)^{\nu}\,\Gamma\left(\nu+\frac{3}{2}\right)}\int\limits_{0}^{\infty}dt\,t^{\nu+1}\,J_{\nu}(\beta t)\,e^{-t\frac{d}{dx}}\,F(x)$$

or

36. $\dfrac{\left(\frac{d}{dx}\right)^{\nu}}{\left(\alpha^2+\frac{d^2}{dx^2}\right)^{\nu+\frac{3}{2}}}F(x)$

$$=\frac{\sqrt{\pi}}{2\alpha\,2^{\nu}\,\Gamma\left(\nu+\frac{3}{2}\right)}\int\limits_{0}^{\infty}dt\,t^{\nu+1}\,e^{-\alpha t}\,J_{\nu}\left(t\frac{d}{dx}\right)F(x),$$

where Re $\nu>-1$.

37. $\left[\sqrt{\dfrac{d^2}{dx^2}+\beta^2}-\dfrac{d}{dx}\right]^{\nu}F(x)=\nu\beta^{\nu}\int\limits_{0}^{\infty}dt\,\dfrac{1}{t}\,J_{\nu}(\beta t)\,e^{-t\frac{d}{dx}}\,F(x)$

or

$$\dfrac{\left[\sqrt{\alpha^2+\frac{d^2}{dx^2}}-\alpha\right]^{\nu}}{\left(\frac{d}{dx}\right)^{\nu}}F(x)=\nu\int\limits_{0}^{\infty}\dfrac{dt}{t}\,e^{-\alpha t}\,J_{\nu}\left(t\frac{d}{dx}\right)F(x),$$

where Re $\nu>0$.

38. $\dfrac{1}{\sqrt{\frac{d}{dx}}}\dfrac{1}{\alpha+\frac{d}{dx}}F(x)=\sqrt{\dfrac{2}{\pi}}\int\limits_{0}^{\infty}dt\sqrt{t}\,e^{-\alpha t}\,K_{\pm\frac{1}{2}}\left(t\frac{d}{dx}\right)F(x)$

or

$$\dfrac{1}{\frac{d}{dx}+\beta}F(x)=\sqrt{\dfrac{2\beta}{\pi}}\int\limits_{0}^{\infty}dt\sqrt{t}\,K_{\pm\frac{1}{2}}(t\beta)\,e^{-t\frac{d}{dx}}\,F(x).$$

39. $\left(\dfrac{d}{dx}\right)^{\nu} \exp\left[-\dfrac{1}{4\alpha}\dfrac{d^2}{dx^2}\right] F(x)$

$$= (2\alpha)^{1+\nu} \int_0^\infty dt\, t^{1+\nu}\, e^{-\alpha t^2}\, J_\nu\left(t\dfrac{d}{dx}\right) F(x),$$

where $\operatorname{Re}\nu > -1$, $\operatorname{Re}\alpha > 0$.

40. $\dfrac{1}{\sqrt{\dfrac{d^2}{dx^2} + \beta^2}}\, F(x) = \dfrac{2}{\pi} \int_0^\infty dt\, K_0(t\beta)\, \cos\left(t\dfrac{d}{dx}\right) F(x).$

41. $\dfrac{d}{dx} \left(\dfrac{d^2}{dx^2} + \beta^2\right)^{-\frac{3}{2}} F(x) = \dfrac{2}{\pi} \int_0^\infty dt\, t\, K_0(\beta t)\, \sin\left(t\dfrac{d}{dx}\right) F(x).$

42. $\dfrac{\dfrac{d}{dx}}{\left[\beta^2 + \dfrac{d^2}{dx^2}\right]^{\nu+\frac{3}{2}}}\, F(x) = \dfrac{1}{\sqrt{\pi}(2\beta)^\nu\, \Gamma\left(\frac{3}{2} + \nu\right)}$

$$\times \int_0^\infty dt\, t^{1+\nu}\, K_\nu(\beta t)\, \sin\left(t\dfrac{d}{dx}\right) F(x),$$

where $\operatorname{Re}\nu > -\dfrac{3}{2}$.

43. $\left(\dfrac{d^2}{dx^2} + \beta^2\right)^{-\frac{1}{2}-\nu} F(x) = \dfrac{2}{\sqrt{\pi}(2\beta)^\nu\, \Gamma\left(\nu + \frac{1}{2}\right)}$

$$\times \int_0^\infty dt\, t^\nu\, K_\nu(\beta t)\, \cos\left(t\dfrac{d}{dx}\right) F(x),$$

where $\operatorname{Re}\beta > 0$, $\operatorname{Re}\nu > -\dfrac{1}{2}$.

44 $\dfrac{1}{\left[2\beta\dfrac{d}{dx} + \dfrac{d^2}{dx^2}\right]^{\nu+\frac{1}{2}}}\, F(x) = \dfrac{\Gamma\left(\frac{1}{2} - \nu\right)}{\sqrt{\pi}(2\beta)^\nu}$

$$\times \int_0^\infty dt\, t^\nu\, [J_\nu(\beta t)\, \cos(\beta t) + N_\nu(\beta t)\, \sin(\beta t)]\, \sin\left(t\dfrac{d}{dx}\right) F(x),$$

where $-1 < \operatorname{Re}\nu < \dfrac{1}{2}$.

45. $\dfrac{1}{\left(\dfrac{d^2}{dx^2} + 2\beta\dfrac{d}{dx}\right)^{\nu+\frac{1}{2}}}\, F(x) = -\dfrac{\Gamma\left(\frac{1}{2} - \nu\right)}{\sqrt{\pi}(2\beta)^\nu}$

$$\times \int_0^\infty dt\, t^\nu\, [N_\nu(\beta t)\, \cos(\beta t) - J_\nu(\beta t)\, \sin(\beta t)]\, \cos\left(t\dfrac{d}{dx}\right) F(x),$$

where $-1 < \operatorname{Re}\nu < \dfrac{1}{2}$.

46. $\dfrac{1}{\sqrt{\frac{d}{dx}}}\ \dfrac{1}{\sqrt{\frac{d^2}{dx^2}+\beta^2}}\ \sqrt{\dfrac{d}{dx}+\sqrt{\dfrac{d^2}{dx^2}+\beta^2}}\ F(x)$

$$=\sqrt{2}\int_{0}^{\infty} dt\ e^{-\frac{1}{2}\beta t}\ I_0\left(\frac{1}{2}\beta t\right)\ \sin\left(t\frac{d}{dx}\right)\ F(x).$$

47. $\dfrac{1}{\sqrt{\frac{d}{dx}}}\ \dfrac{1}{\sqrt{\frac{d^2}{dx^2}+\beta^2}}\ \dfrac{1}{\sqrt{\frac{d}{dx}+\sqrt{\frac{d^2}{dx^2}+\beta^2}}}\ F(x)$

$$=\frac{\sqrt{2}}{\beta}\int_{0}^{\infty} dt\ e^{-\frac{1}{2}\beta t}\ I_0\left(\frac{1}{2}\beta t\right)\ \cos\left(t\frac{d}{dx}\right)\ F(x).$$

Bibliography

Dattoli, G., Quanttromini, M. and Torre, A. (1999). *Nuavo Cimento*, **B 144**, 693.

Dzrbashan, M. M. (1966). *Integer transformations and representations of function in the complex plane*, Moscow (in Russian).

Efimov, G. V. (2008). *Method of functional integrations*, Dubna, JINR, 2008.

Gradshteyn, I. S. and Ryzhik, I. M. (1980). *Table of Integrals, Series and Product*, Academic Press, New York.

Hilfer, R. (ed)(2000). *Applications of fractional calculus in physics*, World Scientific, Singapore.

Namsrai, Kh. (1998). *Square root Klein-Gordon operator and physical interpretation,* Inter, J. Theor. Phys. **37**, 1531.

Samko, S. G., Kilbas, A. A. and Marichev, O. I. (1993). *Fractional integrals and derivatives. Theory and applications*, Gordon and Breach, Amsterdam.

Sveshnikov, A. G. and Tikhonov, A. N. (1967). *Theory of Complex Variable Function*, Fiz-mat publication, Moscow.

Turmetov, B. Kh. and Umarov, S. (1993a). *Fractional integrals in the limiting case*, Doklad RAS (in Russian), Vol.**333**(2), pp. 136-137.

Turmetov, B. Kh. and Umarov, S. (1993b). *About one edge problem for equation with the fractional derivative*, Doklad of RAS (in Russian), Vol.**333**(4), pp. 446-448.

Wheelon, A. D. and Robacker, J. T. (1954). *A short Table of Summable Series* and *A Table of Involving Bessel Functions*, Ramo-Wooldridge.

Whittaker, E. T. and Watson, G. N. (1996). *A course of modern analysis*, Cambridge at the university press.

Zabadal, I., Vilhena, M. and Livotto, P. (2001). *Simulation of chemical reactions using fractional derivatives*, IL.Nuovo Cimento, **116 B**(5), pp. 529-545.

Zel'dovich, Ya. B. and Yaglom, I. M. (1982). *Higher Mathematics for Beginner Physicists and Technicians*, Nauka, Moscow.

Index

CPSIA information can be obtained at www.ICGtesting.com
Printed in the USA
BVOW09*1238120216

436507BV00003B/6/P